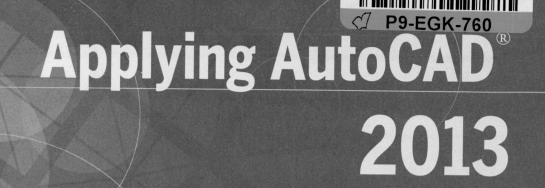

Applying AutoCAD® 2013

Terry T. Wohlers

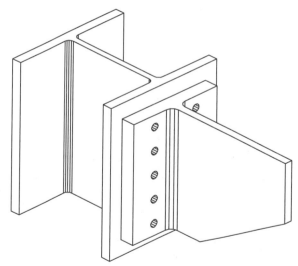

Applying AutoCAD® 2013
is a textbook for those
who wish to learn how to
use AutoCAD software.
AutoCAD is a computer-
aided drafting and design
package produced by
Autodesk, Inc. For information on how to obtain the
AutoCAD software, contact
Autodesk.

This text was created using the AutoCAD® 2013 software.

Mc Graw Hill

Connect
Learn
Succeed™

Acknowledgment

The author wishes to thank Tim Caffrey for his contribution to this book.

The McGraw-Hill Companies

WOHLERS: APPLYING AUTOCAD® 2013

Published by McGraw-Hill, a business unit of The McGraw-Hill Companies, Inc., 1221 Avenue of the Americas, New York, NY 10020. Copyright © 2013 by The McGraw-Hill Companies, Inc. All rights reserved. Printed in the United States of America. Previous editions © 2010, 2011, and 2012. No part of this publication may be reproduced or distributed in any form or by any means, or stored in a database or retrieval system, without the prior written consent of The McGraw-Hill Companies, Inc., including, but not limited to, in any network or other electronic storage or transmission, or broadcast for distance learning.

Some ancillaries, including electronic and print components, may not be available to customers outside the United States.

 This book is printed on recycled, acid-free paper containing 10% postconsumer waste.

1 2 3 4 5 6 7 8 9 0 QDB/QDB 1 0 9 8 7 6 5 4 3 2

ISBN 978-0-07-337551-9
MHID 0-07-337551-9

Vice President & Editor-in-Chief: *Marty Lange*
Editorial Director: *Michael Lange*
Global Publisher: *Raghothaman Srinivasan*
Executive Editor: *Bill Stenquist*
Executive Marketing Manager: *Curt Reynolds*
Development Editor: *Lorraine Buczek*
Lead Project Manager: *Jane Mohr*
Cover Designer: *Studio Montage, St. Louis, MO*
Cover Image: *Courtesy of Autodesk, Inc.*
Buyer II: *Susan K. Culbertson*
Media Project Manager: *Prashanthi Nadipalli*
Compositor: *Fleck's Communications, Inc.*
Typeface: 11/14 Iowan Old Style BT
Printer: Quad/Graphics

All credits appearing on page or at the end of the book are considered to be an extension of the copyright page.

Library of Congress Cataloging-in-Publication Data

Wohlers, Terry T.
  Applying AutoCAD 2013 / Terry T. Wohlers.
    p. cm.
  Includes index.
  ISBN 978-0-07-337551-9 (alk. paper) -- ISBN 0-07-337551-9 (alk. paper)
  1. AutoCAD. 2. Computer graphics. 3. Computer-aided design.
  4. Computer-aided engineering--Computer programs. I. Title.
  T385.W64452 2013
  620'.00420285--dc23
                          2012015275

www.mhhe.com

Contents in Brief

Part 6: Dimensioning and Tolerancing

Part 7: Groups and Details

Part 8: Working in Three Dimensions

Part 9: Solid and Surface Modeling

What's New in 2013

AutoCAD 2013 continues to improve upon the powerful capabilities of the AutoCAD software. You will notice the following differences in *Applying AutoCAD 2013:*

- Restructured online Help system that provides faster results.
- Improvements to the Command line, including making the options for any command selectable by clicking on them in the Command line window.
- New capabilities for viewing objects in Model and Layout viewports and for creating multi-view drawings for model documentation.

About the Author

For more than 30 years, Terry Wohlers has focused his education, research, and practice on design and manufacturing. He has authored 375 books, articles, reports, and technical papers on engineering and manufacturing automation. Also, he has given more than 80 keynote presentations at major industry events in Asia, Europe, the Middle East, the Americas, and Africa.

Wohlers is a renowned CAD educator. He developed and taught the first graduate and undergraduate courses on CAD at Colorado State University in 1983. Since then, he has taught many courses and given countless lectures on CAD and related subjects. In 1986, he founded Wohlers Associates, Inc., an independent consulting firm that provides organizations with technical and strategic advice on new developments and trends in CAD, prototyping, 3D printing, and additive manufacturing.

In May 2004, Wohlers received an Honorary Doctoral Degree of Mechanical Engineering from Central University of Technology (Bloemfontein, South Africa).

In 2007, more than 1,000 industry professionals from around the world selected Wohlers as the #1 most influential person in rapid product development and additive manufacturing.

User's Guide

A Practical Approach

Applying AutoCAD 2013 presents each feature of the AutoCAD® 2013 software in a logical, sequential format. You will build skills as you read about and apply techniques, solve problems, and practice computer-aided drafting and design.

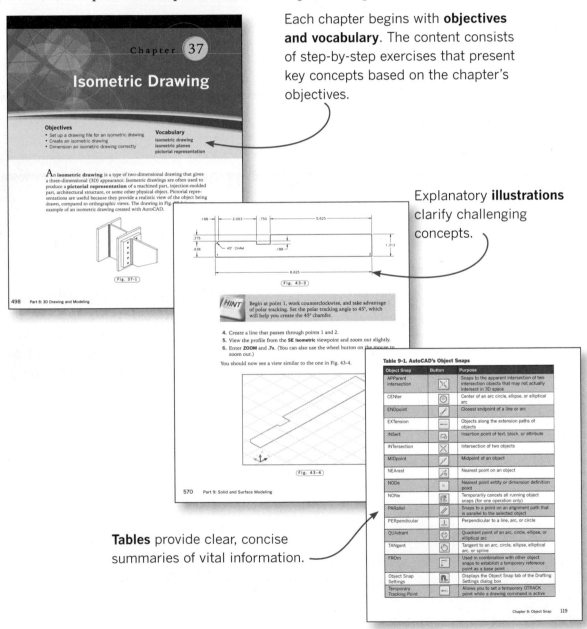

Each chapter begins with **objectives and vocabulary**. The content consists of step-by-step exercises that present key concepts based on the chapter's objectives.

Explanatory **illustrations** clarify challenging concepts.

Tables provide clear, concise summaries of vital information.

Hints suggest efficient ways of accomplishing steps or remind you of procedures you may have forgotten.

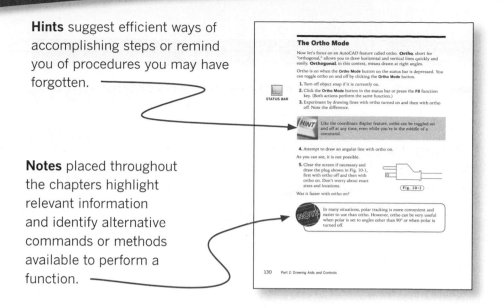

Notes placed throughout the chapters highlight relevant information and identify alternative commands or methods available to perform a function.

Chapter Review & Activities

Each chapter concludes with activities that review, reinforce, and expand learning.

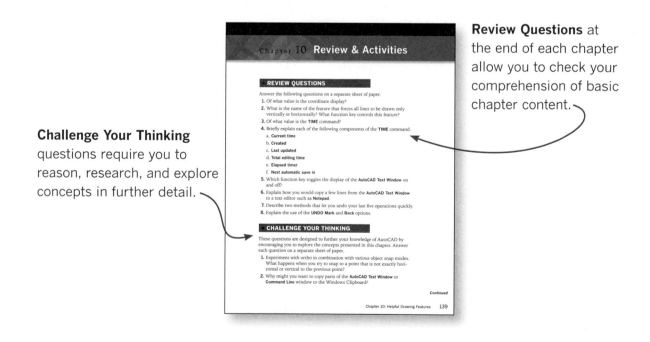

Review Questions at the end of each chapter allow you to check your comprehension of basic chapter content.

Challenge Your Thinking questions require you to reason, research, and explore concepts in further detail.

APPLYING AUTOCAD SKILLS

Problems at the end of each chapter help gauge your understanding of the skills taught in the chapter.

Problems include **real-world objects and concepts** to help you practice and apply chapter skills and concepts in a work-related context.

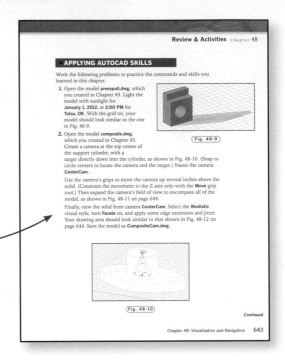

USING PROBLEM-SOLVING SKILLS

Problem-solving activities use representative tasks that you might encounter in industry. These problems require you to synthesize the AutoCAD skills presented throughout the text to arrive at practical solutions.

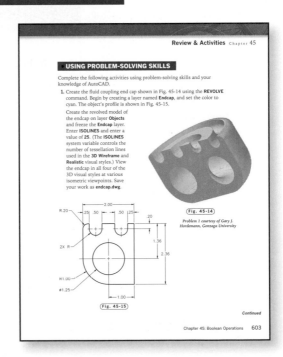

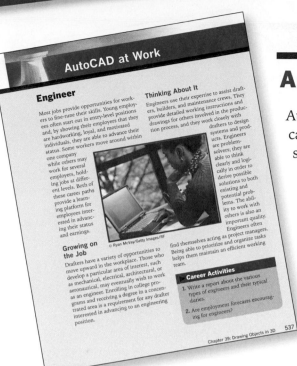

AutoCAD at Work

AutoCAD at Work pages explore the different types of careers open to people with AutoCAD knowledge and skills.

Part Projects

A real-world project is located at the end of each of the nine parts in this book. These projects tie together key concepts and skills and help you apply the skills and techniques you have learned in a realistic manner.

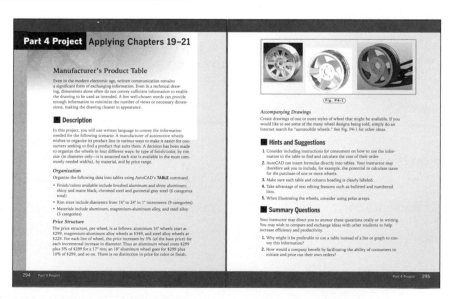

Additional Problems

Additional Problems provide further opportunities to put your AutoCAD skills to work.

The **level of difficulty** of each problem is clearly labeled.

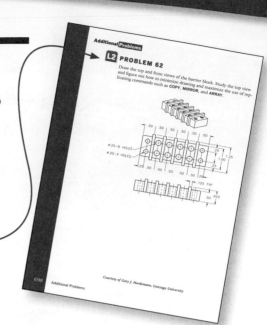

Online Resources

The text features an accompanying website for students and instructors at *www.mhhe.com/wohlers2013*.

The site contains advanced projects for students. Each advanced project presents a problem that you might encounter on the job. Little instruction is given—you are required to plan a solution and then carry it out. This is an opportunity for you to show initiative and creativity on a long-term or group project.

Also included on the site are:

- Instructional Plans
- PowerPoint® Presentations
- Test Bank

- Student Practice Files
- Answer Keys

Contents

Part 1 – Groundwork

Part 2 – Drawing Aids and Controls

Part 3 – Drawing and Editing

Part 4 – Text and Tables

Part 5 – Preparing and Printing a Drawing

Part 6 – Dimensioning and Tolerancing

Part 7 – Groups and Details

Part 8 – Working in Three Dimensions

Part 9 – Solid and Surface Modeling

AutoCAD at Work

Chapter 1

Exploring AutoCAD

Objectives

- Start AutoCAD Vocabulary
- Preview drawing files
- Open drawing files
- View details on an AutoCAD drawing
- Exit AutoCAD

Vocabulary

button
Command line
dialog box
double-clicking
drawing area
icon
resolution
status bar
thumbnail

AutoCAD provides countless methods and tools for producing, viewing, and editing two-dimensional drawings and three-dimensional models. The software permits designers, drafters, engineers, and others to create, revise, model, and document industrial parts and assemblies for prototyping, mold-making, and manufacturing. Around the world, organizations also use AutoCAD for the design of maps, buildings, bridges, factories, and just about every other product imaginable, ranging from car parts and stereo equipment to snow skis and cellular phones.

Starting AutoCAD

The first step in using AutoCAD is to learn how to start the software and open a drawing file. The steps that follow assume a new installation of AutoCAD, with no settings imported from previous versions.

1. Start AutoCAD by **double-clicking** the **AutoCAD 2013 icon** found on the Windows desktop. An icon is a picture on the screen that identifies and activates a tool.

Double-clicking means to position the pointer on the item and press the left button on the pointing device twice very quickly. If the AutoCAD 2013 icon is not present, pick the Windows Start button, move the pointer to Programs (or to All Programs), move it to Autodesk, then to AutoCAD 2013, and pick the AutoCAD 2013 option.

2. If the Startup dialog box appears, click the **Cancel** button.

The Welcome Screen appears, as shown in Fig. 1-1. The Welcome Screen provides access to drawings, learning tools, and online content. It is easily available to you at all times in AutoCAD under the Help command.

Fig. 1-1

3. You will learn more about the Welcome Screen and Help in Chapter 8. Uncheck the box next to Display at Startup in the lower left corner of the window and pick the **Close** button in the lower right corner.

The exact sequence for starting AutoCAD 2013 may vary depending on how the software was left by the last person to use it. If windows do not appear as described, skip to the following step.

The AutoCAD window appears, as shown in Fig. 1-2. Your window may look somewhat different. User commands and settings appear along the top and bottom of the AutoCAD window. The large, blank center portion of the window is the **drawing area**.

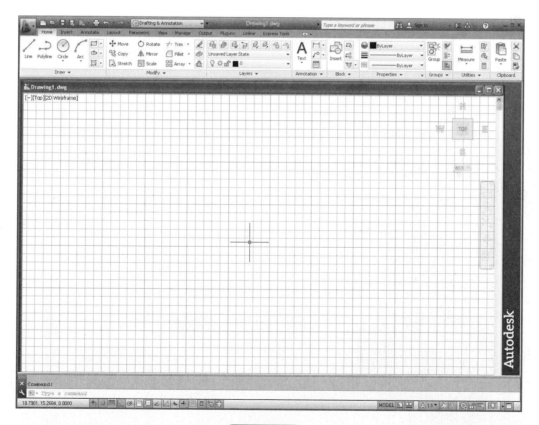

Fig. 1-2

NOTE The exact configuration of the AutoCAD window may vary depending on the screen **resolution** and how it was left by the last person to use the software. The display hardware can cause the appearance of screen elements to vary. Resolution is the number of pixels per inch on a display system; this determines the amount of detail that you can see on the screen.

Previewing Drawing Files

Notice the toolbar in the title bar, just to the right of the large red "A" in the upper left corner of the screen. This is the Quick Access toolbar.

1. On the Quick Access toolbar, click the **Open** button. (The Open button is the small square with a picture of an open file folder.)

Buttons in AutoCAD are the small areas that activate commands or functions. Buttons are identified by an icon, or picture.

This displays the Select File **dialog box**, shown in Fig. 1-3. A dialog box provides information and permits you to make selections and enter informa-

Fig. 1-3

tion. In this case, the dialog box enables you to open a drawing file, as well as select other options. The files and folders that you initially see may differ from those in the figure.

2. Navigate inside the Select File dialog box and find the AutoCAD 2013 folder.

3. Find and double-click the AutoCAD folder named **Sample**.

4. Find and double-click the folder named **Mechanical Sample**.

This displays all of the files contained in the Mechanical Sample folder.

5. To preview a file, rest the pointer on any of the files in the list. If nothing displays in the Preview area of the dialog box, single-click the file. (Your computer may be configured for either of these two methods.)

Notice that a small picture of the drawing, or **thumbnail**, appears at the right in the area labeled Preview. This gives you a quick view of the graphic content of the selected file.

6. Preview each of the remaining drawing files by resting the pointer or single-clicking on each of them.

Opening a Drawing File

7. Double-click the file named **Mechanical-Multileaders.dwg,** or single-click it and pick the **Open** button. (The DWG extension may not be shown as part of the file name. This depends on the configuration of your computer.)

A drawing of a wellhead piping design appears. Notice the tabs at the bottom of the AutoCAD drawing area, as shown in Fig. 1-4. These are the Model and Layout tabs. In this drawing, the Layout tab has been renamed.

Fig. 1-4

8. Pick the **Model** tab.

This displays the model view of the drawing. This is the view for most drafting and design work in AutoCAD.

9. Pick the **Layout** tab, labeled DWG-MB-SS03-PD-19-001 SHT-1.

This displays the drawing as it would appear on a printed or plotted sheet.

Viewing Details

AutoCAD offers powerful tools for viewing details on a drawing. For example, the zoom function allows you to take a closer look at very small sections of the drawing.

In the lower left area of the window, you will see AutoCAD's Command line, which allows you to key in commands and information. AutoCAD also displays important information in this area.

1. Click in the Command line, type **ZOOM** in upper- or lowercase letters, and press **ENTER**.

AutoCAD now asks you to specify the first corner of the zoom window. Imagine a small window surrounding the title block of the drawing, as shown in Fig. 1-5. This will become the zoom window.

2. Using the pointing device, pick a point (by pressing the left button) to specify one of the four corners of the imaginary window. See Fig. 1-5 for the location of the window.

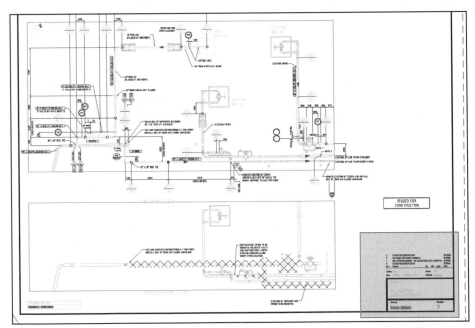

Fig. 1-5

3. Move the pointing device to the opposite corner to create the window and pick another point.

AutoCAD zooms in to fill the drawing area with the area represented by the zoom window.

4. Roll the center wheel button on the mouse, if it has one. This is another way to zoom in and out in the drawing area.

5. Move the pointing device slowly around the drawing. Notice that each item in the drawing becomes highlighted as the cursor touches it.

Exiting AutoCAD

1. Click on the Command line, type the **CLOSE** command using upper- or lowercase letters, and press **ENTER**.

When you press ENTER, AutoCAD asks whether you want to save your changes.

2. Pick the **No** button.

3. In the upper right corner of the AutoCAD window, pick the small box containing an **x**.

It is very important to exit AutoCAD at the end of each chapter. The appearance of some of the buttons may change as you work, depending on your selections. Exiting AutoCAD resets the buttons to their default settings.

Chapter 1 Review & Activities

Chapter 1 Review & Activities

● REVIEW QUESTIONS

Answer the following questions on a separate sheet of paper.

1. What are five examples of products or industries that use AutoCAD?
2. How do you start AutoCAD?
3. What is the process for previewing AutoCAD drawing files prior to opening one?
4. How is it possible to view small details on a complex AutoCAD drawing?
5. How do you exit AutoCAD?

● CHALLENGE YOUR THINKING

These questions are designed to further your knowledge of AutoCAD by encouraging you to explore the concepts presented in this chapter. Answer each question on a separate sheet of paper.

1. Explore the concept behind screen resolution. What is the impact of screen resolution? How can you change your computer's resolution?
2. In AutoCAD's Select File dialog box, explore a second method for previewing thumbnails of AutoCAD drawing files.

● APPLYING AUTOCAD SKILLS

Work the following problems to practice the commands and skills you learned in this chapter.

1. Start AutoCAD.
2. Preview the drawing files located in AutoCAD's Sample folder.
3. Open the drawing file named Mechanical - Multileaders.dwg located in the Mechanical Sample subfolder in the Sample folder. Pick the Model tab. Then pick the Layout tab labeled DWG-MB-SS03-PD-19-001 SHT-1. Pick the Quick View Layouts button on the gray bar at the bottom of the AutoCAD screen and select Model.

4. Open the drawing file named db_samp.dwg in the Sample/Database Connectivity subfolder. Enter the Zoom command and make a window to take a closer look at different areas of the floor plan and other details in the drawing.

5. Exit AutoCAD. Do not save your changes.

• USING PROBLEM-SOLVING SKILLS

Complete the following activities using problem-solving skills and your knowledge of AutoCAD.

1. Start AutoCAD. Open the drawing file named chroma.dwg in the AutoCAD 2013/Support folder. Then open the drawing file named attrib.dwg in the AutoCAD 2013/Sample/ActiveX/ExtAttr folder. Pick the Quick View Drawings button on the gray bar at the bottom of the AutoCAD screen and notice the thumbnails of both open drawings. Rest the pointing device over a drawing thumbnail and notice smaller thumbnails of the Model and Layout views of the drawing appear. Experiment with switching between drawings and between views in this way. Exit AutoCAD without saving.

2. Start AutoCAD. Using AutoCAD's preview feature, find and open each of the following files in AutoCAD's Sample/en-us/DesignCenter folder:

Home - Space Planner.dwg

Landscaping.dwg

Plant Process.dwg

Explore the drawings. What is the purpose of these files? Explain how knowing that these files exist can help you as you use AutoCAD in the future. Exit AutoCAD without saving.

Chapter 2

User Interface

Objectives

- Identify the parts that make up the AutoCAD user interface and describe the function of each part
- Become familiar with AutoCAD's **Ribbon** and learn its display options
- Navigate AutoCAD's **Application Menu**
- Describe the functions of the **Quick Access** toolbar, the **InfoCenter**, the **Command Line** window, and the status bar
- Introduce the **Display** tab on the **Options** dialog box

Vocabulary

context-sensitive
crosshairs
panel
pull-down menu

AutoCAD's user interface includes many parts that are important to the efficient operation of the software. For example, the proper positioning and grouping of buttons improves the speed with which you create, edit, and dimension drawings. Productive users of AutoCAD employ the panels, toolbars, palettes, dialog boxes, and keystrokes to enter commands and interact with AutoCAD. Each of these elements is a part of the AutoCAD user interface.

In this chapter, you will be introduced to the different elements of the user interface. In later chapters, you will explore each element of the user interface further.

Overview of the User Interface

1. Start AutoCAD.

2. If the Startup dialog box appears, click the **Cancel** button.

> As you learned in Chapter 1, the exact configuration of the AutoCAD window may vary depending on the screen resolution and how it was left by the last person to use the software. If windows, buttons, and menus are not present as described in the steps of this chapter, that's okay. You will learn how to display and position them in this chapter and later in the book.

The AutoCAD window appears, as shown in Fig. 2-1. Notice the locations of the title bar, the Ribbon, the Command Line window, and the status bar.

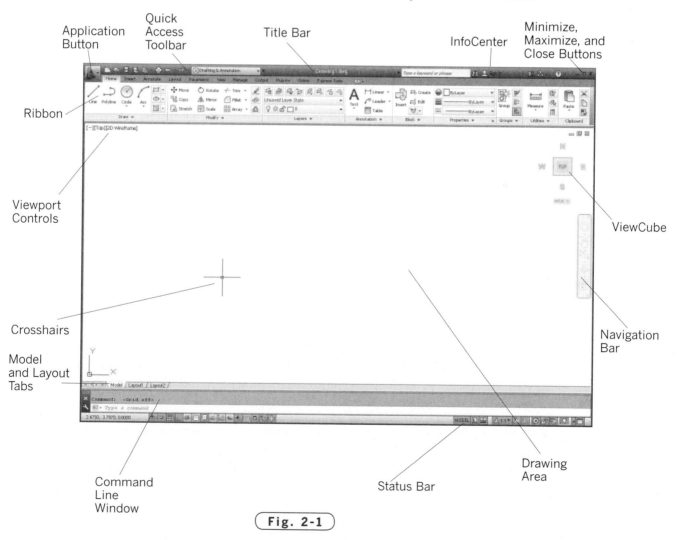

Fig. 2-1

The rest of the screen is the drawing area. As you move the pointing device in the drawing area, notice that the pointer becomes a special cursor called the **crosshairs**. You will use the crosshairs to pick points and select objects in the drawing area. When you move outside the drawing area, the normal Windows pointer appears. You will learn more about the features in the drawing area in future chapters.

The Ribbon

Near the top of the AutoCAD window, just below the title bar, is a wide area called the Ribbon. It provides easy access to many AutoCAD tools with tabs, panels, and buttons. In AutoCAD, a **panel** is a grouping of buttons for similar commands. In the Home tab of the Ribbon, the Draw, Modify, Layers, Annotation, Block, Properties, Groups, Utilities, and Clipboard panels are displayed.

1. Move the pointer to the word Insert in the tab area above the Ribbon and single-click on the **Insert** tab.

A different collection of panels appears on the Ribbon.

2. Select each of the other tabs and notice the panels that appear; then return to the **Home** tab.

Notice that several of the panel title bars contain a small arrow pointing down. This means that when you select them, they will display additional buttons and commands.

3. Click the **Draw** panel title bar.

Two additional rows of buttons appear in the expanded Draw panel.

4. Move the pointer out of the expanded Draw panel and into the drawing area.

The expanded part of the Draw panel disappears.

5. Click inside the **Draw** panel title bar again.

6. Move the pointer to the lower left corner of the expanded panel and select the pin icon.

7. Move the pointer out of the expanded Draw panel and into the drawing area.

Notice the expanded Draw panel remains visible, as shown in Fig. 2-2. AutoCAD allows you to "pin" the expanded panel open. When you no longer need the panel expanded, you can unpin it.

8. Click on the pin icon and move the pointer out of the expanded Draw panel.

The expanded part of the Draw panel disappears.

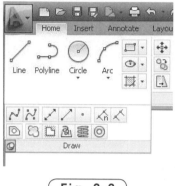

Fig. 2-2

To the right of the last Ribbon tab, Express Tools, is a small button with an arrow. This is the Collapse button.

9. Select the **Collapse** button.

The Ribbon is minimized to the panel titles only. The buttons are now hidden.

10. Move the pointer over the Layers panel title.

The buttons for the Layers panel appear.

11. Select the **Collapse** button again.

The Ribbon is further minimized to display only the panel titles.

12. Select the **Collapse** button once again.

The Ribbon is further minimized to display only the tabs.

13. Move the pointer over the **Home** tab and single click.

The full Ribbon appears, with buttons and panel titles.

14. Move the pointer off of the Ribbon into the drawing area.

The Ribbon disappears.

15. Select the **Collapse** button one last time to return to the Show Full Ribbon display mode.

The Application Menu

Now direct your attention to the red "A" in the upper left corner of the AutoCAD window. This button is called the Application button.

1. Single-click on the **Application** button.

A pull-down menu appears, as shown in Fig. 2-3, although yours may have different drawings listed under Recent Documents. A **pull-down menu** is a vertical list of commands that appears when a button or menu item is selected. This pull-down menu is called the Application Menu. The Application Menu displays options for opening, saving, and other file management commands. It also allows you to search for commands.

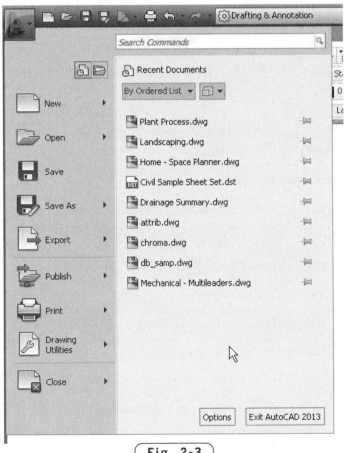

Fig. 2-3

2. Move the pointer into the drawing area and single-click to close the Application Menu.

Other Features

The Title Bar

In addition to housing the Application button, the title bar contains the Quick Access toolbar and the InfoCenter. You were introduced to the Quick Access toolbar in Chapter 1. It contains buttons for the most common functions: New file, Open file, Save, Save As, Print, Undo, and Redo. It also contains the Workspace pull-down menu, which you will learn more about in Chapter 3. The InfoCenter allows you to search for help by typing into the search field or by accessing the AutoCAD Help window.

Command Line Window

Notice the Command Line window near the bottom of the screen. As you learned in Chapter 1, you can enter AutoCAD commands at the Command line. However, in many cases, it is faster and easier to use buttons on the Ribbon and on toolbars, or to use dynamic input. You will learn more about the Command line and dynamic input in Chapter 4.

Status Bar

The Status Bar at the bottom of the screen shows the coordinates of the screen crosshairs, the status of various AutoCAD modes and settings, the Quick View buttons, drawing annotation tools, and several other important commands. The more you work with AutoCAD, the more you will appreciate the availability of the buttons and information on the status bar.

Display Options

1. Single-click the red "A" to display the Application Menu and select the **Options** button near the bottom of the Application Menu.

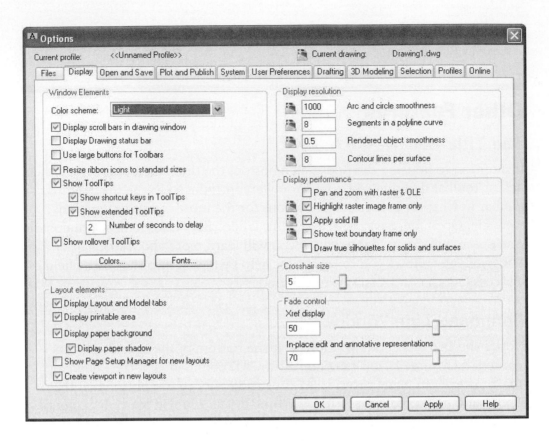

Fig. 2-4

The Options dialog box appears.

2. Select the Display tab in the **Options** dialog box.

The Display tab is shown in Fig. 2-4. In this dialog box you can choose to show, hide, and resize the different elements of the user interface. For example, you can choose to display or hide scroll bars. Another useful tool in the Display tab is the Colors... button, where you can change the color of many elements of the user interface. Some AutoCAD users prefer a dark drawing area over a light drawing area, or choose to personalize parts of the user interface with color.

3. Pick the **Cancel** button to close the Options dialog box.

4. Exit AutoCAD without saving.

• REVIEW QUESTIONS

Answer the following questions on a separate sheet of paper.

1. How do you access the Application Menu?
2. Describe how to display all of the buttons of the Draw panel on the Ribbon. How can you keep them displayed while you work elsewhere in the drawing?
3. What is the quickest way to open a file in AutoCAD?
4. How do you hide all the buttons on the Ribbon?
5. Write the names of items A through N in Fig. 2-5.

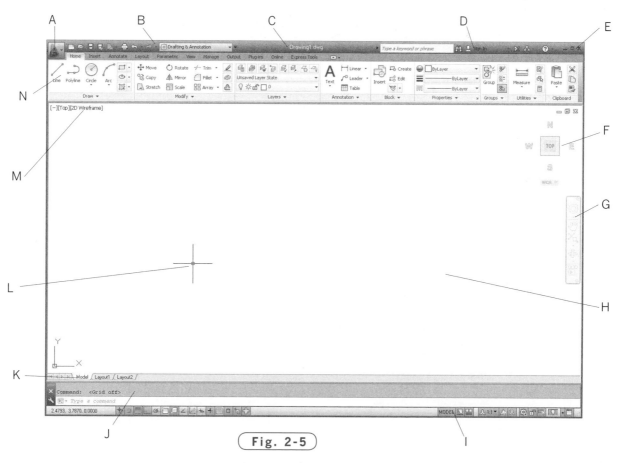

Fig. 2-5

Continued

• CHALLENGE YOUR THINKING

These questions are designed to further your knowledge of AutoCAD by encouraging you to explore the concepts presented in this chapter. Answer each question on a separate sheet of paper.

1. AutoCAD provides ways to change the appearance and colors of the AutoCAD window. For example, if you prefer to work on a blue background, you can change the drawing area to blue. Find out how to customize AutoCAD's appearance and colors. Then write a short paragraph explaining how to customize the display. Tell why it might sometimes be necessary to do so.

2. The AutoCAD window includes two sets of Windows-standard Maximize, Minimize, and Close buttons. Explain why there are two sets. What is the function of each set?

• APPLYING AUTOCAD SKILLS

Work the following problems to practice the commands and skills you learned in this chapter.

1. Navigate to AutoCAD's Sample/Mechanical Sample folder and open and view each of the following drawings. Where possible, use the Zoom button to look closely at the drawings. Close each of the drawings when you are finished.

   Mechanical - Data Links.dwg

   Mechanical - Multileaders.dwg

   Mechanical - Text and Tables.dwg

2. Arrange the panels of the Ribbon so that your AutoCAD screen looks similar to the one in Fig. 2-6.

Fig. 2-6

• USING PROBLEM-SOLVING SKILLS

Complete the following activities using problem-solving skills and your knowledge of AutoCAD.

1. Open the file db_samp.dwg, which is located in the Sample/Database Connectivity folder. Use the Zoom command to zoom in on a small area near the center of the drawing. Notice if scroll bars are available in the default window display. Click on the Application button and select the Options button. In the Options dialog box, pick the Display tab. Then check the box next to Display scroll bars in drawing window and pick OK. (If the box is already checked, just pick OK.) The scroll bars are now displayed. Use the horizontal and vertical scroll bars to navigate the drawing. Pick and drag the movable box within the scroll bars to move quickly around the drawing. Experiment to find other ways to navigate. What other ways does AutoCAD provide? Close the drawing without saving.

2. The following files are located in AutoCAD's Sample/en-us/Dynamic Blocks folder: Architectural - Metric.dwg, Civil - Metric.dwg, and Multileader Tools.dwg. Close all drawings that are open, then open these three drawings. Then click on the View tab of the Ribbon and focus on the User Interface panel. Experiment with the following buttons on the User Interface panel: Tile Horizontally, Tile Vertically, and Cascade. These are some of the ways you can arrange the AutoCAD window when you have more than one drawing open. Describe a situation where it would be preferable to cascade drawings. When would tiling be preferable? When you are finished, quit AutoCAD without saving.

Workspaces, Toolbars, and Palettes

Objectives

- Become familiar with the different workspaces in AutoCAD
- Switch between workspaces
- Display and reorganize docked and floating toolbars
- Describe the functions of palettes

Vocabulary

cascading menu
docked toolbar
floating toolbar
Menu Bar
palette
toolbar
workspace

A **workspace** is a complete user interface, which includes the location and appearance of the drawing area and all menus, buttons, and groupings of buttons. Think of a workspace as a unique, specific arrangement of AutoCAD's features and capabilities that can be retrieved at any time. It is easy to switch between workspaces, modify workspaces, and create custom workspaces to match the specific requirements of a user or a particular task.

AutoCAD features more than one kind of workspace. This allows users to match the workspace to the type of work they are performing or their personal preference. In this chapter, you will be introduced to four workspaces. As you will see, the user interface can be quite different between one workspace and another.

Over the years, AutoCAD has developed several different methods for grouping buttons. The Ribbon is the most recent way to group buttons. Before the Ribbon, there were palettes, and before that, toolbars. You'll learn about toolbars and palettes in this chapter.

Workspaces

1. Start AutoCAD.
2. If the Startup dialog box appears, click the **Cancel** button.
3. If the Welcome Screen appears, click on the **Close** button.

 Depending on how the software was left by the last person to use it, windows, toolbars, or palettes may or may not be present as described in the steps in this chapter. That's okay. You will learn how to display and position them in this chapter. For now, skip to the next step.

Drafting & Annotation Workspace

4. Click on the **Workspace** pull-down menu on the Quick Access toolbar, as shown in Fig. 3-1. If you cannot see the Workspace pull-down menu, make the AutoCAD window wider until it appears.
5. If Drafting & Annotation is not selected in the Workspace menu, select it now.

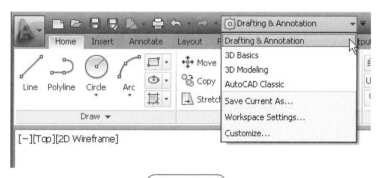

(Fig. 3-1)

AutoCAD displays the Drafting & Annotation workspace. This is AutoCAD's default workspace. It may be the workspace you have been using through the first two chapters of the book. Fig. 2-1 on page 11 describes the parts that make up the Drafting & Annotation workspace.

The 3D Basics Workspace

6. Pick the **Workspace** pull-down menu and select the **3D Basics** workspace, as shown in Fig. 3-2.

AutoCAD displays the 3D Basics workspace. Fig. 3-3 describes the parts that make up the 3D Basics workspace.

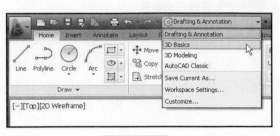

Fig. 3-2

Notice the differences between the workspaces. The 3D Basics workspace contains the Ribbon along the top of the screen, but the tabs and panels on the Ribbon are different.

Application Button
Ribbon
Viewport Controls
Quick Access Toolbar
Title Bar
InfoCenter
Maximize, Minimize, and Close Buttons

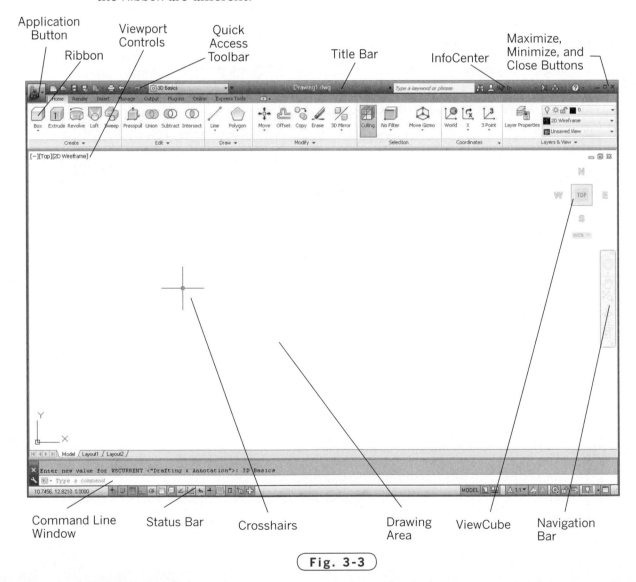

Command Line Window
Status Bar
Crosshairs
Drawing Area
ViewCube
Navigation Bar

Fig. 3-3

The 3D Modeling Workspace

7. Pick the **Workspace** pull-down menu and select the **3D Modeling** workspace, as shown in Fig. 3-4.

AutoCAD displays the 3D Modeling workspace. Fig. 3-5 describes the parts that make up the 3D Modeling workspace. Notice the differences between the workspaces. The tabs and panels on the Ribbon are different in the 3D Modeling workspace.

Fig. 3-4

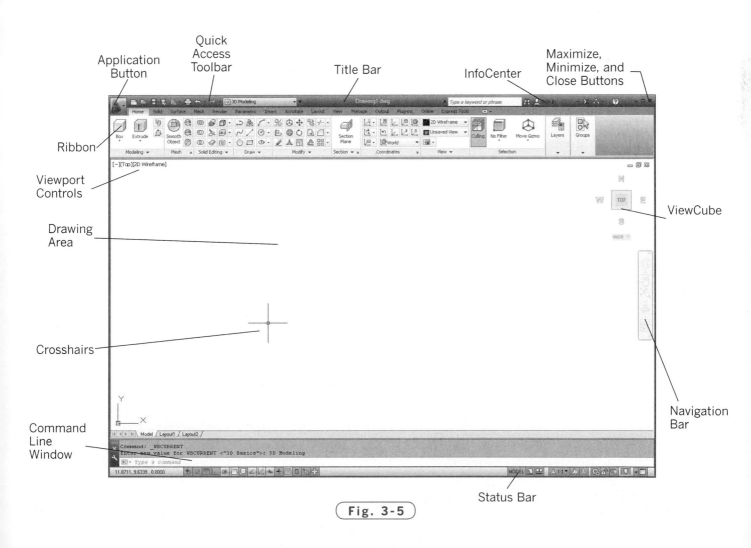

Fig. 3-5

The AutoCAD Classic Workspace

8. Select the **AutoCAD Classic** workspace from the Workspace pull-down menu, as shown in Fig. 3-6.

AutoCAD displays the AutoCAD Classic workspace. Fig. 3-7 describes the parts that make up the AutoCAD Classic workspace.

Fig. 3-6

Again, notice the differences between the workspaces. The AutoCAD Classic workspace contains a row of menu items below the title bar, called the **Menu Bar**, and two rows of buttons below that, as well as a column of buttons on both sides of the drawing area. The Tool Palettes window is floating inside the drawing area.

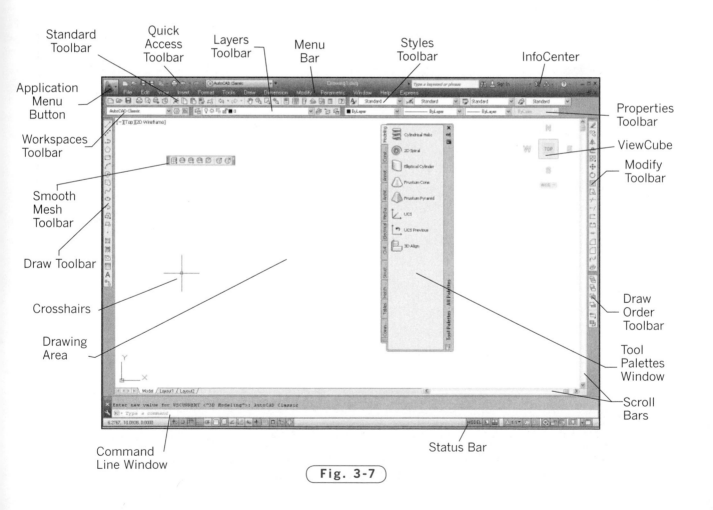

Fig. 3-7

It is possible to modify existing workspaces, rename workspaces, and create new workspaces. You can display the Menu Bar and the scroll bars in any workspace, even though they are hidden in the default settings of the workspaces. You can rearrange the elements on the screen to your liking, then save that arrangement as a workspace. As you learn AutoCAD, keep this in mind.

Toolbars

A **toolbar** is a thin strip that contains a collection of buttons. The AutoCAD Classic workspace includes eight toolbars along the top, left, and right of the AutoCAD window. Under the menu bar (File, Edit, View, etc.), you should see five toolbars in two rows docked across the top of the screen. A **docked toolbar** is one that appears to be part of the AutoCAD screen's border, outside of the drawing area.

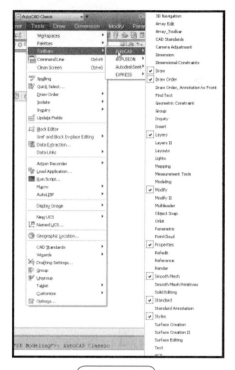

Fig. 3-8

Floating Toolbars

Let's focus on displaying and positioning toolbars. A toolbar that is not docked is called a **floating toolbar**.

1. On the Menu Bar, select the **Tools** pull-down menu, move the pointer to the **Toolbars** option, and then to the **AutoCAD** option on the cascading menu. A **cascading menu** is a second menu that appears when you point to a menu item containing a small arrow pointing to the right.

This displays a list of toolbars that AutoCAD makes available to you, most of which is shown in Fig. 3-8. You can also display the list of toolbars by right-clicking on any docked or floating toolbar on the screen. The ones that are checked are the toolbars that are present on the screen.

To display a toolbar from the list, click its name.

2. Click the word **Lights** to display the Lights toolbar.

3. Close the floating Lights toolbar by picking the **x** (Close button) in the upper right corner of the toolbar.

4. If any other floating toolbars are present, close them by picking the **x** in the upper right corner of each toolbar.

It is easy to move toolbars to a new location. As you work with AutoCAD, you should move floating toolbars to a location that does not interfere with the drawing.

5. Open the **Render** toolbar.

6. Move the **Render** toolbar by clicking and dragging the left or right border of the toolbar.

 AutoCAD allows users to lock toolbars in place. If a toolbar does not move when you try to drag it, right-click any open toolbar, go to the bottom of the pop-up menu, and rest the cursor on Lock Location. From the cascading menu, pick the checkmark next to Floating Toolbars/Panels to unlock the toolbar's position.

Changing a Toolbar's Shape

AutoCAD makes it easy to change the shape of toolbars. You might want to change the shape of a toolbar, for example, to make it fit better in the area of the drawing in which you are working.

1. Move the pointer to the bottom edge of the Render toolbar, slowly and carefully positioning it until a double arrow appears. With the double arrow visible, click and drag downward until the toolbar changes to a vertical shape and then release the button.

2. Move the pointer to the right edge of the Render toolbar, positioning it until a double arrow appears. With the double arrow visible, click and drag to the right until the toolbar changes back to a horizontal shape.

3. Close the Render toolbar.

Docked Toolbars

The AutoCAD Classic workspace has many docked toolbars. The Standard and Styles toolbars fill the first row of toolbars at the top of the screen. The second row includes the Workspaces, Layers, and Properties toolbars. The Draw toolbar is docked vertically on the left side of the screen. Finally, the Modify and Draw Order toolbars are docked on the right side of the screen.

1. Move the pointer to the **Standard** toolbar and click and drag the outer left edge (border) of the toolbar, moving it into the drawing area. The toolbar is undocked and becomes a floating toolbar.

2. Repeat Step 1 for the remaining four toolbars that are docked along the top of the AutoCAD window.

3. Close all five of these floating toolbars.

4. Right-click in any of the three remaining toolbars to display the list of toolbars and select **Layers** to display the Layers toolbar.

5. Repeat Step 4 to display the Workspaces, Properties, Standard, and Styles toolbars.

6. Click and drag the **Standard** toolbar and carefully dock it under the Menu Bar (File, Edit, View, etc.).

The toolbar docks into place.

7. Click and drag the other four toolbars and carefully dock them in place. Refer to Fig. 3-7 on page 24 if you are unsure where they belong.

8. If the Draw, Modify, and Draw Order toolbars are not present and docked in their default locations, position them now as shown in Fig. 3-7.

You should now see eight docked toolbars on the screen.

This textbook assumes that these eight toolbars are present in the AutoCAD Classic workspace in a configuration similar to the one in Fig. 3-7. Instructions are written to work with AutoCAD's default values at installation. It assumes that the screen color, crosshairs, and other elements have not been customized.

Palettes

A **palette** is a special toolbar that contains a large grouping of buttons, tools, or information. Palettes eliminate the need to display a lot of toolbars and reduce the clutter on your computer screen. Like toolbars, palettes can be docked or floating in the drawing area.

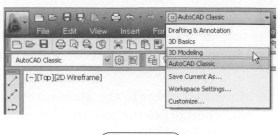

Fig. 3-9

1. In the Workspace pull-down menu, pick the down arrow to the right of AutoCAD Classic and click **3D Modeling**, as shown in Fig. 3-9.

2. Select the **View** tab above the Ribbon, then select the **Tool Palettes** button on the Palettes panel.

The Tool Palettes window is displayed in the drawing area. Fig. 3-10 shows the Tool Palettes window.

3. On the Quick Access toolbar, pick the **Workspace** pull-down menu and select the **AutoCAD Classic** workspace.

In the AutoCAD Classic workspace, the Tool Palettes window and the Smooth Mesh toolbar may be floating.

4. If the Smooth Mesh toolbar is open, close it.

5. If the Tool Palettes window is open, close it by picking the **x** located in the upper right corner.

6. From the **Tools** pull-down menu, move the pointer to the **Palettes** option, and notice the different options available in the cascading menu, as shown in Fig. 3-11.

As you can see, the Ribbon is also a palette.

7. Select **Tool Palettes** from the cascading menu.

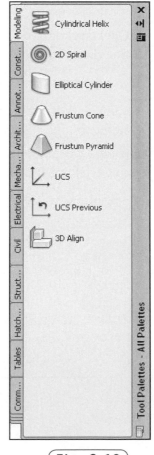

Fig. 3-10

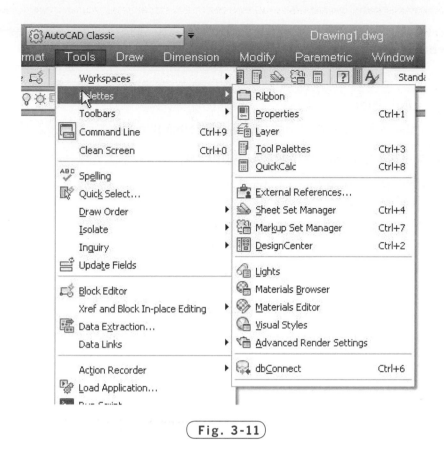

Fig. 3-11

This causes the Tool Palettes window to display. Pressing the CTRL and 3 keys together toggles the display of the Tool Palettes window on and off.

8. Click on the top of the left border of the **Command Line** window and drag it into the drawing area.

As you can see, the Command Line window behaves like a palette. Pressing the CTRL and 9 keys together toggles the display of the Command Line window on and off.

9. Return the **Command Line** window to its docked position by dragging it to the bottom of the AutoCAD screen.

Auto-Hide

AutoCAD's Auto-hide feature increases the size of the drawing area by hiding a palette, except for its title bar, until you place the cursor over the title bar.

1. Rest the pointer over the vertical title bar of the Tool Palettes window and right-click.

This displays a menu of options.

2. Select **Auto-hide**.

3. Move the pointer away from the Tool Palettes window.

The window disappears, except for its title bar.

4. Move the pointer to the title bar.

The window reappears.

5. Move the pointer away from the Tool Palettes window again.

The window disappears. As you can see, Auto-hide helps conserve screen space, which is especially helpful when the AutoCAD window becomes busy.

6. Rest the pointer on the title bar of the Tool Palettes window, right-click, and uncheck **Auto-hide**.

Transparency

Transparency is another feature that helps free up space on the screen.

1. Rest the pointer inside the **Tool Palettes** vertical title bar, right-click, and select the **Transparency...** item.

2. Click and drag the General slider in the dialog box to the far left, click and drag the Rollover slider to the middle position, and pick the **OK** button.

This causes the Tool Palettes window to become nearly transparent. When you move the pointer over it, it becomes semi-transparent.

3. From the Quick Access toolbar, pick the **Open** button and open the file named **db_samp.dwg** in the Sample/Database Connectivity folder.

You should be able to see the drawing through the Tool Palettes window.

4. Right-click inside the Tool Palettes title bar, select **Transparency...**, move the General slider to the middle position, and pick **OK**.

The Tool Palettes window becomes less transparent.

5. Right-click inside the Tool Palettes title bar, select **Transparency...**, move both sliders to the far right, and pick **OK**.

The Tool Palettes window is no longer transparent. Auto-hide and transparency work with all palettes and with the Command Line window.

6. Exit AutoCAD without saving.

Chapter 3 Review & Activities

● REVIEW QUESTIONS

Answer the following questions on a separate sheet of paper.

1. Define workspace in your own words.
2. Briefly describe the differences between the Drafting & Annotation, 3D Basics, 3D Modeling, and AutoCAD Classic workspaces.
3. What is a cascading menu?
4. Describe how to display a toolbar that is not currently displayed on the screen.
5. How do you move a toolbar?
6. How do you make a floating toolbar disappear?
7. Explain how to make the Tool Palettes window semi-transparent.

● CHALLENGE YOUR THINKING

These questions are designed to further your knowledge of AutoCAD by encouraging you to explore the concepts presented in this chapter. Answer each question on a separate sheet of paper.

1. AutoCAD allows users to modify, customize, and create new workspaces. Name two or three different tasks that might best be performed in different workspaces. What would you name each workspace?
2. Workspaces in AutoCAD are similar to physical workspaces in your home or office. What "tools" do you use on the dining room table for a large meal? How would those tools change if you were to use the dining room table to do homework or pay bills?

● APPLYING AUTOCAD SKILLS

1. Starting in the 3D Modeling workspace, rearrange the AutoCAD window so that it is similar to the one in Fig. 3-12 on page 32.
2. Starting in the Drafting & Annotation workspace, rearrange the AutoCAD window so that it is similar to the one in Fig. 3-13 on page 32.

Continued

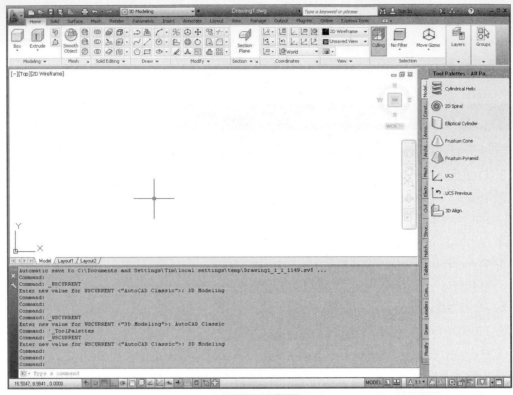

Fig. 3-12

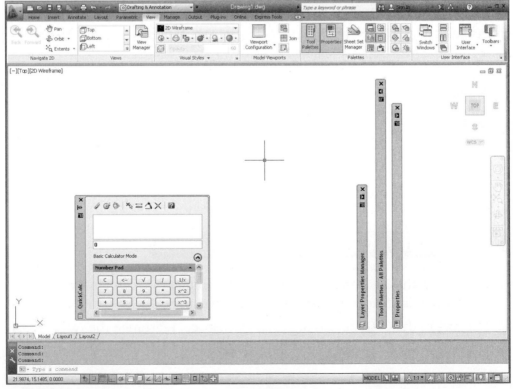

Fig. 3-13

3. Starting in the AutoCAD Classic workspace, rearrange the AutoCAD window so that it is similar to the one in Fig. 3-14. The three floating toolbars are Dimension, UCS, and UCS II.

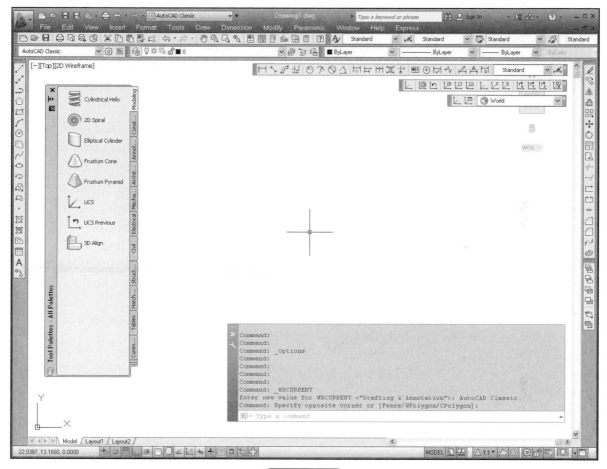

Fig. 3-14

4. Starting in the AutoCAD Classic workspace, display the toolbar list and uncheck a toolbar. Repeat until all toolbars are unchecked. Notice that the drawing area has increased, but you have lost the convenience of toolbar command selection. Click the Tools pull-down menu at the top of the screen, move the pointer over Workspace, and single-click AutoCAD Classic. This restores the toolbars to their default positions in the AutoCAD Classic workspace.

● USING PROBLEM-SOLVING SKILLS

Work the following problems to practice the commands and skills you learned in this chapter.

1. Start AutoCAD. Switch to the 3D Modeling workspace. Modify the workspace by stretching the Command Line window upward and displaying and stretching the Tool Palettes window to the left. The drawing area should cover about a quarter of the screen. Pick the Workspace pull-down menu on the Quick Access toolbar, select Save Current As..., and save the workspace as Workspace1.

2. Delete the workspace you created in problem 1 by picking the Workspace pull-down menu and selecting the Customize... option. This opens the Customize User Interface dialog box. Right click on Workspace1 and pick Delete. Pick the OK button to close the dialog box.

3. Starting in the AutoCAD Classic workspace, open all of the toolbars provided in AutoCAD. Dock them along the sides, top, and bottom of the drawing area. Using the tooltips, investigate the buttons that make up the toolbars. (Tooltips are the words that appear when the pointer rests on the button.) What two problems might you encounter if you left all the toolbars open on the screen? When you are finished, select AutoCAD Classic in the Workspace toolbar at the top left of the screen, and notice that the additional toolbars have all been closed automatically.

Command Entry

Objectives

- Create, save, and open an AutoCAD drawing file
- Prepare AutoCAD for drawing in 2D
- Enter commands using the keyboard and buttons
- Apply shortcut methods to enter commands
- Reenter commands
- Print a drawing

Vocabulary

command aliases
context sensitive
dynamic input
extended tooltips
rubber-band effect
shortcut menu
tooltips

AutoCAD offers several methods of entering and repeating commands. When entering them frequently, as you will practice in this chapter, you will want to use methods that minimize the time it takes. Command aliases and other shortcuts permit you to enter commands with as little effort as possible. If you take advantage of these time-saving features of AutoCAD, you will require less time to complete a drawing. Consequently, you become more attractive to prospective employers looking for highly productive individuals.

In the steps that follow, you will practice entering commands by doing some basic operations such as creating a file and drawing lines. You will apply the concepts you learn in this chapter nearly every time you use AutoCAD.

Creating and Saving a Drawing File

1. Start AutoCAD.
2. If the Startup dialog box appears, click the **Cancel** button.
3. If the Welcome Screen appears, click the **Close** button.

It is important that you start (or restart) AutoCAD at the beginning of each chapter, unless instructed otherwise. This ensures that the buttons on the Ribbon and in the toolbars are set to their default settings.

We will begin this chapter in the 2D Drafting & Annotation workspace.

4. Pick the **Workspace** pull-down menu on the Quick Access toolbar and then select **Drafting & Annotation** from the menu.

 It is important that you select the workspace as instructed, even if that workspace already appears to be displayed. This resets the screen arrangement to the selected workspace.

QUICK ACCESS

5. Pick the **Save** button from the Quick Access toolbar.

 Notice the button printed to the left of Step 5. This has been provided as a visual aid to help you locate and select the button. The screen location of the button appears below the button for reference.

Buttons like this will appear in the margin throughout the chapter in which they are introduced and in subsequent chapters. They will also appear occasionally if the button has not been used recently, to refresh your memory.

The Save Drawing As dialog box appears, as shown in Fig. 4-1. If you use other Windows software, the dialog box should look familiar to you.

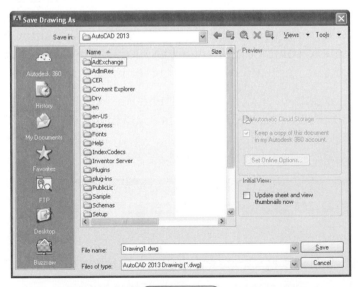

Fig. 4-1

6. Navigate inside the Save Drawing As dialog box and find the AutoCAD 2013 folder.

7. Pick the **Create New Folder** button (located in the upper right area of the dialog box).

This creates a folder named New Folder.

8. Right-click the new folder, select **Rename** from the menu, type your first and last name, and press **ENTER**.

9. Double-click this new folder to make it the current folder.

10. In the File name text box, highlight the name of the drawing file (e.g., **Drawing1.dwg**), type **stuff**, and pick **Save** or press **ENTER**.

 If you do not see the file extension (dwg) after the names of drawing files, extension display may be turned off on your computer. Your work with AutoCAD will be much easier if you turn on the extension display.

Notice that the drawing file name now appears in the title bar of the AutoCAD window.

11. Pick the **x** in the upper right corner of the drawing area to close the stuff.dwg file. Do not pick the **x** in the upper right corner of the AutoCAD window, because this would close AutoCAD.

The stuff.dwg file disappears, and the AutoCAD screen becomes gray because no drawing file is open. Notice also that the Ribbon, status bar, and Command Line window disappear. An abbreviated version of the Quick Access toolbar and a smaller version of the Application button are visible.

Opening a Drawing File

If AutoCAD is already open, you can open a file easily by using the Open button.

1. Pick the **Open** button on the Quick Access toolbar.

The Select File dialog box appears. AutoCAD defaults to the folder you named after yourself.

2. Double-click **stuff.dwg** in the selection window.

You have reopened the stuff.dwg file. Currently, the drawing area is blank.

QUICK ACCESS

Entering Commands

AutoCAD allows you to draw two-dimensional (2D) and three-dimensional (3D) objects. For the next several chapters, you will learn how to enter commands, select objects, and navigate the user interface in the 2D drawing environment.

For the purpose of learning to enter commands, we will use the LINE command. You will learn more about using various AutoCAD commands in the chapters that follow.

Using Dynamic Input

Entering commands at the Command line may force your attention away from the drawing you are creating. **Dynamic input** allows you to enter commands at the cursor instead of moving the pointing device to the Command line. This allows you to keep your focus on the drawing area.

STATUS BAR

1. Find the **Dynamic Input** button on the status bar, and toggle it on (blue) if it is not on already.

2. With the crosshairs anywhere in the drawing area, type **LINE** using upper- or lowercase letters.

Notice that the command appears near the crosshairs as you type, along with a list of commands that begin with the letters LINE.

3. Press **ENTER**.

AutoCAD asks you to specify the first point and shows the coordinates (current position) of the crosshairs. You will learn more about coordinates in Chapter 7.

4. Pick a point in the lower left portion of the drawing area.

AutoCAD's dynamic input prompt now requests the next point. It also displays the distance and angle of the crosshairs from the last point, as shown in Fig. 4-2.

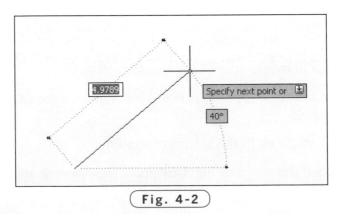

Fig. 4-2

5. Move the crosshairs on the screen and notice how the distance and angle values change dynamically.

6. Instead of picking the next point, type **5** and press the **TAB** key.

Notice that the value you typed appears in the distance box of the dynamic input. Also, a padlock appears to indicate that the value of 5 is locked into place.

7. Type **45** and press **ENTER**.

AutoCAD creates a line segment that is 5 units long at an angle of 45°.

8. Produce the second line segment by picking a point below and to the right of the previous point.

9. Produce the third line segment to create a triangle by moving the crosshairs over the first point and clicking.

You may notice that temporary dotted lines, arcs, and symbols appear on the screen as you complete the triangle in Step 9. These are produced by the AutoCAD features known as object snap and AutoTrack™. You will use these features extensively in the chapters that follow. For now, you can ignore them.

10. Press **ENTER** to terminate the LINE command.

If the entire triangle is not visible in the drawing area, use the mouse wheel, the scroll bars, or other methods of panning and zooming to bring the entire triangle into the drawing area. You will learn more about panning and zooming in Chapters 12 and 13.

Dynamic input can be toggled on and off by clicking the Dynamic Input button on the status bar. You will use dynamic input more in the chapters that follow.

Using the Command Line

In earlier versions of AutoCAD, the only way to enter a command was to enter the command name on the Command line. This method is still available, although other methods have been added in recent years to provide more efficiency.

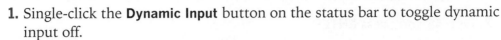

STATUS BAR

1. Single-click the **Dynamic Input** button on the status bar to toggle dynamic input off.
2. With the crosshairs anywhere in the drawing area, type **LINE** in upper- or lowercase letters.

Notice that the command now appears in the Command Line window, along with a list of commands that begin with the letters LINE.

3. Press **ENTER**.

In the Command Line window, notice that AutoCAD asks you to specify the first point.

4. Pick a point anywhere in the drawing area.

AutoCAD requests the next point.

5. Pick a second point anywhere on the screen.

AutoCAD requests another point.

6. Move the crosshairs away from the last point and notice the stretching effect of the line segment.

This is known as the **rubber-band effect**.

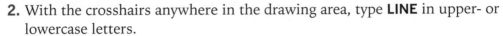

STATUS BAR

7. Produce the second and third line segments to create a triangle, and then press **ENTER** to terminate the LINE command.
8. Click on the **Dynamic Input** button on the status bar to reactivate dynamic input.

You can use the Command line method of entry even when dynamic input is active. Simply single-click at the Command line before typing the command.

Picking Buttons on the Ribbon

It can be faster to pick a button than to enter a command at the keyboard or from a pull-down menu.

DRAW

1. Rest the pointer on the **Line** button located in the Draw panel of the Ribbon, but do not pick the button.

A small window with a brief explanation appears after about one second. Then, after a delay of a few seconds, a larger window with a detailed explanation appears. These windows, called **tooltips**, help you understand the purpose of the button. The larger tooltips are called **extended tooltips**. You can turn tooltips and extended tooltips on or off in the Display tab of the Options dialog box.

2. Slowly position the pointer on top of other buttons and read the information that AutoCAD displays in each tooltip and extended tooltips.

3. Pick the **Line** button and move the pointer into the drawing area.

You have just entered the LINE command. You can see that AutoCAD is again asking where you want the line to begin.

4. Pick a point anywhere in the drawing area; then pick a second point to form a line.

5. Press the **ENTER** key to terminate the LINE command.

6. Pick another button (any one of them) from the Draw panel.

7. Press the **ESC** key to cancel the last entry at the Command line.

As you work with AutoCAD, you will be entering commands by picking buttons, selecting them from menus, or entering them at the keyboard. Occasionally you might accidentally select the wrong command or make a keyboarding error. It is easy to correct such mistakes.

If you catch a keyboarding error before you press ENTER...	use the backspace key to delete the incorrect character(s).
If you select the wrong button or menu item...	press the ESC key to clear the Command line.
If you key in the wrong command...	press the ESC key to clear the Command line.
If you accidentally pick a point or object on the screen...	press the ESC key.

Command Aliases

AutoCAD permits you to issue commands by entering just the first few characters of the command. Command abbreviations such as these are called **command aliases**.

1. Type **L** in upper- or lowercase and press **ENTER**.

This activates the LINE command. You can also enter other commands using aliases. Some of the more common command aliases are listed in Table 4-1.

2. Press the **ESC** key to cancel the LINE command.

3. Enter each of the commands listed in Table 4-1 using the alias method. For now, press **ESC** to cancel each command.

You will learn more about these commands in future chapters.

4. Reenter the **LINE** command using the command alias method and create another line.

5. Press **ENTER** to terminate the LINE command.

Even when you do not know a command alias, AutoCAD can help you enter the command without typing the entire command. We will use the POLYGON command as an example.

6. Type **PO** in upper- or lowercase letters, but do not press **ENTER**.

AutoCAD displays an alphabetical list of all commands that begin with the letters PO.

7. Scroll down the list to the **POL (POLYGON)** command and rest the pointer on it without picking it.

Notice the tool tips for the POLYGON command appear.

8. Pick **POL (POLYGON)** to enter the POLYGON command.

9. Press **ESC** to terminate the command.

Table 4-1. Common Command Aliases

Command	Entry
ARC	A
CIRCLE	C
COPY	CO
DONUT	DO
ELLIPSE	EL
ERASE	E
LINE	L
MOVE	M
PAN	P
POLYGON	POL
SCALE	SC
ZOOM	Z

Command aliases are defined in the AutoCAD file acad.pgp. You can view this file or create additional command aliases by modifying it using a text editor. The file is located in the following folder:

C:\Documents and Settings\username\Application Data\ Autodesk\AutoCAD 2013 - English\R19.0\enu\Support

Substitute your user name for "username" in the file path above.

Reentering the Last Command

After you have entered a command, you have several choices for repeating it.

1. Enter the **LINE** command and create a line segment anywhere in the drawing area. Press **ENTER** to end the command.

2. Press the spacebar or **ENTER**.

DRAW

This enters the last-entered command, in this case the LINE command. You will find that this saves time compared to other methods of entering the command.

3. Create a line segment and then press **ENTER** to end the LINE command.

4. Press the right button on the pointing device.

Right-clicking in AutoCAD displays a **shortcut menu**. The top item on the menu, when selected, repeats the last command you entered.

5. Pick **Repeat LINE**.

6. Draw two of the polygons shown in Fig. 4-3.

The Repeat option is **context sensitive**, which means it changes depending on what the user is doing. In this case, the software remembers the last command you used.

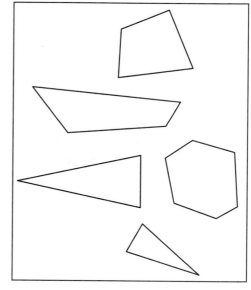

Fig. 4-3

HINT

After you have entered at least two line segments, the prompt changes to include the Close option. This option allows you to close a polygon accurately and automatically. It also terminates the LINE command. To choose the Close option, type the letter C and press ENTER. As you can see, you can use shortcuts or aliases to choose command options as well as to enter commands.

7. Draw the remaining polygons shown in Fig. 4-3 using the LINE command. Be sure to use one of the shortcut methods of reentering the last command.

8. Enter the **LINE** command and draw two connecting line segments, but do not terminate the command.

9. In reply to Specify next point or [Close/Undo], enter **U** for Undo.

AutoCAD backs up one segment, undoing it so that you can recreate it.

10. Practice the **Undo** option with additional polygons.

11. Pick the **Plot** button on the Quick Access toolbar.

QUICK ACCESS

The Plot dialog box appears, as shown in Fig. 4-4. If no printer appears in the Printer/plotter Name field, you must specify a printer.

12. If necessary, ask your instructor to specify a printer, select the printer as shown in Fig. 4-4, and select **OK**.

Your drawing named stuff should print. You will learn more about plotting drawings later in this book.

13. Pick the **Save** button from the Quick Access toolbar to save your work.

QUICK ACCESS

As you work with the different methods of entering commands, you may prefer one method over another. Picking the Line button from the Ribbon may or may not be faster than entering L at the keyboard. Keep in mind that there is no right or wrong method of entering commands. Experienced users of AutoCAD use a combination of methods.

14. Exit AutoCAD.

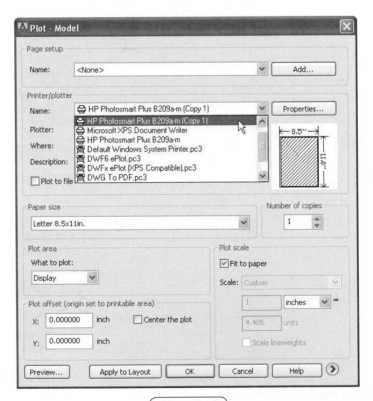

Fig. 4-4

Chapter 4 Review & Activities

● REVIEW QUESTIONS

Answer the following questions on a separate sheet of paper.

1. Describe the purpose of the Open button.
2. When a file has not yet been saved for the first time, what appears when you pick either Save or Save As... from the Application Menu?
3. What appears when you rest the pointer on a button on the Ribbon?
4. What is the fastest and simplest method of reentering the previously entered command?
5. How can you enter commands such as LINE, CIRCLE, and ERASE quickly at the keyboard?
6. What is a fast method of closing a polygon when you are using the LINE command?
7. Explain the use of the LINE Undo option.

● CHALLENGE YOUR THINKING

These questions are designed to further your knowledge of AutoCAD by encouraging you to explore the concepts presented in this chapter. Answer each question on a separate sheet of paper.

1. Discuss the advantages and disadvantages of having more than one way to enter a command. Which methods of entering commands are most efficient? Collect opinions from other AutoCAD users and consider ergonomic factors.
2. Describe a way to open a drawing file in AutoCAD when AutoCAD is not currently running on the computer.

Continued

● APPLYING AUTOCAD SKILLS

Work the following problems to practice the commands and skills you learned in this chapter.

1. Create a new drawing file named prb4-1.dwg and store it in the folder named after yourself. Create the bookcase shown in Fig. 4-5. For practice, enter the LINE command in various ways: using dynamic input, using the Command line, or picking a button. Use the various shortcut methods. Save the drawing.

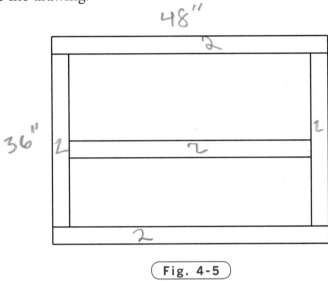

Fig. 4-5

2. Create a new drawing file named prb4-2.dwg and store it in the folder named after yourself. Draw the concrete block shown in Fig. 4-6 using the most efficient method(s) of entering the LINE command.

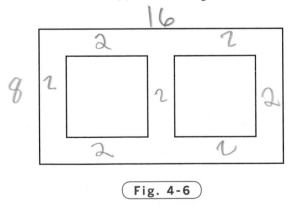

Fig. 4-6

3. Create a new drawing file named prb4-3.dwg and store it in the folder named after yourself. Draw the simple house elevation drawing shown in Fig. 4-7 by picking the Line button from the Draw panel of the Ribbon. When reentering the command, use a shortcut method.

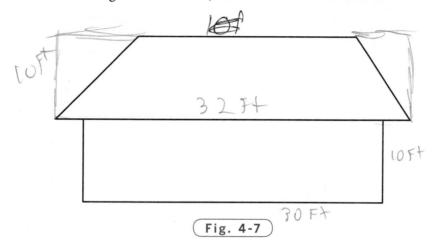

10 Ft

32 Ft

10 Ft

30 Ft

Fig. 4-7

4. Create a new drawing file named prb4-4.dwg and store it in the folder named after yourself. Draw the front and side views of the sawhorse shown in Fig. 4-8 by entering the LINE command's alias at the keyboard. Be sure to use shortcut methods when reentering the command.

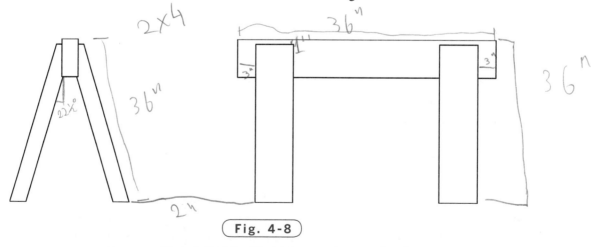

2×4

36"

36"

36"

22½°

36"

1"

3"

3"

2"

Fig. 4-8

Courtesy of Joseph K. Yabu, Ph.D., San Jose State University

Sheet size A

Scale $1\frac{1}{2}" = 1'-0"$

Continued

● USING PROBLEM-SOLVING SKILLS

Complete the following problems using problem-solving skills and your knowledge of AutoCAD.

1. A boat trailer manufacturer needs a new design for the rollers that facilitate raising a boat onto the trailer. The engineering division of your company, which designs the rollers, has proposed the design shown in Fig. 4-9. Create a drawing to be submitted to the trailer manufacturer for approval. Notice that the drawing is dimensioned. Experiment with the buttons on the status bar and see if you can figure out a way to draw the roller approximately to size. Do not dimension the drawing. Save the drawing as roller1.dwg.

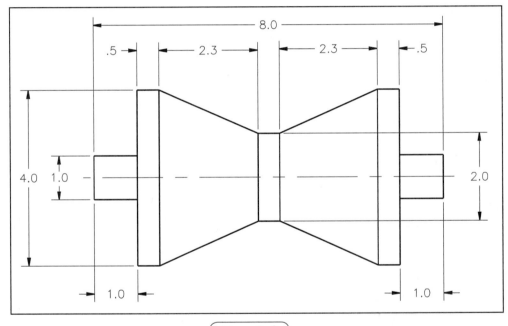

(**Fig. 4-9**)

2. The trailer manufacturer accepted the roller design with the following modification: the 4-inch overall diameter is to be increased to 5 inches. All other dimensions are to remain the same. Open roller1.dwg and make the change. Save the drawing as roller2.dwg.

Chapter 5

Basic Objects

Objectives

- Draw two-dimensional views of holes, cylinders, and rounded and polygonal features
- Create curved objects such as circles, arcs, ellipses, and donuts
- Create rectangles and other types of regular polygons

Vocabulary

concentric
control panels
diameter
donuts
ellipse
radial
radius
regular polygon
tangent
2D representation

Most products, buildings, and maps contain holes, rounded features, and other shapes that are circular in nature. Consequently, when drawing them with AutoCAD, you must use commands that produce circular and radial features. A **radial** feature is one in which every point is the same distance from an imaginary center point. Fig. 5-1 on page 50 shows a race car engine that contains many examples of circular and radial features. Technical drawings also include rectangles and other polygonal shapes, as shown in Fig. 5-1 and other drawings in this book.

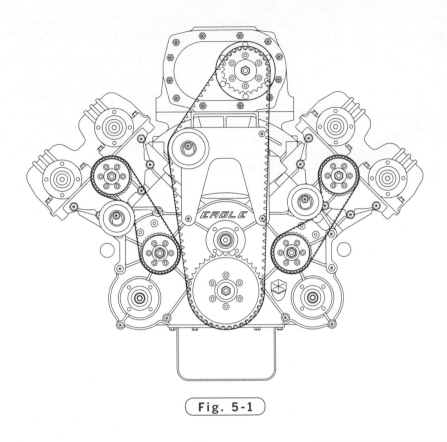

Fig. 5-1

Creating Circles

AutoCAD makes it easy to draw two-dimensional (2D) representations of holes, cylinders, and other round shapes. A **2D representation** is a single profile view of an object seen typically from the top, front, or side.

1. Start AutoCAD. If the AutoCAD window does not appear and you instead see the Startup dialog box, click the **Cancel** button in the dialog box.

2. If the Welcome Screen appears, click the **Close** button.

Remember to exit and restart AutoCAD if it was opened prior to Step 1 to reset all of the buttons to their default appearance. After this chapter, Steps 1 and 2 will be omitted from the procedures. You will be expected to know how to start AutoCAD and reach the main drawing window.

3. Pick the **Workspace** pull-down menu on the Quick Access toolbar and select **Drafting & Annotation**, even if it already appears as the current workspace.

4. Pick the **Save** or **Save As** button from the Quick Access toolbar, or select **Save** or **Save As...** from the Application Menu.

This causes the Save Drawing As dialog box to appear.

5. Open the folder with your name and highlight **Drawing1.dwg**, located to the right of File name.

6. Type **engine** in upper- or lowercase letters and pick **Save** or press **ENTER**.

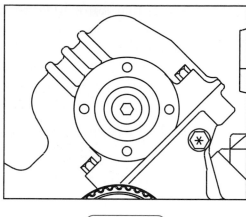

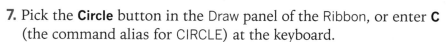
Fig. 5-2

This creates and stores a new drawing file named engine.dwg in your named folder.

The detail drawing in Fig. 5-2 shows one of the engine's cylinder heads. Let's create the center portion of the cylinder head. (For now, disregard the bolt at the center of the head. You will add it later in this chapter.)

First, turn your attention to the Ribbon. In the Drafting & Annotation workspace, the Ribbon contains **control panels**, or groupings, of buttons specifically for creating two-dimensional drawings.

DRAW

7. Pick the **Circle** button in the Draw panel of the Ribbon, or enter **C** (the command alias for CIRCLE) at the keyboard.

Notice the instruction on the Command line: Specify center point for circle. Other options are in brackets. These same instructions are displayed on the dynamic input prompt. The options are also displayed if you press the down arrow key.

8. Use the pointing device to pick a center point near the center of the drawing area.

9. Move the crosshairs and notice that you can "drag" the radius of the circle.

If the radius of the circle does not change as you move the crosshairs, a feature called DRAGMODE may be turned off. To turn it back on, enter DRAGMODE and then A (for Auto).

Notice that AutoCAD is requesting the radius or diameter of the circle. The **radius** of a circle is the length of a line extending from its center to any point on the circle. The **diameter** of a circle is the length of a line that extends from one side of a circle to the other and passes through its center.

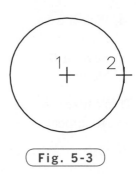

Fig. 5-3

10. Pick a point a short distance from the center, as shown in Fig. 5-3.

This completes the command and forms the innermost circular feature of the cylinder head.

DRAW

Concentric Circles

Now let's draw a concentric circle to represent the next feature. **Concentric** circles are circles that share a common center.

1. Reenter the **CIRCLE** command.

HINT Use one of the shortcuts to reenter the command.

2. Move the crosshairs to the center of the circle you created.

A small circle appears at the center of the circle. This is the Center object snap. Object snaps are like magnetic points that lock onto a specific point on an object easily and accurately. You will learn more about object snaps in Chapter 9.

3. While the small circle is present, pick a point.

AutoCAD snaps to the center point of the existing circle.

4. Drag the radius into place to create a second circle, as shown in Fig. 5-4. Review the detail of the cylinder head to approximate the size of the circle.

Fig. 5-4

5. Repeat Steps 1 through 4 to place the remaining concentric circles, as shown in the detail of the cylinder head in Fig. 5-2 (page 51).

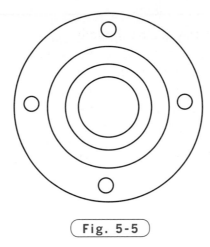

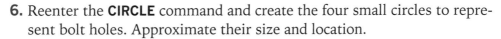
Fig. 5-5

Circle

DRAW

6. Reenter the **CIRCLE** command and create the four small circles to represent bolt holes. Approximate their size and location.

Your drawing should now look similar to the one in Fig. 5-5.

7. Save the drawing file.

Tangent Circles

Circles that touch at a single point are said to be **tangent**. The Ttr option of the CIRCLE command allows you to create a circle with a specified radius that is tangent to two other circles or arcs.

1. Select the **New** button from the Quick Access toolbar to begin a new drawing file.

QUICK ACCESS

This displays the Select template dialog box. This dialog box permits you to select from a number of template files. A template file contains pre-established settings and values that speed the creation of new drawings. You will learn more about them later in the book.

2. Select the **acad.dwt** file from the list of options and pick the **Open** button.

3. Create two circles and position them relatively close to one another.

4. Pick the down-facing arrow at the bottom of the **Circle** button, and select the **Tan, Tan, Radius** button from the pull-down menu.

Circle

DRAW

This enters the Tan, Tan, Radius option of the CIRCLE command. Other ways to enter this option is to reenter the CIRCLE command and type T at the keyboard or click on the Ttr (tan, tan, radius) option in brackets on the Command line.

5. Read the instructions on the Command line and then move the crosshairs over one of the two circles.

A small tangent symbol appears, showing that your pick will "snap" to a point that is tangent to the circle.

If the small tangent symbol does not appear, pick the Object Snap button on the status bar to turn object snap on and then try Step 5 again.

6. Pick a point on the first circle.

7. Move the crosshairs to the other circle and pick a point.

8. For the radius of the new circle, enter **2**.

AutoCAD creates a new circle that is tangent to the two circles. It's okay if the circle extends off the screen. If you see the message Circle does not exist on the Command line, repeat Steps 4 through 8, but use a larger radius value.

Creating Arcs

DRAW

The person who created the engine drawing shown in Fig. 5-1 used arcs to produce the radial features on some of the individual parts. In AutoCAD, the ARC command is often used to produce arcs.

1. Pick the **Arc** button on the Draw panel of the Ribbon or enter **A** at the keyboard.

This activates the ARC command. The default option for the ARC command is 3-Point.

2. Pick three consecutive points anywhere on the screen.

3. Reenter the **ARC** command and produce two additional arcs using the same method.

Options for Creating Arcs

DRAW

AutoCAD offers several options for creating arcs. The option you choose depends on the information you know about the arc. Suppose you want to create an arc for an engine part and you know the start point, the endpoint, and the radius of the arc.

1. Pick the down-facing arrow at the bottom of the **Arc** button and select the **Start, End, Radius** button from the pull-down menu.

2. Pick a start point, then pick an end point a short distance away.

3. Move the crosshairs and watch the arc form dynamically on the screen.

4. Enter **3** for the **Radius**.

An arc forms on the screen.

You can also create this arc using the 3-Point Arc command: before selecting the second point, enter E or pick the End option on the Command line for

End, and before selecting the third point, enter R or pick the Radius option on the Command line for Radius.

Suppose you don't know the radius of the arc or where the endpoint should be. However, you know where the center point of the arc should be, and you know how long the arc must be.

DRAW

5. Pick the down-facing arrow at the bottom of the **Arc** button and select the **Start, Center, Angle** option from the pull-down menu.

6. Pick a start point, then pick a center point for the arc.

7. Enter a number (positive or negative) up to 360. (The number specifies the angle in degrees.)

An arc forms on the screen. You can also create this arc using the 3-Point Arc command and the Command line prompt by entering C or picking the Center option on the Command line for Center before selecting the second point, and entering A or picking the Angle option on the Command line for Angle before selecting the third point.

8. Experiment with the other options on your own.

As you can see, AutoCAD's options allow you to create an arc accurately using the information that is available to you. Understanding these options can save time when you need to produce accurate engineering drawings that contain circular or radial features.

Series of Tangent Arcs

In some cases, you may need to draw a series of arcs that are connected to one another. The Continue option of the ARC command provides an efficient way to draw such a series.

The line in Fig. 5-6 is really a series of tangent arcs. This line was developed for a map. It represents a stretch of Highway 34 through Rocky Mountain National Park, west of Estes Park, Colorado. Let's reproduce the map line to practice creating tangent arcs.

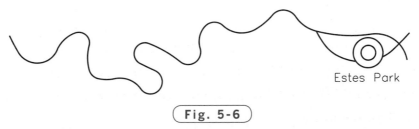

Estes Park

Fig. 5-6

1. Enter the **ARC** command and create an arc using the **3-Point** option.

2. Reenter the **ARC** command by pressing the spacebar or **ENTER**; press the spacebar or **ENTER** a second time.

DRAW

Notice how another arc segment develops as you move the crosshairs.

3. Create the next arc segment.

Notice that the new arc is tangent to the first one. (The point of tangency is the point at which the two arcs join.)

4. Repeat Steps 2 and 3 until you have finished the road.

5. Create the concentric circles that mark Estes Park, but omit the text.

You can also enter the Continue option of the Arc command by selecting the Continue button on the Arc pull-down menu.

Drawing Ellipses

The ELLIPSE command enables you to create mathematically correct ellipses. An **ellipse** is a regular oval shape that has two centers of equal radius. Ellipses are used to construct shapes that become a part of engineering drawings. For example, the race car engine includes a partial ellipse, as shown in Fig. 5-7.

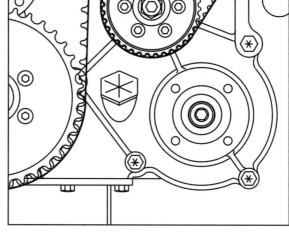

Fig. 5-7

DRAW

1. Pick the down-facing arrow to the right of the **Ellipse** button on the Ribbon and select the **Axis, End** option from the pull-down menu. Or, enter the **ELLIPSE** command at the keyboard.

2. Pick a point for an endpoint of the first axis, as shown in Fig. 5-8.

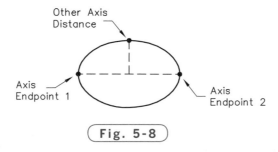

Fig. 5-8

3. Pick a second point for the other endpoint of the first axis directly to the right or left of the first point.

4. Move the crosshairs and watch the ellipse develop. Pick a point or enter a numeric value, such as **.3**, and the ellipse will appear.

5. Experiment with the **Center** option on your own. How might this option be useful?

Elliptical Arcs

Many parts that include elliptical shapes do not use the entire ellipse. To draw these shapes, you can use the ELLIPSE Arc option.

1. Pick the down-facing arrow to the right of the **Ellipse Axis, End** button and select the **Elliptical Arc** button.

2. Pick three points similar to those shown in Fig. 5-8.

A temporary ellipse forms.

DRAW

3. In reply to Specify start angle, pick a point anywhere on the ellipse.

4. As you slowly move the crosshairs counterclockwise, notice that an elliptical arc forms.

5. Pick a point to form an elliptical arc.

Donuts

The DONUT command allows you to create thick-walled or solid circles, known in AutoCAD as **donuts**. Drafters commonly use donuts to represent features on a machine part, architectural drawing, or map.

1. Click on the Draw panel title bar on the Ribbon to display the expanded panel, then select the **Donut** button.

DRAW

2. Specify an inside diameter of **.4** …

3. … and an outside diameter of **.75**.

The outline of a small donut locks onto the crosshairs and is ready to be dragged and positioned by its center.

4. Place the donut anywhere in the drawing by picking a point.

5. Move the crosshairs away from the new solid-filled donut and notice the Command line at the bottom of the screen.

6. Place several additional donuts in the drawing.

7. Press **ENTER** to terminate the command.

Rectangles

One of the most basic shapes used by drafters and designers is the rectangle. You can create rectangles using the LINE command, but doing so has some disadvantages. For example, you would have to take the time to make sure that the corner angles are exactly 90°. Also, each line segment would be a separate object, which would make it difficult to modify the rectangle at a later time. Therefore, AutoCAD provides the RECTANG command, which allows you to create a rectangle with perfect corners and as a single object. Clicking anywhere on a rectangle created with the RECTANG command selects the entire rectangle.

DRAW

1. On the Draw panel of the Ribbon, pick the **Rectangle** button.

As you can see, AutoCAD is asking for the first corner of the rectangle. Other options appear in brackets at the Command line.

2. In reply to Specify first corner point, pick a point at any location.

3. Move the crosshairs in any direction and notice that a rectangle begins to form.

4. Pick a second point at any location to create the rectangle.

5. Create a second rectangle. Since the **RECTANG** command was just entered, reenter it by pressing the spacebar or **ENTER**, or right-click and pick **Repeat RECTANG** from the shortcut menu.

6. Create a third rectangle.

7. Close the drawing without saving it.

Polygons

The POLYGON command enables you to create regular polygons with 3 to 1024 sides. A **regular polygon** is one with all sides of equal length. Using the POLYGON command, let's insert the bolt head into the engine drawing you started earlier in this chapter.

1. If the engine.dwg file is not open, pick the **Open** button on the abbreviated Quick Access toolbar and open it.

2. Pick the down-facing arrow to the right of the **Rectangle** button and select the **Polygon** button. You can also enter the **POL** alias at the keyboard.

DRAW

Notice the <4> at the end of the AutoCAD prompt. This is the default value, meaning that if you were to press ENTER now, AutoCAD would enter 4 in reply to Enter number of sides. To represent the bolt head shown in Fig. 5-2 on page 51, you will need a hexagon (six-sided polygon). ·

3. Enter **6**.

AutoCAD now needs to know if you want to select the center point or define an edge of the polygon. Let's specify a center.

4. Move the pointing device over any of the larger circles in the drawing. When the small circle symbol appears at the center of the cylinder head, pick a point to select the center of the circle as the center point of the hexagon. (The Object Snap button on the status bar must be depressed.)

AutoCAD allows you to create a polygon by inscribing it inside a circle of a specified diameter or by circumscribing it around the specified circle. Fig. 5-9 shows the difference.

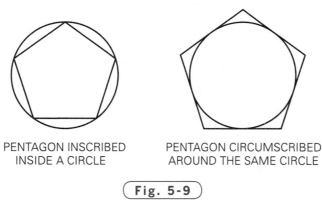

PENTAGON INSCRIBED
INSIDE A CIRCLE

PENTAGON CIRCUMSCRIBED
AROUND THE SAME CIRCLE

Fig. 5-9

5. Press **ENTER** to select the **I** (**Inscribed**) default value.

AutoCAD now wants to know the radius of the circle within which the polygon will appear.

6. With the pointing device, move the crosshairs from the center of the polygon and notice that a hexagon begins to form.

7. Pick a point to create the hexagonal bolt head at an appropriate size relative to the cylinder head. Refer to Fig. 5-2, if necessary.

NOTE To create the polygon with a more accurate size, you could have entered a specific numeric value, such as .5, at the keyboard. (Entering a 0 before the decimal point is optional.)

QUICK ACCESS

The drawing of the cylinder head component is now complete. Your drawing should look similar to the one in Fig. 5-10. The relative diameters of the circles, the sizes of the bolt heads relative to the circles, and the location and size of the four small bolt holes may be different in your drawing because we did not use exact sizes.

8. Save your work and exit AutoCAD.

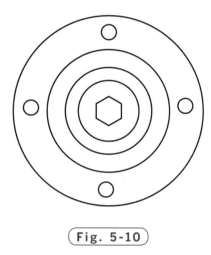

(Fig. 5-10)

HINT Back up your files regularly. The minute or two required to make a backup copy may save you hours of work. Experienced users back up regularly because they understand the possible consequences if they do not.

● REVIEW QUESTIONS

Answer the following questions on a separate sheet of paper.

1. Briefly describe the following methods of producing circles.

   a. 2 points

   b. 3 points

   c. tan tan radius

2. In which Ribbon control panel are the Arc and Circle buttons found?

3. What function does the Arc Continue option serve?

4. Explain the purpose of the DONUT command.

5. When you create a polygon using the Inscribed in circle option, does the polygon appear inside or outside the imaginary circle?

6. What are the practical differences between creating a rectangle using the LINE command and creating it using the RECTANG command? Why might you choose one command instead of the other?

7. In addition to the LINE and RECTANG commands, which other command is capable of creating a rectangle? When might you choose this method?

● CHALLENGE YOUR THINKING

These questions are designed to further your knowledge of AutoCAD by encouraging you to explore the concepts presented in this chapter. Answer each question on a separate sheet of paper.

1. Review the information in this chapter about specifying an angle in degrees. How might you be able to create an arc in a clockwise direction? Try your method to see if it works, and then write a paragraph describing the method you used.

2. Experiment further with the ARC command. Is it possible to create a single arc that has a noncircular curve using the options available for this command? (A noncircular curve is one in which not all the points are exactly the same distance from a common center point.) Explain your answer.

Continued

• APPLYING AUTOCAD SKILLS

Work the following problems to practice the commands and skills you learned in this chapter.

1. Complete the drawings shown in Figs. 5-11 and 5-12. Don't worry about text matter or exact shapes, sizes, or locations, but do try to make your drawings look similar to the ones in the figures. Save the drawings as prb5-1a.dwg and prb5-1b.dwg.

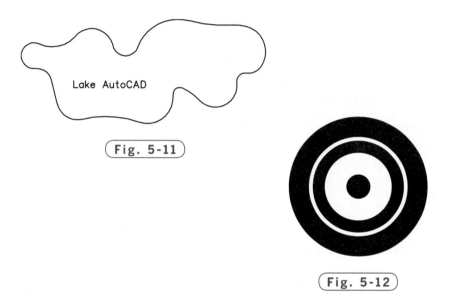

Lake AutoCAD

Fig. 5-11

Fig. 5-12

2. Use the LINE, POLYGON, CIRCLE, and ARC commands to create the hex bolt and nut shown in Fig. 5-13. Save the drawing as prb5-2.dwg.

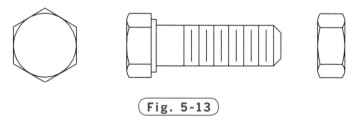

Fig. 5-13

Courtesy of Joseph K. Yabu, Ph.D., San Jose State University

3. Create the object shown in Fig. 5-14 by first drawing a six-sided polygon with the POLYGON command. Then draw the six-pointed star using the LINE command. Save the drawing as prb5-3.dwg.

4. In the same drawing, create the same object again, but draw the star using two three-sided polygons. Save your work.

Fig. 5-14

5. How would you draw a five-pointed star? Draw it in the same drawing file and save your work.

6. Draw a block with a rectangular cavity like the one shown in Fig. 5-15 using the RECTANG and LINE commands. Save the drawing file as prb5-6.dwg.

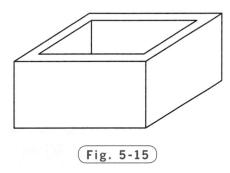

Fig. 5-15

7. The two objects shown in Fig. 5-16 are composed entirely of equal-sided and equal-sized polygons with common edges surrounding a central polygon. Can this be done with equal-sided and equal-sized polygons of any number of sides? Answer this question by trying to draw such objects using the POLYGON command with polygons of five, six, seven, and eight sides.

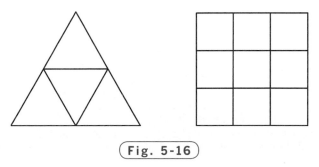

Fig. 5-16

Problems 3–7 courtesy of Gary J. Hordemann, Gonzaga University

Continued

8. Draw the screw heads shown in Fig. 5-17. Save the drawing as prb5-8.dwg.

9. Draw the eyebolt shown in Fig. 5-18 using all of the commands you have learned. Save the drawing as prb5-9.dwg.

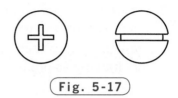

Fig. 5-17

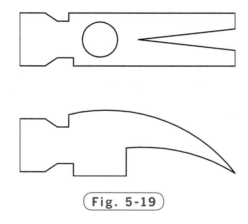

Fig. 5-18

10. Using the LINE, ARC, and CIRCLE commands, draw the two views of a hammer head shown in Fig. 5-19. Save the drawing as prb5-10.dwg.

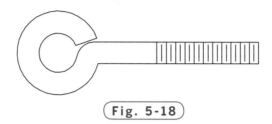

Fig. 5-19

11. Use the ARC and LINE commands to create the screwdriver shown in Fig. 5-20. Save the drawing as prb5-11.dwg.

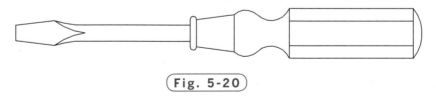

Fig. 5-20

Problems 9–11 courtesy of Joseph K. Yabu, Ph.D., San Jose State University

• USING PROBLEM-SOLVING SKILLS

Complete the following activities using problem-solving skills and your knowledge of AutoCAD.

1. Draw the ski-lift rocker arm shown in Fig. 5-21 for a design study by a gondola manufacturer. Do not include dimensions. Save your drawing as rockerarm.dwg.

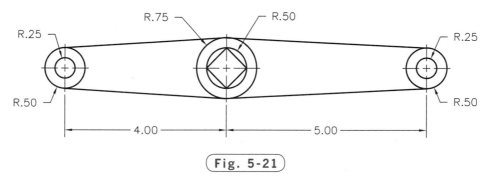

Fig. 5-21

2. The traditional mantle clock body drawing shown in Fig. 5-22 is necessary for the reconstruction of antique and collectible clocks. Draw the clock body. Do not include dimensions. Save your drawing as mantle.dwg.

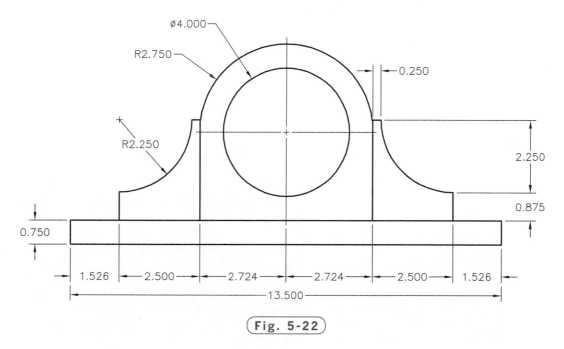

Fig. 5-22

3D Print Shop Coordinator

With more manufacturers accepting 3D data to drive computer-controlled equipment, drafters and engineers are moving rapidly toward designing in 3D. Designing in 3D has an important advantage over traditional methods. The 3D files can be sent to a 3D printer or additive manufacturing device to create a physical prototype. This prototype can undergo testing before the manufacturer goes to the trouble and expense of creating tooling for the product. Design changes can be made relatively inexpensively at this stage.

Photo Courtesy of Z Corporation

The equipment required to produce a 3D prototype is still expensive, although costs are coming down. As a result, many companies choose to outsource their prototyping needs. They create the 3D CAD files and send them to a 3D print shop for processing.

Types of 3D Printing

Several types of 3D printing devices are available. In addition to 3D printers, 3D scanners and freeform modeling systems are now being used. Many 3D print shops invest in several different types of printing devices. This gives the shop the flexibility to serve a wide range of customers, but it requires a skilled coordinator.

Coordinating a Print Shop

To coordinate a 3D print shop successfully, you will need to be familiar not only with the equipment in your shop, but also with the various CAD software programs used to create the data used by these machines. You will need to know which machines can handle which software. The best way to prepare for this career is to take drafting and modeling classes offered at area high schools, community colleges, and vocational schools. You will also need management experience and the ability to work well both with customers and with technical personnel.

▶ Career Activities

1. Find out if there are any 3D print shops in your area. What services do they offer?

2. Research on the Internet to learn more about the various types of 3D printing and the capabilities of each.

Chapter 6

Object Selection

Objectives

- Identify AutoCAD objects
- Select objects to create a selection set
- Add and remove objects to and from selection sets
- Erase and restore objects
- Resize objects parametrically

Vocabulary

entity
grips
noun/verb selection
object
pickbox
Quick Properties panel
selection previewing
selection set
selection window
verb/noun selection

W hen using AutoCAD, you will find yourself selecting objects so that you can move, copy, erase, or perform some other operation. Because drawings can become very dense with lines, dimensions, and text, it is important to use the most efficient methods of selecting these objects. Usually, the quickest way to move or copy an object is to click and drag. This is true whether you are using an inexpensive shareware program on a low-cost computer or sophisticated CAD software on an expensive computer. This chapter covers this and alternative methods of selecting objects and performing basic editing operations, such as resizing objects and erasing.

Objects

An **object**, also called an **entity**, is an individual predefined element in AutoCAD. The smallest element that you can add to or erase from a drawing is an object.

The following list gives examples of object types in AutoCAD.

3Dface	leader	solid
3Dsolid	line	spline
arc	mline	text
attribute	multiline text	tolerance
body	point	trace
circle	polyline	vertex
dimension	ray	viewport
ellipse	region	xline
image	shape	

You will learn more about these objects and how to use them in future chapters.

Selecting Objects

Basic object selection in AutoCAD is straightforward—you select the object by picking it with the left button on the pointing device. However, AutoCAD also offers more advanced object selection techniques. This chapter illustrates many of these techniques.

DRAW

DRAW

1. Start AutoCAD, pick the **Workspace** pull-down menu on the Quick Access toolbar, and select (or re-select) the **Drafting & Annotation** workspace.

2. Using the **LINE** and **CIRCLE** commands, draw the triangle and circle shown in Fig. 6-1.

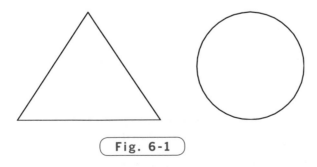

Fig. 6-1

 In this book, all new drawings use Imperial (English) units unless otherwise specified.

3. Enter the ERASE command by picking the **Erase** button from the Modify panel on the Ribbon or by entering **E** at the keyboard.

MODIFY

The AutoCAD prompt changes to Select objects. Notice that the crosshairs have changed to a small box called the **pickbox**. The pickbox is used to pick objects. At this point, you can also create a window to select points.

4. Pick a point anywhere to the lower left of the triangle.

5. Move the pickbox to the right and notice that a blue box forms.

This box defines the **selection window**.

6. Move the pickbox so that the triangle fits completely inside the window and pick a point.

The objects to be erased are highlighted with broken lines, as shown in Fig. 6-2. The highlighted objects make up the **selection set**.

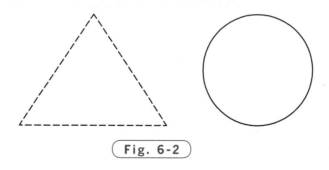

Fig. 6-2

7. Press **ENTER** to make the triangle disappear. (This also terminates the ERASE command.)

Restoring Erased Objects

What if you have erased an object by mistake and you want to restore it? Let's restore the triangle.

8. Pick the **Undo** button on the Quick Access toolbar.

QUICK ACCESS

The triangle reappears. The Undo button conforms to standard Windows functionality, so you may be familiar with it from other Windows applications. Picking this button repeatedly undoes your previous actions, one at a time, beginning with the most recent.

Using a Crossing Window

The Crossing object selection option selects all objects that cross the window boundary as well as those that lie completely within it.

MODIFY

1. Enter the **ERASE** command.
2. Type **C**, press **ENTER**, and pick a point to the left of the triangle.
3. Move the pickbox to the right to form a green box and notice that the box's border is made up of broken lines.
4. Form the box so that it crosses over at least one side of the triangle and press the pick (left) button in reply to Specify opposite corner.

Notice that AutoCAD selected those objects that cross over the box as well as any objects that lie completely within it.

> **HINT** You can also select objects using a crossing window without formally entering the Crossing option. When the Select objects prompt appears, simply pick two points to form a window, being sure to pick the *right-most* point first.

Polygonal Selection Options

The WPolygon and CPolygon object selection options are similar to the Window and Crossing options. However, they offer more flexibility because they are not restricted to the rectangular window shape used by the Window and Crossing options.

MODIFY

1. Fill the screen with several objects, such as lines, circles, arcs, rectangles, polygons, ellipses, and donuts.
2. Enter the **ERASE** command.
3. In reply to Select objects, enter **WP** for WPolygon. (W is for Window).
4. Pick a series of points that form a polygon of any shape around one or more objects. As you create the polygon, notice that the area inside the polygon is blue.
5. Press **ENTER** to close the polygon.

 When forming the polygon, AutoCAD automatically connects the last point to the first point.

AutoCAD selects those objects that lie entirely within the polygon.

6. Press **ENTER** to erase the objects.

The CPolygon option is similar to WPolygon.

MODIFY

7. Enter the **ERASE** command or press the spacebar.

8. In response to Select objects, enter **CP** for CPolygon. (C is for Crossing.)

9. Pick a series of points to form a polygon. Make part of the polygon cross over at least one object and press **ENTER**.

As you create the polygon, notice that its border is made up of broken lines, and that the area inside the polygon (the selection window) is green. The broken lines and green selection window indicate that the Crossing option is in effect. Solid lines around a blue selection window indicate a normal (not crossing) Window selection.

AutoCAD selects those objects that cross the polygon, as well as those that lie completely within it. As you can see, CPolygon is similar to the Crossing option, whereas WPolygon is similar to the basic Window option.

Using a Fence

The Fence option is similar to the CPolygon option, except that you do not close a fence as you do a polygon. When you select objects using the Fence option, AutoCAD looks for objects that touch the fence. Objects "inside" the fence are not selected unless they actually touch the fence.

MODIFY

1. Enter **ERASE**.

2. Enter **F** for Fence.

3. Draw a line that crosses over one or more objects and press **ENTER**.

4. Press **ENTER** to complete the erasure.

Selecting Single Objects

Suppose you want to select a single object, such as a line segment.

MODIFY

1. Recreate the triangle and circle you drew previously.

2. Pick the **Erase** button or enter **E** at the keyboard.

3. Using the pickbox, pick two of the line segments in the triangle.

AutoCAD automatically highlights objects as the pickbox rolls over them. This is called **selection previewing**. This feature is helpful when you need to select individual objects in a crowded area of the drawing. It allows you to see in advance which object will be selected when you click the button on the pointing device.

4. Press **ENTER** to complete the command and to make the selected objects disappear.

If you are using a pointing device that has more than one button, pressing one of them, usually the one on the right, is the same as pressing the ENTER key. This is normally faster than pressing ENTER at the keyboard.

Selecting the Last Object Drawn

The Last object selection option automatically selects the last object you drew. This option works whenever the Select objects prompt is present.

QUICK ACCESS

1. Pick the **Undo** button to restore the line segments.

Notice that because you erased both line segments in a single ERASE operation, you only have to pick the Undo button once to make both lines reappear.

MODIFY

2. Enter the **ERASE** command.

3. Type **L** (for Last) and press **ENTER**.

AutoCAD highlights the last object you drew.

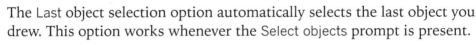

The last object you drew may be different from the last object you selected in the drawing. If you want to reselect the last object you selected, enter P (for Previous) instead of Last. The Previous option is especially useful when you need to reselect an entire group of objects.

4. Press **ENTER**.

If you continue to enter ERASE and Last, you will erase objects in the reverse order from which you created them. However, it is usually faster to use the Undo button for this purpose.

5. Pick the **Undo** button to restore the object you deleted using the Last option.

Selecting the Entire Drawing

The All object selection option permits you to select all of the objects in the drawing quickly.

1. Enter the **ERASE** command.

2. In reply to Select objects, type **All** and press **ENTER**.

AutoCAD selects all objects in the drawing.

3. Press **ENTER** again to erase all of the objects.

4. Pick **Undo** to restore the objects.

Editing Selection Sets

AutoCAD allows you to add and remove objects from a selection set while you are in the process of creating it. For example, if you accidentally pick an object that you did not mean to select, you can remove it without affecting the rest of the selected objects. This is especially helpful when you have many objects in a selection set on a complex drawing.

Removing Objects from a Selection Set

To remove an unwanted object from a selection set, you must activate the Remove objects mode.

1. Enter the **ERASE** command.

2. Create a selection window to select the triangle and the circle.

Both the triangle and the circle should now be highlighted. Suppose you have decided not to erase the circle.

3. Type **R** (for Remove) and press **ENTER**.

Notice that the AutoCAD prompt changes to Remove objects. You can now remove one or more objects from the selection set. The object(s) you remove from the selection set will not be erased.

4. Remove the circle from the selection set by picking it.

Note that it is no longer highlighted.

5. Press **ENTER** to complete the ERASE operation.

6. Pick the **Undo** button to restore the triangle.

Adding to the Selection Set

After you have used the Remove objects option to remove objects from a selection set, you can select additional objects to include in the set.

1. Repeat Steps 1 through 4 in the previous section ("Removing Objects from a Selection Set").

2. Instead of pressing ENTER, type **A** for Add and then press **ENTER**.

Notice that the AutoCAD prompt changes back to **Select objects**.

3. Select the circle.

4. Press **ENTER** to complete the ERASE operation.

So you see, you can add and remove objects as you wish until you are ready to perform the operation. The objects selected are indicated by broken lines. These selection procedures work not only with the ERASE command, but also with other commands that require object selection, such as MOVE, COPY, MIRROR, ARRAY, and many others.

5. Pick the **Undo** button to restore the objects to the screen.

QUICK ACCESS

Using Grips

Another method of selecting objects is to pick them using the pickbox located at the center of the crosshairs. When you pick an object without first entering a command, small boxes called **grips** appear at key points on the object. Using these grips, you can perform several basic operations such as moving, copying, or changing the shape or size of an object.

DRAW

1. Enter the **RECTANG** command and draw a rectangle of any size and at any location.

STATUS BAR

2. Find the Quick Properties button on the status bar. If it is off, pick it once to toggle **Quick Properties** on (blue).

3. Select any point on the rectangle. (No command should be entered.)

Notice that blue grips appear at midpoint of every side and at every corner of the rectangle. Also notice the floating panel that appears to the right of the rectangle. This is the **Quick Properties panel**. When turned on, Quick Properties displays some of the properties of selected objects. For this chapter, when the Quick Properties panel appears, close it by picking the x in the upper right corner.

Editing Size and Shape with Grips

4. Move the crosshairs until it locks over the lower right grip, but do not pick the grip yet.

AutoCAD displays the current length of the two lines that meet at the lower right grip and options associated with the STRETCH command, as shown in Fig. 6-3.

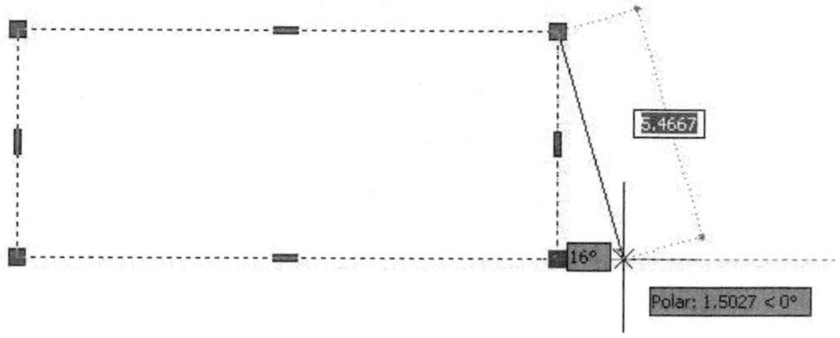

Fig. 6-3

5. Move the crosshairs to each of the other corner grips to see the current length of other sides of the rectangle.

6. Pick the lower right grip to select it.

The grip turns red, showing that you have selected it. The length of the right side of the rectangle is now displayed in an editable text box, as shown in Fig. 6-4.

Fig. 6-4

7. Move the crosshairs horizontally on the screen and notice how the lengths update dynamically as you move the grip around. See Fig. 6-5.

8. Pick a point to the right of the original grip.

The length of the right side of the rectangle changes and the shape is no longer rectangular.

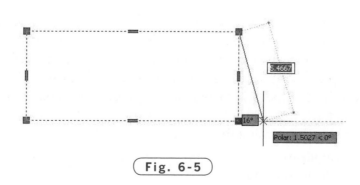

Fig. 6-5

Modifying Objects with Grips

1. Create another rectangle and select any one of the four corner grips.
2. Press the right button on the pointing device (right-click) to display a shortcut menu.
3. Pick **Move**, drag the rectangle to a new location, and pick a point.
4. Display the shortcut menu again and select other editing options, such as **Copy**, **Scale**, and **Rotate**, and complete these operations. Follow the instructions at the AutoCAD prompt.

Edge Grips

You can also modify objects using edge grips.

1. Create a new rectangle, select it, and move the crosshairs until it locks over one of the four edge grips that lie on the midpoints of its sides, but do not pick the grip yet.

AutoCAD displays a short menu of options: Stretch, Add Vertex, and Convert to Arc.

2. Pick the **Stretch** option, and move the crosshairs to stretch the rectangle. Click once to end the Stretch command.
3. Move the crosshairs over the edge vertex again, and select the **Convert to Arc** option.
4. Move the crosshairs and notice that the arc forms dynamically on the screen. Click once to define the arc's radius.

You can pick the radius of the arc with the pointing device, or type a value at the keyboard.

Modifying the Grips Feature

For most AutoCAD users, the standard grips characteristics, such as color and size of the grip boxes, are adequate. In some companies or circumstances, however, it may be necessary to adjust the grips so that you can work with them more easily. For example, if your company has customized the background of the drawing area to dark gray, you may want to change the color of unselected grip boxes from blue to yellow so that you can see them more easily. You may need to adjust the size of the grip boxes depending on your display resolution. Grip boxes viewed at 1600×1200 resolution are twice as small as those viewed at 800×600. AutoCAD allows you to make changes such as these using the Options dialog box.

1. Pick the **Application** button; then select the **Options** button at the bottom of the Application Menu.

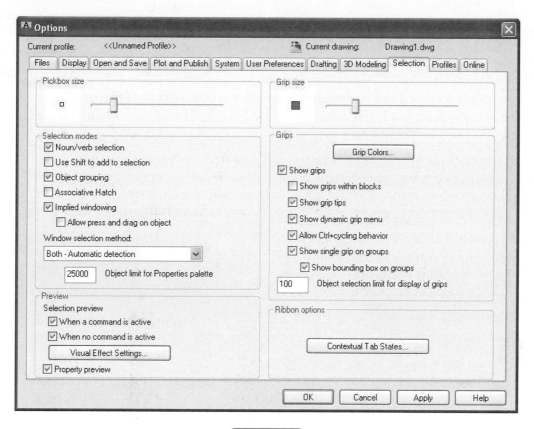

Fig. 6-6

The Options dialog box includes ten tabs with important information and settings in each one.

2. Pick the **Selection** tab.

Information and settings related to the grips feature appear in the right half of the dialog box, as shown in Fig. 6-6.

3. In the area labeled Grip Size, move the slider bar and watch how it changes the size of the box to the left of it.

4. Increase the size of the grips box and pick the **OK** button.

5. Select one of the objects and notice the increased size of the grips.

6. Pick one of the grips and perform an editing operation.

7. Display the Options dialog box and adjust the size of the grips box to its original size.

The Selection tab also permits you to enable and disable the grips feature, as well as control the assignment of grips within blocks. (You will learn more about blocks in Chapter 32.) You can also make changes to the color of selected and unselected grips. For now, do not change these settings.

8. Close the Options dialog box.

Using Grips with Commands

The traditional AutoCAD method of entering a command and then selecting the object to be edited is sometimes called **verb/noun selection**. First you tell the software what to do (using a verb such as ERASE) and then you select the object to be acted upon (using a noun such as LINE).

When you use grips with editing commands, you are using **noun/verb selection**. See Fig. 6-7. First you select the object, and then you tell the software what to do with the object. You may find the grips (noun/verb) method convenient to use, especially if you are familiar with other graphics programs that use this technique. However, either method is acceptable. The method you choose is a matter of personal preference.

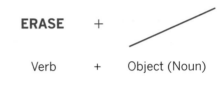

ERASE   +

Verb   +   Object (Noun)

VERB/NOUN SELECTION

MODIFY

1. Select any object in your drawing.
2. Pick the **Erase** button or enter **E** for ERASE.

As you can see, AutoCAD erased the object without first asking you to select objects. You have just used the noun/verb technique.

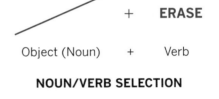

+   **ERASE**

Object (Noun)   +   Verb

NOUN/VERB SELECTION

(Fig. 6-7)

3. Pick another object and press the **Delete** key on the keyboard.

As you can see, pressing the Delete key is the same as entering the ERASE command.

QUICK ACCESS

4. Pick the **Undo** button from the Quick Access toolbar twice to undo the deletions.
5. Exit AutoCAD without saving.

Chapter 6 Review & Activities

• REVIEW QUESTIONS

Answer the following questions on a separate sheet of paper.

1. In AutoCAD, what is an object? Give three examples.

2. After you enter the ERASE command, what does AutoCAD ask you to do?

3. Experiment with and describe each of these object selection options.

   a. Last

   b. Previous

   c. Crossing

   d. Add

   e. Remove

   f. Undo

   g. WPolygon

   h. CPolygon

   i. Fence

   j. All

4. How do you place a window around one or more objects during object selection?

5. If you erase an object by mistake, how can you restore it?

6. How can you retain part of what has been selected for erasure while remaining in the ERASE command?

7. What is the fastest way of erasing the last object you drew?

8. When you use a graphics program, such as AutoCAD, what is usually the quickest way to edit an object?

9. Describe how you would remove an object from a selection set.

10. Explain the primary benefit of using the grips feature.

11. What editing functions become available to you when you use grips?

12. Explain how to copy an object using grips.

13. Explain the difference between the noun/verb and verb/noun selection techniques.

Continued

● CHALLENGE YOUR THINKING

These questions are designed to further your knowledge of AutoCAD by encouraging you to explore the concepts presented in this chapter. Answer each question on a separate sheet of paper.

1. As you have seen, AutoCAD's grips feature is both easy and convenient to use. With that in mind, discuss possible reasons AutoCAD also includes specific commands that you must enter to perform some of the same functions you can do easily with grips.

2. Describe a situation in which the Remove option can be useful for more than just removing an object that was added to a selection set by mistake.

● APPLYING AUTOCAD SKILLS

Work the following problems to practice the commands and skills you learned in this chapter.

1. Create a new drawing. Draw a circle and a rectangle as illustrated in Fig. 6-8. Use the RECTANG command to create the rectangle. Then enter the ERASE command and try to erase one line of the rectangle. At another location in the drawing, create another rectangle, but this time use the LINE command. Try to erase one line of the second rectangle. What is the difference? Do not save the drawing.

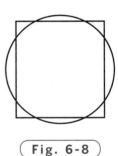

Fig. 6-8

2. Create a new drawing. Use the LINE command to draw polygons *a* through *e* as shown in Fig. 6-9. Enter the ERASE command and use the various object selection options to accomplish the following operations without pressing ENTER.

   • Place a window around polygon *a*.

   • Pick two of polygon *b*'s lines for erasure.

   • Use the Crossing option to select polygons *c* and *d*.

   • Remove one line selection from polygon *c* and one from polygon *d*.

   • Pick two lines from polygon *e* for erasure, but then remove one of the lines so it won't be erased.

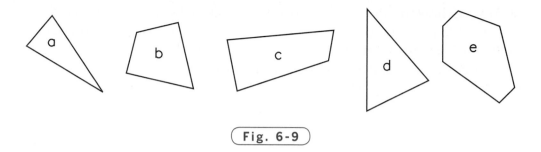

Fig. 6-9

After you have completed the five operations, press ENTER. You should have nine objects (line segments) left on the screen. Save the drawing as prb6-2.dwg in the folder with your name, but do not close the file.

3. Select Save As... in the Application Menu and enter prb6-3.dwg for the file name. Restore the line segments that you deleted in problem 2. Then use grips to rearrange the polygons in order of size from smallest to largest. Save your work.

● USING PROBLEM-SOLVING SKILLS

Complete the following activities using problem-solving skills and your knowledge of AutoCAD.

1. The computer sketch in Fig. 6-10 was sent to the project engineer for comments and has now been routed to your desk with the attached note. Use the ELLIPSE, ARC, and LINE commands to create the sketch of the bushing. Because this is only a rough sketch, you may approximate the dimensions. Make the change requested by the engineering department. Save the drawing as ch6-bushing.dwg.

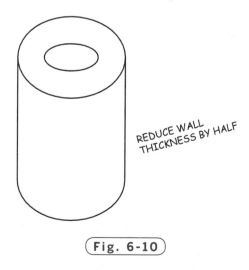

REDUCE WALL THICKNESS BY HALF

Fig. 6-10

Continued

2. Your architectural firm's customer wants the gable-end roof changed to a hip roof, as shown in Fig. 6-11. Draw the front elevation with both roof lines. Since this is only a roof line change, approximate the size and location of the windows and door. Save the drawing as ch6-roof.dwg.

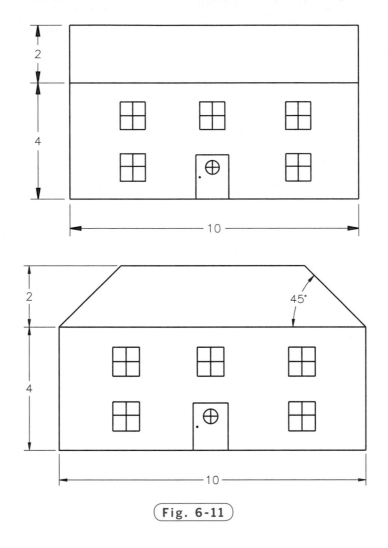

Fig. 6-11

Chapter 7

Entering Coordinates

Objectives

- Describe several methods for entering coordinates
- Locate points and draw objects using the Cartesian coordinate system
- Apply the absolute, relative, polar, and polar tracking methods for entering coordinates

Vocabulary

absolute points
alignment paths
Cartesian coordinate system
coordinate pair
coordinates
origin
polar method
polar tracking method
relative method
world coordinate system (WCS)

Sets of numbers used to describe specific locations on a grid are called **coordinates**. In AutoCAD, coordinates are used to specify locations of objects in a drawing. In previous chapters, when you used the pointing device to pick the endpoints of lines, etc., AutoCAD recorded the coordinates of the points that you picked on the screen.

You can also use the keyboard to enter coordinates. The keyboard method allows you to specify points and draw lines of any specific length and angle. This method also applies to creating arcs, circles, and other objects. By specifying the coordinates of objects, you can achieve the accuracy that is required for engineering drawings.

AutoCAD uses a **Cartesian coordinate system** similar to that used in geometry. In AutoCAD, this system is called the **world coordinate system (WCS)**. In the two-dimensional version of this system, two axes are used. See Fig. 7-1. This is the version used for most drafting work. The X axis is an infinite, imaginary line that runs horizontally. The Y axis is an infinite, imaginary line that runs vertically. Each axis is numbered with sequential positive and negative numbers. Each combination of one x value and one y value (a **coordinate pair**) specifies a unique point in the coordinate system.

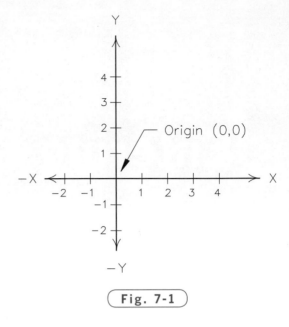

Fig. 7-1

The point at which the axes cross is known as the **origin**. At the origin, the values of x and y are both zero. The values to the right and above the origin are positive. Those below and to the left of the origin are negative.

In three-dimensional modeling, a third axis—the Z axis—is added at right angles to the other two axes to allow you to specify the third dimension, depth.

Methods of Entering Coordinates

The Cartesian system does not limit you to entering coordinate pairs to specify points. Consider the following ways to specify points using the LINE command. Note that dynamic input must be off for the following methods to work as described.

Absolute Method

When you enter specific x and y values, AutoCAD places the points according to the Cartesian coordinate system. Points entered in this way are considered to be **absolute points**. See Fig. 7-2.

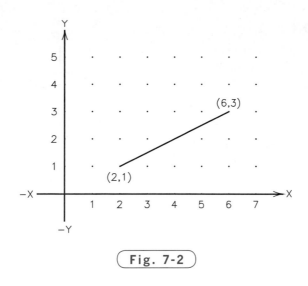

Fig. 7-2

Example:

    LINE Specify first point: 2,1
    Specify next point or [Undo]: 6,3

This begins the line at absolute point 2,1 and ends it at absolute point 6,3.

Relative Method

The **relative method** allows you to enter points based on the position of a point that has already been defined. AutoCAD determines the position of the new point relative to a previous point, as shown in Fig. 7-3. Notice the use of the @ symbol in the following example. Using this symbol tells AutoCAD that the following coordinate pair should be read relative to the previous point.

Example:

    LINE Specify first point: 2,1
    Specify next point or [Undo]: @2,0

This draws a line 2 units in the positive X direction and 0 units in the Y direction from absolute point 2,1. In other words, the distances 2,0 are relative to the location of the first point.

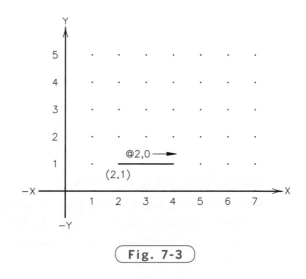

Fig. 7-3

Polar Method

Another form of relative positioning is the **polar method**. Using this method, you can produce lines at precise angles. The area surrounding a point is divided into 360 degrees, as shown in Fig. 7-4. If you specify an angle of 60 degrees, for example, the line extends upward from the last point, as shown in Fig. 7-5. Notice the use of the < symbol to specify the angle.

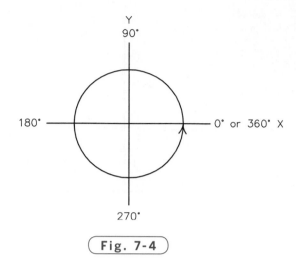

Fig. 7-4

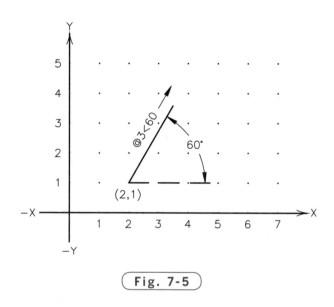

Fig. 7-5

Example:

```
LINE Specify first point: 2,1
Specify next point or [Undo]: @3<60
```

This produces a line segment 3 units long at a 60° angle. The line begins at absolute point 2,1 and extends for 3 units at a 60° angle.

Polar Tracking Method

AutoCAD also provides the **polar tracking method**, which is similar to the polar method, except that it's faster and easier. With this method, you can produce lines at precise angles and lengths with minimal keyboard entry. See Fig. 7-6.

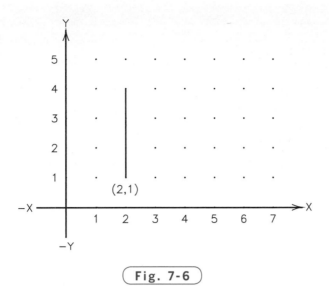

Fig. 7-6

Polar tracking causes AutoCAD to display temporary **alignment paths** at prespecified angles to help you create objects at precise positions and angles. When the cursor follows an alignment path, AutoCAD assumes that the alignment path defines the direction of the line. Polar tracking is on when the Polar Tracking button on the status bar is toggled on.

Example:

LINE Specify first point: 2,1
Specify next point or [Undo]: Use
the 90° alignment path that
appears automatically on the
screen and enter 3.

This produces a line segment
3 units long at a 90° angle. The
line begins at absolute point 2,1
and extends 3 units upward
vertically.

The alignment paths appear at 90° angles by default. You can change the angle of the alignment paths to meet your needs for specific drawing tasks. The following steps guide you through this process.

1. Start AutoCAD, pick the **Workspace** pull-down menu on the Quick Access toolbar, and select (or re-select) the **Drafting & Annotation** workspace.

2. Find the **Polar Tracking** button on the left side of the status bar, right-click, and select **Settings...** from the shortcut menu.

STATUS BAR

Fig. 7-7

The Drafting Settings dialog box appears, as shown in Fig. 7-7.

Line

DRAW

3. In the Polar Angle Settings area of the dialog box, pick the down arrow and change the increment angle to **30**.

4. Pick the **OK** button to close the dialog box.

5. Enter the **LINE** command and pick a point anywhere in the drawing area.

6. Move the crosshairs away from and around the point.

At 30° increments around the point, AutoCAD displays an alignment path and a small box showing the distance and angle from the point.

7. Position the crosshairs so that they are at a 60° (1 o'clock) position from the point and enter **2**.

AutoCAD produces a line 2 units long at a 60° angle.

8. Use the Polar Tracking feature to produce additional line segments.

9. Erase everything in the drawing file.

10. Display the Drafting Settings dialog box, set the polar tracking increment angle to **90**, and pick the **OK** button.

Entering Coordinates with Dynamic Input

When dynamic input is on, entering coordinates at the keyboard produces different results than the methods we have just described. This is because

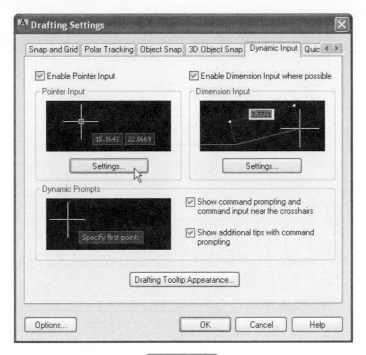

Fig. 7-8

dynamic input uses preset default formats for coordinate entry. Dynamic input is on when the Dynamic Input button on the status bar is toggled on.

1. Display the Drafting Settings dialog box once again and pick the **Dynamic Input** tab.

2. Select the **Settings...** button in the Pointer Input area, as shown in Fig. 7-8.

This displays the Pointer Input Settings dialog box. As you can see, when using dynamic input, AutoCAD forces you to choose between polar and Cartesian for the default format and between absolute and relative methods for coordinate entry.

When dynamic input is on, AutoCAD uses these default settings during coordinate entry. Using the @ and < symbols with dynamic input on will produce unpredictable results.

3. Pick the **Cancel** button in the Pointer Input Settings dialog box.

4. Pick the **Cancel** button in the Drafting Settings dialog box.

Applying Methods of Coordinate Entry

Some methods of coordinate entry work better than others in a given situation. The best way to become familiar with the various methods of coordinate entry, and which to use when, is to practice using them.

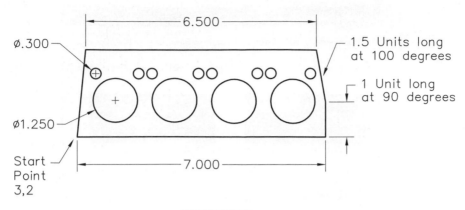

6.500

Ø.300

1.5 Units long
at 100 degrees

1 Unit long
at 90 degrees

Ø1.250

Start
Point
3,2

7.000

Fig. 7-9

STATUS BAR

1. Turn dynamic input off by releasing the **Dynamic Input** toggle button on the status bar.

2. Enter the **LINE** command.

3. Using the methods presented in the chapter, create the drawing of the gasket shown in Fig. 7-9. Don't worry about the exact locations of the holes, and don't dimension the drawing. However, do make the drawing precisely the size shown. The top and bottom edges of the gasket are perfectly horizontal. (You may need to enter **Z** and **A** to ZOOM All. This makes the gasket larger on the screen. You will learn more about ZOOM All later in the book.)

As you draw the holes in the gasket, keep in mind that the Ø (*theta*) symbol indicates that the dimensions are given as diameters.

You can enter negative values for line lengths and angles.

    LINE Specify first point: 5,5
    Specify next point or [Undo]: @2<–90

This polar point specification produces a line segment 2 units long downward vertically from absolute point 5,5. This is the same as entering @2<270. Try it and then erase the line.

ve your work in a file named gasket.dwg and exit AutoCAD.

rk

● REVIEW QUESTIONS

Answer the following questions on a separate sheet of paper.

1. What is the relationship between AutoCAD's world coordinate system and the Cartesian coordinate system?

2. Briefly describe the differences among the absolute, relative, polar, and polar tracking methods of point specification.

3. Explain the advantage of specifying endpoints at the keyboard rather than with the pointing device.

4. What is an advantage of specifying endpoints with the pointing device rather than at the keyboard?

5. What is the effect on the direction of the new line segment when you specify a negative number for polar coordinate entry?

6. In your own words, describe the ways dynamic input might change coordinate entry.

● CHALLENGE YOUR THINKING

These questions are designed to further your knowledge of AutoCAD by encouraging you to explore the concepts presented in this chapter. Answer each question on a separate sheet of paper.

1. Why may entering absolute points be impractical much of the time when you are completing drawings?

2. Consider the following command sequence:

   LINE Specify first point: 4,3
   Specify next point or [Undo]: 5,1

   What polar coordinates could you enter at the Specify next point or [Undo] prompt to achieve exactly the same line?

3. When might the polar tracking feature be useful?

4. When might the default settings of dynamic input be useful?

Continued

• APPLYING AUTOCAD SKILLS

Work the following problems to practice the commands and skills you learned in this chapter. Turn dynamic input off and ZOOM All at the beginning of each drawing.

1. Fig. 7-10 shows the end view of a structural member for an architectural drawing. Create the rectangle using the LINE command and the following keyboard entries.

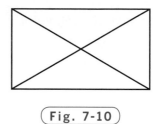

Fig. 7-10

Specify first point: 1,1

Specify next point or [Undo]: 10.5, 1

Specify next point or [Undo]: @0,5.5

Specify next point or [Close/Undo]: @9.5<180

Specify next point or [Close/Undo]: c

Draw a line from the intersection at the lower left corner of the rectangle to the upper right corner of the rectangle. Use coordinate pairs to place the endpoints of the diagonal line exactly at the corners of the triangle. The new line defines the rectangle's diagonal distance, or the hypotenuse of each of the two triangles you have created. Then draw a line from the lower right corner to the upper left corner of the rectangle to create the other diagonal. Save the drawing as prb7-1.dwg.

2. Draw the $3^1/_2''$ disk shown in Fig. 7-11 with dynamic input on. If necessary, temporarily change the default settings for dynamic input in the Pointer Input Settings dialog box. Save the drawing as prb7-2.dwg.

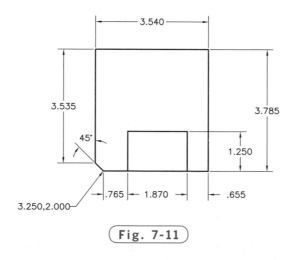

Fig. 7-11

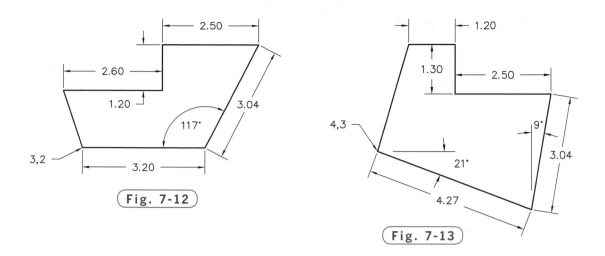

Fig. 7-12

Fig. 7-13

3-4. List on a separate sheet of paper exactly what you would enter when using the LINE command to produce the drawings of sheet metal parts shown in Figs. 7-12 and 7-13. Then step through the sequence in AutoCAD. Try to use four methods—absolute, relative, polar, and polar tracking—to enter the points. Note that the horizontal lines in the drawings are perfectly horizontal. Be sure to ZOOM All at the beginning of each drawing. Save the drawings as prb7-3.dwg and prb7-4.dwg.

5. Draw the front view of the spacer plate shown in Fig. 7-14. Start the drawing at the lower left corner by entering a first point using absolute coordinates of 3,3. Determine and use the best methods of coordinate entry to draw the spacer plate. Save the drawing as prb7-5.dwg.

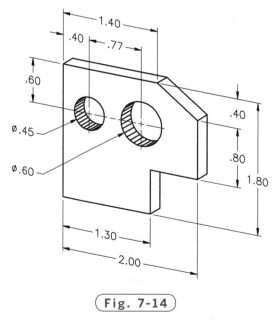

Fig. 7-14

Problem 5 courtesy of Gary J. Hordemann, Gonzaga University

Continued

6. Draw the front view of the locking end cap shown in Fig. 7-15. Start the drawing at the center of the object using absolute coordinates of 5,5. Use absolute coordinates to locate the centers of the four holes, and use polar tracking with the RECTANG command to draw the rectangular piece. Save the drawing as prb7-6.dwg.

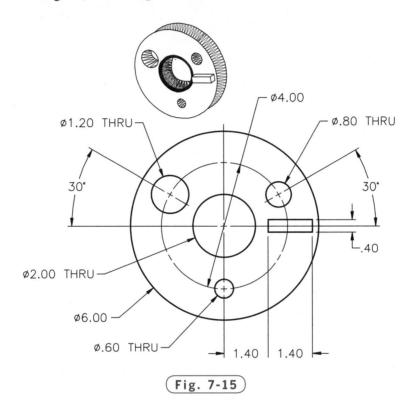

Fig. 7-15

• USING PROBLEM-SOLVING SKILLS

Complete the following activities using problem-solving skills and your knowledge of AutoCAD.

1. Draw the shim shown in Fig. 7-16 with dynamic input off. Decide on the best method of coordinate entry for each point, based on the dimensions shown and your knowledge of the various methods. Position the drawing so that the lower left corner of the shim is at absolute coordinates 1,1. Do not include the dimensions. Save the drawing as ch7-shim.dwg.

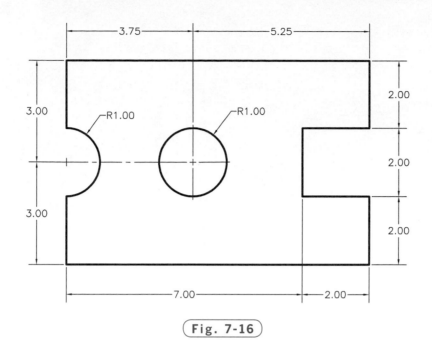

Fig. 7-16

2. Draw the front view of the slider block shown in Fig. 7-17. Assume the spacing between dots is .125. Start the drawing at the lower left corner using absolute coordinates of 1,3. Use the RECTANG command to draw the rectangular slot. Use relative polar coordinates to locate the endpoints of the lines, and absolute rectangular coordinates to locate the centers of the three holes. Save the drawing as ch7-block.dwg.

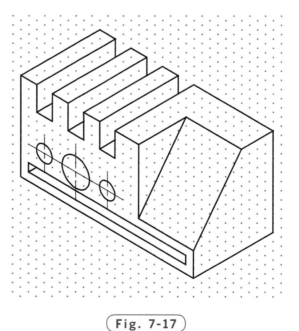

Fig. 7-17

Help and File Maintenance

Objectives

- Obtain help on AutoCAD commands and topics
- Use AutoCAD's content-sensitive help features
- Copy, rename, move, and delete drawing files
- Create and delete folders
- Audit and recover damaged drawing files

Vocabulary

audits
context-sensitive help
file attributes

AutoCAD provides a rich mix of online help and file-searching capabilities. Its *Learning Resources, Tutorials,* and *Sample Files* are no more than a few clicks away no matter what you are doing in AutoCAD. Autodesk, the maker of AutoCAD, also allows users to download AutoCAD Help so it can be accessed offline. Online assistance on how you can obtain consulting, training, and support is also available.

When you work with AutoCAD, there will be times when you want to browse and review the contents of a folder or delete, rename, copy, or move files. This chapter gives you practice in using these functions. The chapter also discusses commands that attempt to repair drawing files containing errors.

Obtaining Help

The AutoCAD software offers many forms of help. Most of these can be obtained by entering the HELP command.

1. Start AutoCAD and reselect the **Drafting & Annotation** workspace.
2. Find and open the **engine.dwg** file.

The cylinder head component that you created in Chapter 5 should now be visible on the screen.

INFO CENTER

3. Pick the **Help** button on the right side of the InfoCenter, or enter **HELP** or **?** at the keyboard.

 You can also obtain help by pressing the F1 function key.

The AutoCAD Help window opens as shown in Fig. 8-1.

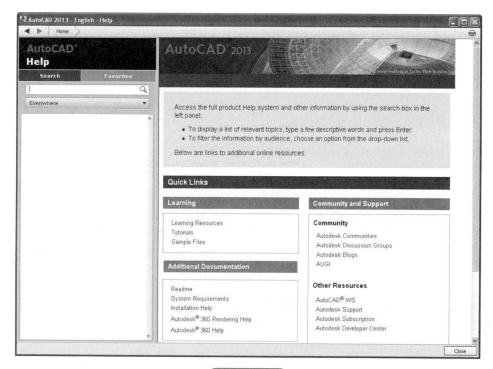

Fig. 8-1

Using the Help Window

 This section assumes that you have Internet access on the computer that is running AutoCAD. If you do not have Internet access, talk with your instructor about connecting to the Internet or downloading AutoCAD Offline Help.

4. Focusing your attention on the left side of the Help window, click in the Search edit box in the upper area of the window, type **move objects**, and press **ENTER**.

AutoCAD lists all help topics that contain the words *move* or *objects*.

5. Use the scroll bar to move through the topics.

6. Locate and click on the item named **About Moving Objects**.

7. Scroll down the right side of the screen to see the entire discussion. Note the links at the bottom for related tasks, references, and concepts.

8. Pick **To Move an Object Using a Displacement** in the Related Tasks section.

9. Pick **Commands for Moving and Rotating Objects** in the Related References section.

Notice the navigation aids in the upper left corner of the Help window, including the back and forward arrows.

10. Click the back arrow twice to return to the About Moving Objects topic.

As you can see, you can easily navigate and search the valuable information available in the Help window.

11. Close the Help window.

Using the Help Quick Links

Additional resources are available in the Quick Links area of the Help window.

1. Pick the **Help** button, enter **HELP** or **?** at the keyboard, or press the **F1** function key to open the Help window again.

The Quick Links area contains links for Learning, Additional Documentation, and Community and Support. It also provides a link to download Help so it can be accessed offline.

2. Click on the **Learning Resources** link under Learning.

A browser window titled AutoCAD Services & Support appears.

3. Click on the **Tutorials** link under Learn the Basics and review the video tutorials available.

As you can see, Autodesk offers many tutorials to install, learn, and understand the features of AutoCAD.

4. Click on the **Support** link on the left side of the AutoCAD Services & Support browser window.

The Support window provides access to additional resources, including product information, troubleshooting, discussion groups, error reporting, and training videos.

5. Explore the AutoCAD Services & Support window on your own, including the Documentation, Training, Community, and Consulting areas.

6. Close the AutoCAD Services & Support browser window.

7. Close the Help window.

The InfoCenter

The InfoCenter is located in the title bar in the upper right corner of the AutoCAD window. In addition to containing the Help button, it allows you to access additional resources and information.

1. Type **erase** in the edit box in the InfoCenter and press **ENTER** or click on the **Search** button.

As you can see, the Help window appears with a list of topics related to erase. The edit box allows you to search AutoCAD Help from the main AutoCAD window.

2. Close the Help window.

The InfoCenter provides access to Autodesk 360, a cloud-based platform that provides access to storage, a collaboration workspace, and other cloud services. Some of the Autodesk 360 resources are free. Users must create an account to access this service, and sign in to use it.

The Autodesk Exchange Apps website can be accessed from the InfoCenter. This site helps users personalize their software with apps developed within the community of users. The InfoCenter allows users to check for product updates, send feedback to Autodesk, and participate in the Customer Involvement Program.

You can also open the Welcome Screen from the InfoCenter. You may remember seeing the Welcome Screen when you first launched AutoCAD in Chapter 1.

3. Pick the down-facing arrow to the right of the Help button on the InfoCenter.

4. Select the **Welcome Screen...** option from the pull-down menu.

The Welcome Screen appears. The Work panel allows you to select a new drawing, open recent files, or open sample files. The Learn panel provides access to tutorial videos and other online resources. The Extend panel provides another way to access the Autodesk Exchange Apps and Autodesk 360 resources.

5. Close the Welcome Screen.

Context-Sensitive Help

AutoCAD allows you to obtain help in the middle of a command, when you are most likely to need it. If another command is active when you enter the HELP command, AutoCAD automatically displays help for the active command. This is known as **context-sensitive help**. When you close the help window, the command resumes.

1. Enter the **GRID** command at the keyboard.

2. Press the **F1** function key to display help information on the GRID command.

3. Close the Help window and press the **ESC** key to cancel the GRID command.

File and Folder Navigation

AutoCAD's Select File dialog box can help you organize and manage AutoCAD-related files.

1. Pick the **Open** button from the Quick Access toolbar.

AutoCAD displays the Select File dialog box, as shown in Fig. 8-2 on page 101.

2. Find and open the folder with your name—the one you created in Chapter 4.

Fig. 8-2

Viewing File Details

3. Pick the **Views** drop-down menu and select **Details**, unless it is grayed out, which means it is the current selection.

This provides a listing of each file, along with its size, type, the author, and the date and time it was created and last modified. These items are known as **file attributes**. In folders that contain many drawing files, it is sometimes easier to find files if they are listed in a certain order. By default, files are listed in alphabetical order by name. You can change the order in which the files and folders are listed by clicking the column names, which are actually buttons.

4. Click the bar labeled Name at the top of the list box.

This reverses the order of the files. They are now in reverse alphabetical order (from Z to A).

5. Click the same Name bar again.

The files are once again listed in alphabetical order.

6. Click the bar labeled Size.

This places the files in order of size from smallest to largest. Clicking it again would reorder the files from largest to smallest. Clicking the bars labeled Type and Date Modified organizes the files according to type and date, respectively. You may need to resize the columns or use the scroll bar to view these options.

7. Click the bar labeled Date Modified; click it again.

8. From the **Views** drop-down menu, pick **List** to produce a basic listing of the files.

Finding Drawing Files

Within the Select File dialog box, AutoCAD provides other options for reviewing and finding files.

1. From the **Tools** pull-down menu, located in the upper right area of the dialog box, select the **Find...** item.

This displays the Find dialog box.

2. In the text box after Named, enter **db_samp** and pick the **Find Now** button.

AutoCAD searches for the file named db_samp.dwg. If AutoCAD successfully finds it and displays the file details in the lower area of the dialog box, go to Step 4. If the search is unsuccessful, proceed to Step 3.

3. Click the down arrow located at the right of Look in, select the disk drive on which AutoCAD was installed, and pick the **Find Now** button.

4. Pick the **OK** button to open the folder containing this file.

5. Select **Find...** again from the **Tools** pull-down menu.

6. Pick the down arrow located at the right of Type.

This permits you to specify other file types to search.

7. Click outside the pull-down menu to close it.

8. Pick the **Browse** button.

The Browse button enables you to select a specific folder or disk drive to search.

9. Pick the **Cancel** button.

10. Pick the **Date Modified** tab.

This gives you the option of specifying dates within which the file you are seeking was created or modified.

11. Pick the **Cancel** button.

Copying Files

The Select File dialog box permits you to copy files easily from within AutoCAD.

1. If necessary, pick the **Back to** (left-facing arrow) button to return to the folder with your name on it.
2. Right-click and drag the **stuff.dwg** file to an empty area within the list box area and release the right button on the pointing device.
3. Pick **Copy Here** from the shortcut menu that appears.

A new file named Copy of stuff.dwg appears.

Renaming Files

You can also rename files using the Select File dialog box.

1. Right-click the new file named **Copy of stuff.dwg**.
2. Select **Rename** from the shortcut menu.

You should now see a blinking cursor at the end of the file name with a box around it.

3. Type the name **junk.dwg** and press **ENTER**.

The new file named junk.dwg has the same contents as stuff.dwg.

Creating New Folders

It's also fast and easy to create new folders.

1. Pick the **Create New Folder** button.

A new folder named New Folder appears.

2. To rename the folder, right-click the new folder, select **Rename** from the menu, type **Junk Files**, and press **ENTER**.

Moving Files

Let's move a file to the new folder.

1. Right-click **junk.dwg** and drag and drop it into the **Junk Files** folder.

A shortcut menu appears.

2. Pick **Move Here**.

The junk.dwg file is now located in the Junk Files folder.

Deleting Files and Folders

AutoCAD also allows you to delete unwanted files and folders using the Select File dialog box.

1. Open the **Junk Files** folder.

2. Right-click **junk.dwg** and select **Delete** from the shortcut menu.

3. Pick the **Yes** button in the Confirm File Delete dialog box if you are sure you have selected junk.dwg.

4. Pick the **Up one level** button.

5. Right-click the **Junk Files** folder and select **Delete** from the shortcut menu.

6. If you are sure that you selected the **Junk Files** folder, pick the **Yes** button in the **Confirm Folder Delete** dialog box to send this folder to the Windows Recycle Bin.

7. Close the Select File dialog box.

 You can use methods similar to those discussed in this section even when AutoCAD is not open. The Microsoft Windows environment allows you to copy, rename, move, and delete files of all types, including AutoCAD drawing files.

Diagnosing, Repairing, and Recovering Files

Occasionally, AutoCAD drawing files may become corrupted. AutoCAD supplies two commands to help fix corrupted files.

Auditing a Drawing File

The AUDIT command is available as a diagnostic tool. It **audits**, or examines the validity of, the current drawing file. It can also correct some of the errors it finds. The AUDIT command automatically creates an audit report file containing an ADT file extension when the AUDITCTL system variable is set to 1 (On). The default setting is 0.

Probable causes of damaged files include:

- An AutoCAD system crash
- A power surge or a disk write error while AutoCAD is writing the file to disk

When a file becomes damaged, AutoCAD may refuse to edit or plot the drawing. In some cases, a damaged file may cause an AutoCAD internal error or fatal error.

1. Open the file named **stuff.dwg**.
2. On the **Application Menu**, select **Drawing Utilities** and **Audit**.

This enters the AUDIT command.

3. In reply to Fix any errors detected? press **ENTER** to accept the No default.
4. Press the **F2** function key to view the AutoCAD Text Window and read all of the message.

The AutoCAD Text Window displays information that is similar but not necessarily identical to that shown in Fig. 8-3.

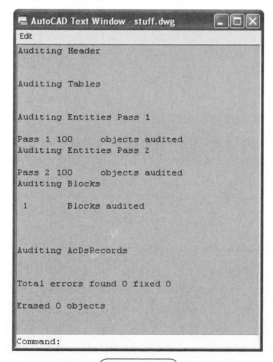

Fig. 8-3

If you had entered Yes in reply to Fix any errors detected? the report would have been the same, as long as the file contained no errors.

NOTE AutoCAD offers a command similar to AUDIT named RECOVER. This command attempts to open and repair damaged drawing files. The recovery process has been embedded in the OPEN command, so when you try to open a damaged file, AutoCAD automatically executes the RECOVER command and attempts to repair it.

If you know or suspect that a file has been damaged, you can enter the RECOVER command directly at the AutoCAD prompt. This produces the Select File dialog box, from which you can choose the file to be recovered.

Permanently Damaged Files

A drawing file may be damaged beyond repair. If so, drawing recovery with the AUDIT and RECOVER commands will not be successful.

Each time you save an AutoCAD file, AutoCAD saves the changes to the current DWG file. AutoCAD also creates a second file with a BAK (backup) file extension. This file contains the previous version of the file—the version prior to saving. If the DWG file becomes damaged beyond repair, you can rename the BAK file to a DWG file and use it instead. To prevent the loss of data, save often and produce a backup copy of your DWG files frequently.

The Drawing Recovery Manager

When a hardware problem, power failure, or software problem causes AutoCAD to terminate unexpectedly, the Drawing Recovery Manager opens the next time you start AutoCAD. It displays a list of the drawing files that were open at the time the program terminated.

If you pick a file, AutoCAD lists all the available versions of the drawing that may be recovered, including the DWG file, the BAK file, and the SV$ (auto-save) file. These file types are listed in the order of their time stamps. You can double-click the most recent file, and AutoCAD will attempt to repair and open it. This allows you to recover the most recently saved version of the drawing that is available.

   5. Exit AutoCAD without saving unless the file was damaged and recovered. In that case, save the changes to the file.

Chapter 8 Review & Activities

● REVIEW QUESTIONS

Answer the following questions on a separate sheet of paper.

1. When you pick the HELP button, what does AutoCAD display?
2. Suppose you have entered the MIRROR command. At this point, what is the fastest way of obtaining help on this command?
3. Name and briefly describe each of the options available on the InfoCenter toolbar.
4. What is context-sensitive help? Explain how to use it in AutoCAD.
5. What is the purpose of each of the five items located in the upper right area of the Select File dialog box?
6. In the Select File dialog box, what information can you display by picking Details from the Views drop-down menu?
7. What is the primary purpose of the AUDIT command?

● CHALLENGE YOUR THINKING

These questions are designed to further your knowledge of AutoCAD by encouraging you to explore the concepts presented in this chapter. Answer each question on a separate sheet of paper.

1. Explore the benefit of typing questions in the Search tab of the AutoCAD Help screen.
2. AutoCAD gives you the option of creating a report file when you run the AUDIT command. Explain why such a report might be useful.

● APPLYING AUTOCAD SKILLS

Using the features described in this chapter, complete the following activities. Complete problems 1 through 5 from within the Select File dialog box.

1. Display a list of drawing files from one of AutoCAD's folders.
2. Display a list of template files found in one of AutoCAD's folders.
3. Rename one of your drawing files. Then change it back to its original name.

Continued

4. Make a copy of the file Welding.dwg in the AutoCAD Sample/en-us/ DesignCenter folder. Rename the copy MyWelding.dwg and move it to your named folder.

5. Delete MyWelding.dwg from your named folder.

6. You know you have a drawing of a house somewhere on your hard drive, but you don't remember the full name of the drawing. You recall that it includes "Textstyles," and because it's a drawing file, you know that it has a DWG file extension. Find the drawing. What is its name and what does it contain?

• USING PROBLEM-SOLVING SKILLS

Complete the following activities using problem-solving skills and your knowledge of AutoCAD.

1. You need to draw two concentric circles. You have already drawn the inner circle. The outer circle needs to be .5 larger. Enter the OFFSET command from the Modify panel of the Ribbon and access context-sensitive help to offset the diameter of the first circle by .5.

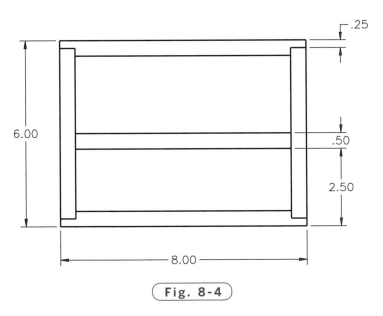

Fig. 8-4

2. It's time to build the bookcase you designed in Chapter 4. Make a copy of the bookcase you saved in your named folder as prb4-1.dwg. Modify the drawing as shown in Fig. 8-4 and save the modification as bookcase.dwg. Delete the copy of prb4-1.dwg.

Mechanical Drafter

Mechanical devices are all around you. At the grocery store, conveyer belts help move shoppers' selections from their carts to the cashier. Mechanics use hydraulic lifts to raise cars in order to work on parts underneath. Military personnel use gauges, tracking devices, and other equipment. Have you ever taken a mechanical object apart just to see how it works? If so, you might be interested in learning more about mechanical drafting.

On the Job

Mechanical drafters make up the largest category of drafters in the United States. They design machinery and mechanical tools. Accurate, detailed drawings of mechanical equipment must be provided for manufacturers to use as guides during the building process. Today, most mechanical drafters use AutoCAD or a similar CAD program to assist with the overall design process.

A Mechanical Mindframe

Curiosity about how mechanical devices operate and the ability to visualize how

© Monty Rakusen/Getty Images/RF

an object might look from all angles are necessary skills for a mechanical drafter. The ability to communicate and work well with others is also important.

Mechanical drafters often consult with the individuals who initially thought up product ideas; they also confer regularly with other drafters and mechanical engineers. Perhaps most importantly, mechanical drafters must have a knack for understanding exactly how mechanical items work. Attention to detail is important both when studying existing devices to improve future designs and when creating the detail drawings for new products.

► Career Activities

1. Look around the room. Make a list of the mechanical devices you see.

2. Try to sketch a rough example of how you think a design drawing might look. Share your drawing with a classmate and compare ideas.

Applying Chapters 1–8

Rocker Arm

Engineering drawing refers to the creation of highly accurate drawings using drafting instruments or, more commonly today, CAD software. Engineering drawings are used, among other things, to describe a part so that the manufacturer can build it correctly.

▶ Description

The rocker arm shown in Fig. P1-1 was submitted to your manufacturing company as a hand sketch. The basic design has been approved, and now the marketing department needs an electronic version of the drawing for use in a new product brochure. The drawing must reflect the actual dimensions of the rocker arm, although for this use, it is not necessary to show the dimensions on the drawing.

Your task is to draw the rocker arm accurately using the AutoCAD commands and procedures you learned in Part 1. Be sure to read the "Hints and Suggestions" for this project before you begin. Save the drawing as rockerarm.dwg in your named folder.

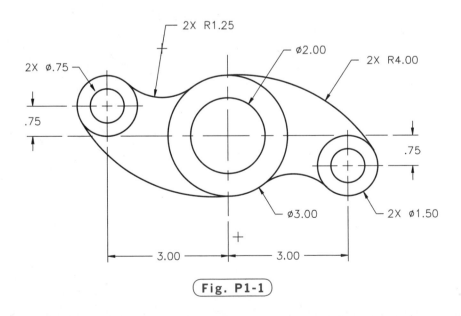

Fig. P1-1

▶ Hints and Suggestions

1. The lines made up of short and long dashes are center lines (lines that intersect at the exact center of a feature). Use these lines and your knowledge of coordinates to draw the rocker arm accurately.

2. Study the drawing to determine the best order in which to draw the parts of the rocker arm. Keep in mind that it is *not* necessary to construct the part linearly, so that each line connects to the next. For example, you may choose to create all of the holes and cylindrical features before you draw the other parts of the rocker arm.

3. If you have trouble visualizing how to draw the part in AutoCAD, try sketching it first on graph paper.

▶ Summary Questions

Your instructor may direct you to answer these questions orally or in writing. You may wish to compare and exchange ideas with other students to help increase efficiency and productivity.

1. Pinpoint any difficulties you had in drawing the rocker arm and explain how you solved them.

2. Describe the order in which you drew the parts to make up the rocker arm. Why did you choose this order?

3. Study the drawing in P1-1 once more. Does this drawing provide everything needed to manufacture the actual part? If not, what is missing?

4. In this project, you have created an electronic drawing file to be used by the marketing department because marketing needed the file in electronic format. However, there are other advantages to creating the file electronically. How else might this file be used? How would it need to be modified to be useful to other departments? Explain your answer.

Chapter 9

Object Snap

Objectives

- Set and use running object snap modes
- Apply object snaps that are not currently turned on
- Change object snap settings to increase productivity
- Use the **Quick Setup** wizard

Vocabulary

aperture box
object snap tracking
object snaps
perpendicular
quadrant
quadrant points
running object snap modes
running object snaps
system variable
wizard

One of the most useful features of AutoCAD is the **object snap** feature. Object snaps are like magnets that permit you to pick endpoints, midpoints, center points, points of tangency, and other specific points easily and accurately. **Object snap tracking**, which is used in conjunction with object snaps, provides alignment paths that help you produce objects at precise positions and angles. Combined with polar tracking, these features speed the precision drawing process.

Running Object Snaps

You can use object snaps in different ways to make a drawing task easier. When you know that you will be needing one or more specific object snaps several times, you can preset them to "run" in the background as you are working. Object snaps that have been preset in this way are known as **running object snaps**.

Quick Setup Wizard

In preparation for using object snaps, we will use the Quick Setup wizard to establish basic settings quickly in AutoCAD. Quick Setup is based on the acad.dwt template file.

1. Start AutoCAD, and re-select the **Drafting & Annotation** workspace.

2. Enter **STARTUP** at the keyboard and enter **1**.

STARTUP is a system variable. A **system variable** is a setting that controls a specific function in AutoCAD. The default value of the STARTUP system variable is 0. When it is set at 1, AutoCAD displays the Startup dialog box when you start the software.

3. Enter the **NEW** command.

The Create New Drawing dialog box appears. The four buttons at the top permit you to open a drawing, start a drawing from scratch, use a template file, or use a wizard. A **wizard** is a prompted sequence of steps that streamlines a task or set of tasks. Using the Start from Scratch button is equivalent to beginning a new drawing with STARTUP set at 0. In any case, all of the settings come from the acad.dwt template file.

4. Pick the **Use a Wizard** (fourth) button.

5. Select **Quick Setup** and pick **OK**.

AutoCAD displays a dialog box that focuses on the unit of measurement. Decimal is the default selection. Notice the sample in the right area of the box.

6. Accept the **Decimal** default selection and pick the **Next** button.

The dialog box concentrates on the drawing area. As you can see, the default drawing area is 12 × 9 units.

7. For Width, enter **95**, and for Length, enter **70**.

8. Pick the **Finish** button and save your work in a file named **bike.dwg**.

The drawing area now represents 95 × 70 units. These units can represent millimeters, kilometers, feet, miles, or whatever we want. We will use them as inches.

9. Enter **Z** for ZOOM and **A** for All.

This zooms the drawing area to the full 95 × 70 inches. You will learn more about the ZOOM command in Chapter 12.

Specifying Object Snap Modes

1. Click on the **Polar Tracking, Object Snap, Object Snap Tracking**, and **Dynamic Input** toggle buttons on the status bar to enable these features, unless they are already enabled.

When status bar toggle buttons are enabled, they have a blue background. When they are disabled, they have a grey background. The Snap Mode, Grid Display, and Ortho Mode buttons should be disabled.

2. Enter the **OSNAP** command.

This displays the Drafting Settings dialog box, as shown in Fig. 9-1. Entering the OSNAP command is equivalent to right-clicking the Object Snap button on the status bar and selecting Settings... from the shortcut menu.

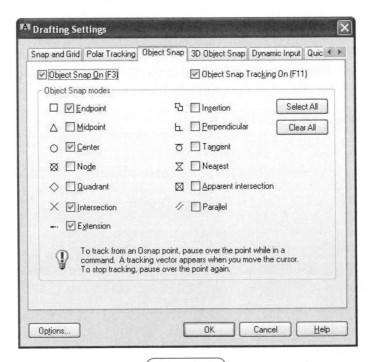

Fig. 9-1

3. Review the object snap modes.

The ones that are checked are called **running object snap modes**. These object snaps are "on" and function automatically as you work with drawing and editing commands in AutoCAD.

4. Check **Endpoint, Center, Intersection,** and **Extension,** and uncheck all the others. (Some or all of them may be checked already.)

Notice that Object Snap On and Object Snap Tracking On are checked. This is because the OSNAP and OTRACK buttons on the status bar are on (depressed). A third way to turn them on and off is to press the F3 and F11 function keys, as indicated in the dialog box.

5. Pick the **OK** button.

Using Preset Object Snaps

Let's use AutoCAD's object snaps to create a basic drawing of the new mountain bike frame design shown in Fig. 9-2. Be sure to save your work every few minutes as you work through this chapter.

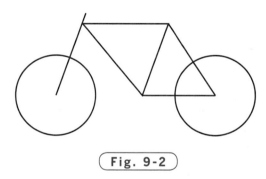

Fig. 9-2

1. In the lower left area of the screen, draw a 22"-diameter circle to represent the front wheel, as shown in Fig. 9-2.

The Center object snap selects the exact center of a circle or arc. Let's use it to start a 24" line at the center of the bike's front wheel. This line will begin the outline of the bike frame.

2. Enter the **LINE** command.

3. At the Specify first point prompt, move the crosshairs so that it touches the circle or move the crosshairs to the circle's center.

A small circular marker appears at the circle's center. The marker indicates that AutoCAD has selected the point at the center of the circle.

DRAW

DRAW

4. While the marker is present, pick a point.

AutoCAD snaps to the circle's center.

5. For the next endpoint, use polar coordinates to extend the line **24″** up and to the right at a **70°** angle.

 Type 24, press the TAB key, type 70, and press ENTER, or enter @24<70 at the keyboard.

6. Press **ENTER** to terminate the LINE command.

Object Snap Tracking

Next, we are going to draw the horizontal line using AutoCAD's **object snap tracking** feature.

1. Reenter the **LINE** command and move the crosshairs to the second endpoint of the line until the Endpoint object snap marker appears.

The purpose of doing this is to "acquire" the Endpoint object snap. When you acquire an object snap, a small + (plus sign) appears at the snap point. An acquired object snap also displays a small x near the crosshairs. To remove an acquired object snap, pass over the snap point a second time.

2. Slowly move the crosshairs around the endpoint and notice the temporary alignment paths that appear.

As you may recall, alignment paths are temporary lines that display at specific angles from an acquired object snap. Object snap tracking is active when the alignment paths appear from one or more acquired object snaps. This feature is a part of AutoCAD's AutoTrack™. You can toggle AutoTrack on and off with the Polar Tracking and Object Snap Tracking buttons on the status bar. Because object snap tracking works in conjunction with object snaps, one or more object snaps must be set before you can track from an object's snap point.

3. Move the crosshairs down the line that you created earlier so that a small x appears on the line.

4. Enter **3**.

This starts a new line 3″ from the endpoint of the first line.

5. Draw a horizontal line segment 24″ to the right.

HINT When creating the second point, use polar tracking to force the line horizontal and enter 24.

6. In reply to Specify next point, click at the Command line and enter **@21<250** to create the next line segment.

The dynamic input prompt can also be used, but the result will depend on the position of the crosshairs at the time you type the angle value of 250. This is because AutoCAD measures the angle either clockwise or counterclockwise from the horizontal, depending on where you position the crosshairs.

7. In reply to Specify next point, move the crosshairs to the intersection of the first and second line segments.

The intersection of the two lines is also an endpoint of the second line. Therefore, either the Intersection or Endpoint object snap marker will appear. An x-shaped marker depicts the Intersection object snap, and a square marker indicates the Endpoint object snap.

8. Snap to this point and press **ENTER** to terminate the LINE command.

9. Using the **Endpoint** object snap and polar tracking, create the second horizontal line. Make it 20″ long.

10. Create the final segment of the bike frame, using Fig. 9-2 on page 115 as a guide.

11. Create the second wheel for the bike. Use the **Endpoint** object snap to place the center of the wheel at the right endpoint of the lower horizontal line. Make it is the same size as the front wheel.

The drawing of the new bike design is now complete.

12. Save your work.

Specifying Object Snaps Individually

Another way to use object snaps is to specify them individually as you need them. You can do this either at the keyboard or by displaying and using the Object Snap toolbar. Using this toolbar streamlines the selection of object snaps.

WINDOWS

1. From the **View** tab of the **Ribbon**, select the **Toolbars** button on the **User Interface** panel.
2. Rest the pointer over the **AutoCAD** item in the pull-down menu, then select **Object Snap** from the cascading menu.

The Object Snap toolbar appears, as shown in Fig. 9-3.

Fig. 9-3

3. Rest the pointer on each of the buttons and read the tooltips.

Table 9-1 contains a list of object snap modes and their functions. To enter an object snap mode at the keyboard, enter only the capitalized letters.

Snapping to Quadrant Points

OBJECT SNAP

1. Enter the **LINE** command.
2. From the **Object Snap** toolbar, pick the **Snap to Quadrant** button.

Quadrant points on a circle are the points at 0°, 90°, 180°, and 270°. If you were to connect these points with horizontal and vertical lines, as shown in Fig. 9-4, you would divide the circle into four equal parts. Each part is a **quadrant** of the circle.

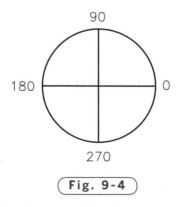

Fig. 9-4

Table 9-1. AutoCAD's Object Snaps

Object Snap	Button	Purpose
APParent intersection		Snaps to the apparent intersection of two intersection objects that may not actually intersect in 3D space
CENter		Center of an arc circle, ellipse, or elliptical arc
ENDpoint		Closest endpoint of a line or arc
EXTension		Objects along the extension paths of objects
INSert		Insertion point of text, block, or attribute
INTersection		Intersection of two objects
MIDpoint		Midpoint of an object
NEArest		Nearest point on an object
NODe		Nearest point entity or dimension definition point
NONe		Temporarily cancels all running object snaps (for one operation only)
PARallel		Snaps to a point on an alignment path that is parallel to the selected object
PERpendicular		Perpendicular to a line, arc, or circle
QUAdrant		Quadrant point of an arc, circle, ellipse, or elliptical arc
TANgent		Tangent to an arc, circle, ellipse, elliptical arc, or spline
FROm		Used in combination with other object snaps to establish a temporary reference point as a base point
Object Snap Settings		Displays the Object Snap tab of the Drafting Settings dialog box
Temporary Tracking Point		Allows you to set a temporary OTRACK point while a drawing command is active

3. Move the crosshairs over one of the two wheels until a small diamond marker appears, along with the tooltip Quadrant.

4. Move to each of the four quadrant points and pick one of them.

5. Enter **ESC** to cancel the LINE command.

Snapping Perpendicularly

The Perpendicular object snap allows you to snap to a point perpendicular to an object such as a line, multiline, polyline, ray, spline, construction line, or even a point on a circle or arc. (A line is **perpendicular** to another line when the two form a right angle.)

1. Enter the **LINE** command and pick a point anywhere above the top horizontal segment of the bike.

2. Pick the **Snap to Perpendicular** button from the **Object Snap** toolbar.

3. Move the crosshairs along the top horizontal segment of the frame until a right angle marker appears, along with the tooltip Deferred Perpendicular, and pick a point.

OBJECT SNAP

AutoCAD defers the point where the perpendicular line will intersect until after you have picked a point, then snaps the line perfectly perpendicular to the horizontal segment of the frame.

QUICK ACCESS

4. Press **ENTER** to terminate the LINE command and pick the **Undo** button to remove the line.

Snapping Parallel Lines

You can use the parallel object snap to produce parallel line segments.

OBJECT SNAP

1. Reenter the **LINE** command, pick another point above the bike, and pick the **Snap to Parallel** object snap from the toolbar.

2. Move the crosshairs over the fork of the frame (the first line you drew) and then move the crosshairs until a parallel marker appears, and then back and forth until a temporary line is parallel to the fork.

3. Pick a point anywhere to form the parallel line.

4. Press **ENTER** and pick the **Undo** button.

Snapping to Points of Tangency

The Tangent object snap permits you to snap to a point on an arc or circle that, when connected to the last point, forms a line tangent to that object.

OBJECT SNAP

1. Enter the **LINE** command and pick the **Snap to Tangent** button on the Object Snap toolbar.

2. Move the crosshairs along the bottom half of the front wheel until a Snap to Tangent marker appears, along with the tooltip Deferred Tangent.

AutoCAD defers the point of tangency until after you have picked both end-points of the line segment.

3. Pick a point anywhere along the bottom half of the front wheel.

4. Pick the **Snap to Tangent** button again.

5. Move the crosshairs along the bottom half of the rear wheel until the Snap to Tangent marker appears and pick a point.

OBJECT SNAP

AutoCAD draws a line tangent to both wheels.

6. Using polar tracking, extend the line 12″ to the right and press **ENTER** to end the LINE command.

Snapping to the Nearest Point

The Nearest Point object snap mode snaps to a location on the object that is closest to the crosshairs. This is particularly useful when you want to make certain that the point you pick lies precisely on the object and not a short distance away from it.

1. Pick the **Osnap Settings...** button from the Object Snap toolbar.

This also displays the Object Snap tab of the Drafting Settings dialog box. (You entered the OSNAP command to display it the last time.)

OBJECT SNAP

2. Pick the **Select All** button and pick **OK**.

All object snaps are now on.

3. Enter the **LINE** command and slowly move the crosshairs over the snap points of the objects you have drawn.

An hourglass-shaped marker depicts the Nearest object snap.

4. Snap to a **Nearest** point on one of the objects.

Cycling Through Snap Modes

Pressing the TAB key cycles through all of the running snap modes associated with an object. This is useful when an area is densely covered with objects, making it difficult to snap to a particular point.

1. Rest the crosshairs on the rear wheel.

2. Press the **TAB** key.

3. Press **TAB** several more times, stopping for a moment between each.

As you can see, AutoCAD moves from one object snap to the next.

4. Press **ESC** to cancel the LINE command.

5. Pick the **Osnap Settings...** button on the Object Snap toolbar.

6. Pick the **Clear All** button, check **Endpoint**, **Center**, **Intersection**, and **Extension**, and pick **OK**.

OBJECT SNAP

Object Snap Settings

Other settings allow you to tailor the behavior of object snaps to meet your individual needs.

OBJECT SNAP

Changing the Settings

1. Pick the **Osnap Settings...** button on the Object Snap toolbar.
2. Pick the **Options...** button in the dialog box.

Marker, Magnet, and Display AutoSnap tooltip should be checked. When Marker is checked, geometric markers appear at each of the snap points. (These are the markers you have used throughout this chapter.) When Magnet is checked, AutoCAD locks the crosshairs onto the snap target. When Display AutoSnap tooltip is checked, AutoCAD displays a tooltip that describes the snap location.

3. Check the **Display AutoSnap aperture box** check box.

This causes an **aperture box** to appear at the center of the crosshairs when you snap to objects. The aperture box defines an area around the center of the crosshairs within which an object or point will be selected when you use the pointing device to pick it.

4. Pick the **Colors...** button below the AutoSnap Settings to display the Drawing Window Colors dialog box.
5. With 2d AutoSnap marker highlighted under Interface element, pick the down arrow under Colors and select green.
6. Pick the **Apply & Close** button.
7. In the **AutoSnap Marker Size** area, move the slider bar slightly to the right, increasing the size of the marker.

In the AutoTrack Settings area of the dialog box, all three check boxes should be checked. When Display polar tracking vector is checked, AutoCAD displays an alignment path for polar tracking. When Display full-screen tracking vector is checked, AutoCAD displays alignment vectors as infinite lines. When Display AutoTrack tooltip is checked, AutoCAD displays AutoTrack tooltips.

The Alignment Point Acquisition area controls the method of displaying alignment vectors. When Automatic is picked, AutoCAD displays tracking vectors automatically when the aperture moves over an object snap. When Shift to acquire is picked, AutoCAD displays the tracking vectors only if you press the SHIFT key while you move the aperture over an object snap.

8. In the **Aperture Size** area, slightly increase the size of the aperture using the slider bar.

9. Pick the **Apply** and **OK** buttons.

10. Pick the **OK** button in the Drafting Settings dialog box.

Applying the New Settings

Now let's examine how the changes made in the dialog box affect the drawing process.

DRAW

1. Enter the **ARC** command and move the crosshairs over the drawing.

Notice that the object snap markers appear when the aperture box contacts the object.

2. Press **ESC** to cancel the **ARC** command.

The AutoSnap settings are stored with AutoCAD, not with the drawing. This means that AutoCAD remembers the changes. Likewise, running object snap modes are stored with AutoCAD.

3. Right-click the **Object Snap** button on the status bar and select **Settings...** from the shortcut menu.

OBJECT SNAP

This is a third way to display the Object Snap tab of the Drafting Settings dialog box.

4. Pick the **Options...** button.

5. Uncheck **Display AutoSnap aperture box**, change the marker color back to its original color by selecting the **Restore current element** in the Drawing Window Colors dialog box, and reduce the marker and aperture box to their original sizes.

6. Pick **Apply** and **OK**.

7. Pick **OK** again to close the Drafting Settings dialog box.

As you work with polar tracking, object snap, and object snap tracking, keep in mind that you can toggle them on and off at any time using the Polar Tracking, Object Snap, and Object Snap Tracking buttons on the status bar. Most of the time, you will want to use these features, but there will be times when you will want to disable them.

8. Practice using the remaining object snaps on your own.

9. Close the Object Snap toolbar.

10. Enter **STARTUP** at the AutoCAD prompt and enter **0**.

11. Save your work and exit AutoCAD.

● REVIEW QUESTIONS

Answer the following questions on a separate sheet of paper.

1. How is the Quick Setup wizard useful?

2. Explain the purpose of the object snaps.

3. Describe two methods for using AutoCAD's object snap modes.

4. In order to snap a line to the center of a circle, what part of the circle must the crosshairs touch?

5. What is the benefit of using running object snap modes?

6. Describe a situation in which you would want to change the aperture box size.

7. Briefly describe the use of each of the following object snap modes.
   a. Apparent Intersection
   b. Center
   c. Endpoint
   d. Extension
   e. Insertion
   f. Intersection
   g. Midpoint
   h. Nearest
   i. Node
   j. None
   k. Parallel
   l. Perpendicular
   m. Quadrant
   n. Tangent

8. Explain the benefit of object snap tracking.

9. Explain the relationship between polar tracking, object snap, and object snap tracking. Can you use object snap tracking when object snap is turned off?

• CHALLENGE YOUR THINKING

These questions are designed to further your knowledge of AutoCAD by encouraging you to explore the concepts presented in this chapter. Answer each question on a separate sheet of paper.

1. Devise a new object snap that does not currently exist in AutoCAD. The new snap mode must provide a useful service. Write a short paragraph describing the new object snap. Explain what it does and why it is useful.

2. Investigate the Apparent Intersection object snap. Of what use is an object snap that allows you to snap to a point where two objects only *seem* to intersect? Give an example.

• APPLYING AUTOCAD SKILLS

Work the following problems to practice the commands and skills you learned in this chapter.

1. Draw the square on the left in Fig. 9-5. Then use object snaps to make the additions shown on the right to produce a top view of a jeweler's new design for a cut gemstone.

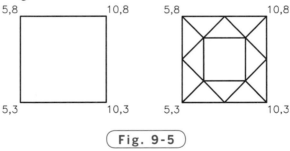

Fig. 9-5

2. Create a new drawing and set running snap modes for Center and Endpoint. Then create the shaft bearing in Fig. 9-6 using the dimensions shown. Start the lower left corner of the drawing at absolute coordinates 1,1 and work counterclockwise. Save the file as shaftbearing.dwg.

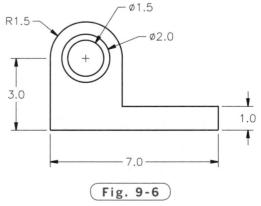

Fig. 9-6

Continued

3. Draw the top view of the video game part shown in Fig. 9-7. Start by drawing a circle of any size at any location. Then use the Quadrant and Center object snaps to draw the four lines to divide the circle into quadrants, as shown on the left in Fig. 9-8. (You must draw four lines, not two.) Use the ARC command with the Start, Center, End option and Quadrant, Midpoint, and Center object snap modes to draw the arcs. Erase the quadrant lines. Save the drawing as gamepart.dwg.

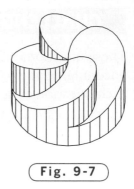

Fig. 9-7

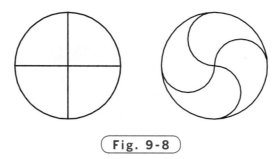

Fig. 9-8

Courtesy of Gary J. Hordemann, Gonzaga University

● USING PROBLEM-SOLVING SKILLS

Complete the following activities using problem-solving skills and your knowledge of AutoCAD.

1. The drawing in Fig. 9-9 represents a variation of a magnet for a new motor design. Draw the magnet. Select the appropriate drawing unit of measurement, drawing area, and object snap modes. Do not include dimensions, but save the drawing as magnet.dwg.

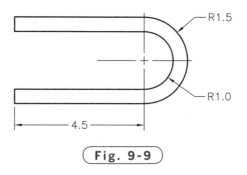

Fig. 9-9

2. Suppose your company supplies special machine parts for a plastic injection molding company. Draw the new design for a pivot arm to swing the mold halves apart, as shown in Fig. 9-10. Use the Quick Setup wizard and select the appropriate drawing unit of measurement and drawing area. Set the appropriate object snap modes. Do not include dimensions. Save the drawing as injection.dwg.

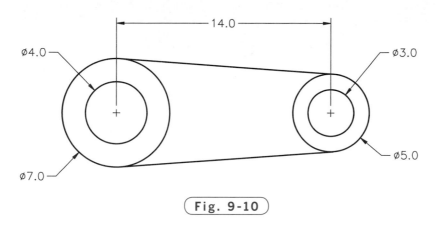

Fig. 9-10

3. The three figures in Fig. 9-11 are constructed of polygons according to the following rules:

a. There are as many polygons as there are sides to the polygons.

b. All polygons, except the two on the ends, share a single edge with each of two other polygons.

If we imagine these figures to be groups of blocks sitting on a horizontal surface, then all three groups are stable (assuming a fair amount of friction between the surfaces). Using the Edge option of the POLYGON command with either the Intersection or the Endpoint object snap, draw similar groupings of hexagons and heptagons. Draw them as though they were sitting on a horizontal surface, and be sure to follow the two rules stated above. Is either grouping unstable? Can you formulate a general rule about the stability of such groupings that would apply to polygons with any number of sides?

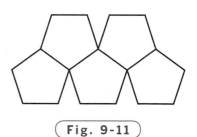

Fig. 9-11

Courtesy of Gary J. Hordemann, Gonzaga University

Chapter 10

Helpful Drawing Features

Objectives

- Use and change the display of coordinate information
- Restrict lines to vertical or horizontal orientations
- Track time spent on a drawing file
- Use the **AutoCAD Text Window** and the **Command Line** window
- Undo and redo multiple operations

Vocabulary

ortho
orthogonal

Several features in AutoCAD provide information to help users become more productive. The coordinate display, for example, gives size information on the drawing area. The ortho feature helps you draw horizontal and vertical lines quickly and accurately. The TIME command tracks the time you spend working with AutoCAD. The AutoCAD Text and Command Line windows provide historical information on your interactions with the software. With AutoCAD's undo and redo features, you don't have to worry about making a wrong move.

Coordinate Display

Coordinate information, in digital form, is located in the lower left corner of the AutoCAD window and is part of the status bar. The coordinate display tracks the current position (or coordinate position) of the crosshairs as you move the pointing device. It also gives the length and angle of line segments as you draw.

1. Start AutoCAD and re-select the **Drafting & Annotation** workspace.

2. Move the crosshairs in the drawing area by moving the pointing device and note how the coordinate display on the left side of the status bar changes with the movement of the crosshairs.

3. Enter the **LINE** command and draw a line segment of a polygon. Note the coordinate display as you draw the line segment.

 AutoCAD provides shortcut keys for commands that are used frequently. Shortcut keys allow you to enter commands very quickly. AutoCAD's shortcut keys include the function keys (F1, F2, F3, and so on) and the combination of the Control (CTRL) key and another key pressed at the same time. The following steps use this shortcut method.

4. Press the **CTRL** key and the letter **I** at the same time, draw another line segment, and notice the coordinate display.

As you can see, polar coordinate information is now displayed.

5. Press the **CTRL** and **I** keys again, draw another line segment or two, and watch the coordinate display.

The coordinate display should now be off.

6. While in the LINE command, click the coordinate display and note the change in the coordinate display as you move the crosshairs.

7. Create additional line segments, clicking the coordinate display between each.

As you can see, clicking the coordinate display serves the same function as pressing the CTRL + I shortcut combination.

8. Press **ENTER** to terminate the LINE command.

The Ortho Mode

Now let's focus on an AutoCAD feature called ortho. **Ortho**, short for "orthogonal," allows you to draw horizontal and vertical lines quickly and easily. **Orthogonal**, in this context, means drawn at right angles.

Ortho is on when the Ortho Mode button on the status bar is toggled on. You can turn ortho on and off by clicking the Ortho Mode button.

STATUS BAR

1. Turn off object snap if it is currently on.

2. Click the **Ortho Mode** button in the status bar or press the **F8** function key. (Both actions perform the same function.)

3. Experiment by drawing lines with ortho turned on and then with ortho off. Note the difference.

> **HINT** Like the coordinate display feature, ortho can be toggled on and off at any time, even while you're in the middle of a command.

4. Attempt to draw an angular line with ortho on.

As you can see, it is not possible.

5. Clear the screen if necessary and draw the plug shown in Fig. 10-1, first with ortho off and then with ortho on. Don't worry about exact sizes and locations.

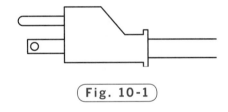

(Fig. 10-1)

Was it faster with ortho on?

>  In many situations, polar tracking is more convenient and easier to use than ortho. However, ortho can be very useful when polar is set to angles other than 90° or when polar is turned off.

6. Turn ortho mode off and turn object snap on.

Tracking Time

AutoCAD keeps track of time while you work. With the TIME command, you can review this information.

1. Enter **TIME** at the keyboard.

The AutoCAD Text Window appears, providing information similar to that shown in Fig. 10-2. Of course, the dates and times will be different. Here's what this information means.

Current time:	Current date and time
Created:	Date and time the drawing was created
Last updated:	Date and time the drawing was last updated
Total editing time:	Total time spent editing the current drawing
Elapsed timer (on):	A timer you can reset or turn on or off
Next automatic save in:	Time before the next automatic save occurs

If the current date and time are not displayed on the screen, they were not set correctly in the computer, or the computer's battery may be weak or dead.

2. If the date and time are incorrect, right-click the time located in the lower right corner of the Windows taskbar on your computer screen, pick **Adjust Date/Time**, make the necessary changes, and pick the **OK** button.

3. Press **ESC** to end the TIME command, but do not close the AutoCAD Text Window.

```
AutoCAD Text Window - Drawing1.dwg                    _ □ X
Edit

Command: '_.coords
Enter new value for COORDS <2>: 0
Command:
Command:
Command: _line
Specify first point:
Specify next point or [Undo]:  <Coords on>
Specify next point or [Undo]:  <Coords on>
Specify next point or [Close/Undo]:

Command:  <Ortho on>
Command:  <Osnap off>
Command: time

Current time:              Saturday, April 14, 2012  8:27:28:593 AM
Times for this drawing:
  Created:                 Saturday, April 14, 2012  7:29:01:827 AM
  Last updated:            Saturday, April 14, 2012  7:29:01:827 AM
  Total editing time:      0 days 00:58:26:781
  Elapsed timer (on):      0 days 00:58:26:781
  Next automatic save in: 0 days 00:05:26:969

Enter option [Display/ON/OFF/Reset]:
```

Fig. 10-2

4. Reenter the **TIME** command.

What information is different? Check the current time, total editing time, and elapsed time.

5. With AutoCAD's TIME command entered, type **D** for **Display** and notice what displays on the screen.

This provides updated time information.

6. Enter the **OFF** option and display the time information again.

The elapsed timer should now be off.

 An example of when you might specify OFF is when you want to leave the computer to take a break. When you return, you turn the time back ON. This keeps an accurate record of the actual time (elapsed time) you spend working on a project.

7. Reset the timer by entering **R** and display the time information once again.

The elapsed timer should show 0 days 00:00:00.000.

8. Last, turn on the timer and display the time information.

Notice that the elapsed timer keeps track of the time only while the timer is turned on.

Why is all of this time information important? In a work environment, the TIME command can track the amount of time spent on each project or job, making it easier to charge time to clients.

9. Press **ESC** to end the TIME command, but do not close the AutoCAD Text Window.

AutoCAD Text Window

As you are aware, AutoCAD displays the time information in the AutoCAD Text Window. AutoCAD uses this window for various purposes, as you will see in future chapters. A basic understanding of how to use this window will help you use AutoCAD more efficiently.

Displaying the Window

1. Pick a point in the drawing area outside the AutoCAD Text Window.

This causes the graphics screen to appear in front of the AutoCAD Text Window, hiding it from view.

2. Press the **F2** function key to make the AutoCAD Text Window come to the front.

You can also make it reappear by picking the AutoCAD Text Window button on the Windows taskbar.

3. In the upper right corner of the AutoCAD Text Window, pick the **Minimize** button (the dash).

4. Press **F2** or pick the **AutoCAD Text Window** button on the Windows taskbar to make it reappear.

 You can press F2 at any time to display the AutoCAD Text Window. F2 serves as a toggle switch to toggle the display of the window.

5. Using the vertical scrollbar, scroll up the list of text in the AutoCAD Text Window.

Observe that AutoCAD maintains a complete history of your activity.

Copying and Pasting

Notice that the AutoCAD Text Window offers an Edit pull-down menu. This menu allows you to perform basic copying and pasting functions within the window.

1. Select the **Edit** pull-down menu from the **AutoCAD Text Window.**

Recent Commands provides a cascading menu of the commands you have used recently. The list of recent commands is also available by clicking on the Recent Commands button in the Command Line window.

Copy permits you to copy a selected portion of the text to the Windows Clipboard. The Clipboard is a memory space in the computer that temporarily stores information (text and graphics). After you have copied information to the Clipboard, it is easy to paste it into another software application. The Copy option is grayed out and not available because text has not been highlighted.

Copy History copies the entire contents of the AutoCAD Text Window to the Windows Clipboard. Paste enables you to paste the contents of the

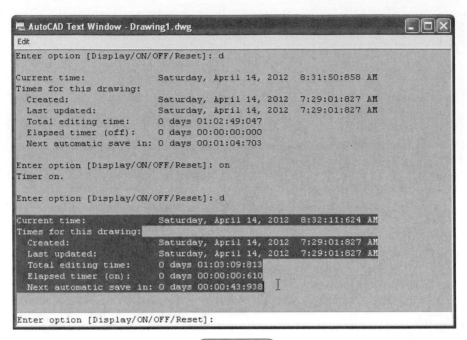

```
AutoCAD Text Window - Drawing1.dwg                          [_][□][X]
 Edit
Enter option [Display/ON/OFF/Reset]: d

Current time:                Saturday, April 14, 2012  8:31:50:858 AM
Times for this drawing:
  Created:                   Saturday, April 14, 2012  7:29:01:827 AM
  Last updated:              Saturday, April 14, 2012  7:29:01:827 AM
  Total editing time:        0 days 01:02:49:047
  Elapsed timer (off):       0 days 00:00:00:000
  Next automatic save in:    0 days 00:01:04:703

Enter option [Display/ON/OFF/Reset]: on
Timer on.

Enter option [Display/ON/OFF/Reset]: d

Current time:                Saturday, April 14, 2012  8:32:11:624 AM
Times for this drawing:
  Created:                   Saturday, April 14, 2012  7:29:01:827 AM
  Last updated:              Saturday, April 14, 2012  7:29:01:827 AM
  Total editing time:        0 days 01:03:09:813
  Elapsed timer (on):        0 days 00:00:00:610
  Next automatic save in:    0 days 00:00:43:938        I

Enter option [Display/ON/OFF/Reset]:
```

Fig. 10-3

Clipboard to the Command line. Paste to CmdLine pastes highlighted text to the Command line. The Options... menu item displays the Options dialog box, which allows you to change various settings to meet individual needs.

2. Using the pointer, highlight the most recent time information, as shown in Fig. 10-3.

3. Pick **Copy** from the **Edit** pull-down menu in the AutoCAD Text Window or press the **CTRL** and **C** keys to copy the text.

4. Close the AutoCAD Text Window by clicking the **Close** button (the **x**) in the upper right corner of the window.

5. Minimize AutoCAD by picking the **Minimize** button (the dash) located in the upper right corner.

6. Launch a text editor such as **Notepad**. (Notepad is a standard Windows program and is part of the Accessories group of Windows utilities.)

7. In the text editor, select **Paste** from the **Edit** menu or press the **CTRL** and **V** keys.

This pastes the time information from the Windows Clipboard. At this point, you could name and save this file and/or print this information.

8. Exit the text editor without saving the file.

9. Click the **AutoCAD** button on the Windows taskbar to make the AutoCAD window reappear.

Command Line Window

Now focus your attention on the Command Line window (the window that contains the Command prompt).

1. Position the pointer so that its tip is touching any part of the Command Line window's left border.

2. With the Windows pointer, click and drag the Command Line window upward into the drawing area until a floating window forms.

You can move the window anywhere on the screen, and you can resize its width and height.

3. Pick the **Customize** button in the title bar of the Command Line window and select the **Transparency...** option.

This displays the Transparency dialog box that was discussed in Chapter 3.

COMMAND LINE WINDOW

4. Move the **General** slider to the far left and pick **OK**.

The Command Line window is now transparent.

5. Pick the **Customize** button again, select the **Transparency...** option, move the **General** slider to the far right, and pick **OK**.

6. Drag the Command Line window downward and dock it into its original position.

COMMAND LINE WINDOW

Copying and Pasting

It is possible to copy and paste text into the Command Line window. You may find this feature useful as you work with AutoLISP, an AutoCAD programming language.

1. Use the scroll bar at the right of the Command Line window to scroll up. The scroll bar is invisible until you move the pointer to the right edge of the Command Line window.

2. Identify a command (any command) and highlight it.

3. With the pointer inside the **Command Line** window, right-click and pick **Paste to CmdLine**.

AutoCAD pastes the highlighted text at the Command line.

4. Pick the **Recent Commands** button in the command prompt area on the Command Line window.

COMMAND LINE WINDOW

5. Pick one of the commands from the list.

AutoCAD enters the command.

6. Press the **ESC** key.

Editing Command Entries

AutoCAD allows you to correct misspellings at the Command line.

1. Type **TME** (a misspelled version of the **TIME** command) at the **Command** line, but do not press **ENTER**.

2. Using the left arrow key, back up and stop between the **T** and **M**.

3. Type the letter **I**.

 The Insert key serves as a toggle to turn on and off the insert mode. If the Insert function is not active, press the Insert key to turn it on.

4. Assuming that **TIME** is now spelled correctly, press **ENTER**.

5. Press **ESC** to cancel.

Table 10-1 briefly describes the keys that you can use to edit text on the Command line.

6. Close the AutoCAD Text Window.

7. On your own, experiment with the Command line navigation options shown in Table 10-1.

Table 10-1. Editing Keys

Key	Action
Left arrow	Moves the cursor back (to the left)
Right arrow	Moves the cursor forward (to the right)
Up arrow	Displays the previous line in the command history
Down arrow	Displays the next line in the command history
HOME	Places the cursor at the beginning of the line
END	Places the cursor at the end of the line
INSERT	Turns on and off the insertion mode
DEL	Deletes the character to the right of the cursor
Backspace key	Deletes the character to the left of the cursor
CTRL + V	Pastes text from the clipboard

Undoing Your Work

As you may recall from earlier chapters, you can undo or redo one or several actions by picking the Undo and Redo buttons in the Quick Access toolbar. AutoCAD also offers an UNDO command that provides further options.

Reversing Multiple Operations

The basic UNDO command is similar to the Undo button. It undoes, or reverses, one or more operations, beginning with the most recent.

1. Draw the objects shown in Fig. 10-4 at any size and location.

2. Enter the **UNDO** command at the keyboard.

3. Review the list of command options. Then enter **2** for the number of operations.

By entering 2, you told AutoCAD to back up two steps.

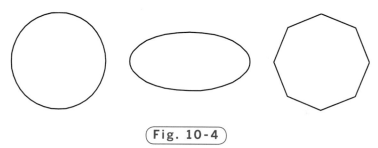

Fig. 10-4

 AutoCAD also includes the U command, which allows you to back up one step at a time. The action of the U command can be reversed by the REDO command.

Returning to a Previous Drawing Status

Suppose you want to proceed with drawing and editing, but you would like the option of returning to this point in the session at a later time. The UNDO command's Mark and Back options allow you to do this efficiently.

1. Draw a small rectangle.

2. Enter the **UNDO** command.

3. Enter the **Mark** option by picking **Mark** on the Command line or by typing **M** and pressing **ENTER**.

DRAW

AutoCAD has (internally) marked this point in the session.

4. Perform several operations, such as drawing and erasing objects and changing the buttons on the status bar.

Now suppose you decide to back up to the point where you drew the rectangle in Step 1.

5. Pick the **Recent Commands** button on the Command line, and then select the **Back** option.

You have just practiced two of the most common uses of the UNDO command. Other UNDO options exist, such as BEgin and End. Refer to AutoCAD's online help for more information on these options.

Controlling the Undo Feature

The undo feature can use a large amount of disk. You may want to disable the UNDO command partially or entirely by using the UNDO Control option.

Let's experiment with the UNDO Control.

1. Enter the **UNDO** command and then the **Control** option.
2. Enter the **One** option.
3. Draw two arcs anywhere in the drawing area.
4. Enter the **UNDO** command.

UNDO now permits you to undo only your last operation.

5. Press **ENTER**.
6. Enter **UNDO**, **Control**, and **All**.

The undo feature is once again fully enabled.

7. Enter **UNDO** and note the complete list of options.
8. Enter **ESC** to cancel the command.
9. Exit AutoCAD without saving.

• REVIEW QUESTIONS

Answer the following questions on a separate sheet of paper.

1. Of what value is the coordinate display?
2. What is the name of the feature that forces all lines to be drawn only vertically or horizontally? What function key controls this feature?
3. Of what value is the TIME command?
4. Briefly explain each of the following components of the TIME command.
   a. Current time
   b. Created
   c. Last updated
   d. Total editing time
   e. Elapsed timer
   f. Next automatic save in
5. Which function key toggles the display of the AutoCAD Text Window on and off?
6. Explain how you would copy a few lines from the AutoCAD Text Window to a text editor such as Notepad.
7. Describe two methods that let you undo your last five operations quickly.
8. Explain the use of the UNDO Mark and Back options.

• CHALLENGE YOUR THINKING

These questions are designed to further your knowledge of AutoCAD by encouraging you to explore the concepts presented in this chapter. Answer each question on a separate sheet of paper.

1. Experiment with ortho in combination with various object snap modes. What happens when you try to snap to a point that is not exactly horizontal or vertical to the previous point?
2. Why might you want to copy parts of the AutoCAD Text Window or Command Line window to the Windows Clipboard?

Continued

• APPLYING AUTOCAD SKILLS

Work the following problems to practice the commands and skills you learned in this chapter.

1. Set the elapsed timer to ON. Complete the design shown in Fig. 10-5 for an over-the-hood storage area for a garage. Use ortho to keep the lines perfectly horizontal and vertical. When you have completed the drawing, set the elapsed timer to OFF and write down the total time it took you to complete the drawing.

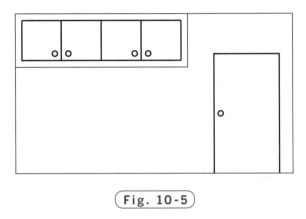

Fig. 10-5

2–4. Compare the use of ortho with object snap tracking and polar tracking by drawing the objects shown in Figs. 10-6 through 10-8. Create a new drawing to hold the objects. Turn on the coordinate display and note the display as you construct each object.

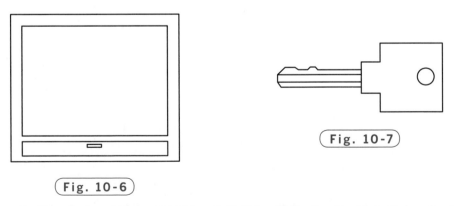

Fig. 10-6

Fig. 10-7

Problems 3, 4, and 5 courtesy of Joseph K. Yabu, Ph.D., San Jose State University

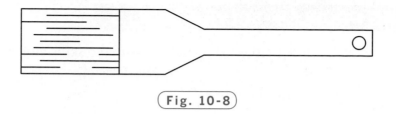

(Fig. 10-8)

5. Create a new drawing file and draw the lamp head shown in Fig. 10-9. Copy everything entered at the Command Line window to the Clipboard and paste it into a text editor such as Notepad. Then print it. (Read problem 6 before completing this problem.)

6. Use AutoCAD's timer to review the time you spent completing problem 5. Print this information or record it on a separate sheet of paper.

7. Draw the simple house elevation shown in Fig. 10-10 at any size with ortho or object snap tracking and polar tracking. Prior to drawing the roof, use UNDO to mark the current location in the drawing. Then draw the roof.

8. With UNDO Back, return to the point prior to drawing the roof. Draw the roof shown in Fig. 10-11 in place of the old roof. Use the UNDO command as necessary as you complete the drawing.

(Fig. 10-9)

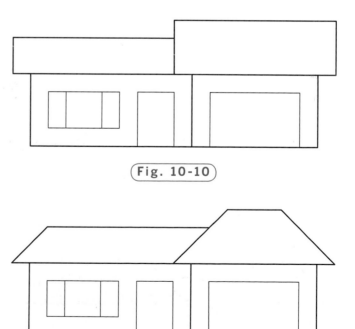

(Fig. 10-10)

(Fig. 10-11)

Continued

● USING PROBLEM-SOLVING SKILLS

Complete the following activities using problem-solving skills and your knowledge of AutoCAD.

1. As the packaging engineer for a major manufacturer, you have been asked to design a support for a new product to hold it in place within the container. The drawing (Fig. 10-12) is not to scale. Create a drawing to scale according to the dimensions given. Keep track of your hours, because the time will be billed to the customer. Do not dimension, but save the drawing.

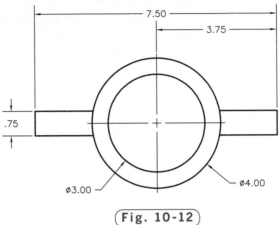

Fig. 10-12

2. The core of an electromagnet is made up of a series of thin "slices," or laminations. The laminations concentrate the electromagnetic field, adding strength. Draw the lamination shown in Fig. 10-13. Keep track of your time. Do not dimension, but save the drawing.

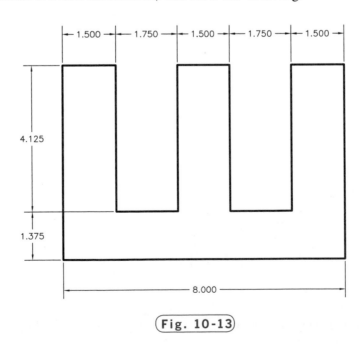

Fig. 10-13

3. Draw the front view of the tube bulk-head shown in Fig. 10-14. Before drawing the holes, use the UNDO command to mark the current location in the drawing. After completing the bulkhead, use UNDO to replace the two holes with two new ones. Draw both of the new holes with a diameter of 1.00. Change the center-to-center distance from 1.40 to 1.60.

4. The two identical hubs shown in Fig. 10-15 are meant to be mounted back-to-back in an assembly. Draw the front view of one hub. Before drawing the four small holes, use the UNDO command to mark the current location in the drawing. After completing the hub, use UNDO to replace the four holes with eight new ones, equally spaced around the circumference. Use a new hole diameter of .284 and a new center line diameter of 2.50.

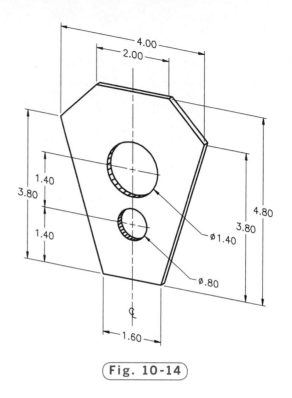

Fig. 10-14

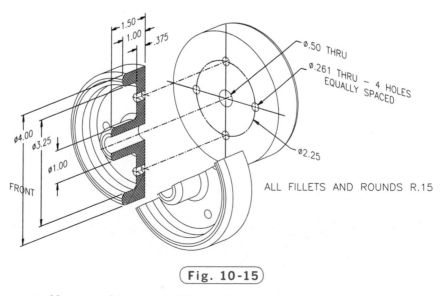

Fig. 10-15

Problems 3 and 4 courtesy of Gary J. Hordemann, Gonzaga University

Chapter 11

Construction Aids

Objectives

- Set and use AutoCAD's visual grid system
- Establish a snap grid that is appropriate to the current drawing task
- Use construction lines to project views in a multiview drawing
- Become more familiar with the **AutoCAD Classic** workspace

Vocabulary

alignment grid
aspect ratio
construction lines
hidden lines
multiview drawing
offsetting
orthographic projection
plane
rays
snap grid
xlines

AutoCAD permits you to create a nonprinting grid system for visual layout of drawings. The snap grid complements the visual grid, assisting with the layout process. Construction lines provide a means for projecting lines from one view to another using orthographic projection. All of these construction aids make it easier to create accurate multiview drawings in the shortest amount of time possible.

Since you were introduced to workspaces in Chapter 3, you've worked exclusively in the Drafting & Annotation workspace. In this chapter, you will learn to use the AutoCAD Classic workspace.

Alignment Grid

The GRID command allows you to set an **alignment grid** of lines of any desired spacing, making it easier to visualize distances and drawing sizes. You can use the grid with both English (Imperial) and metric units and in both 2D and 3D modeling. In this chapter, we will create a drawing of a plastic extrusion using metric units.

1. Start AutoCAD and set the **STARTUP** system variable to **1**.
2. Enter the **NEW** command at the keyboard.

The Create New Drawing dialog box appears.

3. Pick **Start from Scratch**, pick the **Metric** radio button, and pick **OK**.
4. Click on the Workspace pull-down menu on the Quick Access toolbar and select the **AutoCAD Classic** workspace.
5. Close all floating toolbars and palettes that are open on the screen.
6. Move the crosshairs to the upper right corner of the drawing area and read the coordinate display.

Since you are now working in the metric system, the unit of measure is millimeters (mm). The drawing area of 12 × 9" is now approximately 500 × 350 mm, depending on the location of the origin and the size of the window representing the drawing area on your computer.

STATUS BAR

7. Pick the **Grid Display** button located in the status bar.

A grid appears on the screen.

8. Enter the **GRID** command at the keyboard.

Notice that the default value for the grid spacing is 10. This means that the grid lines are currently spaced 10 mm apart.

9. Change the spacing to 20 mm by entering **20**.
10. Reenter the **GRID** command and change the spacing back to 10 mm.
11. Turn the grid off by picking the **Grid Display** button.
12. Make the grid visible again.

STATUS BAR

You can also turn the grid on and off by pressing CTRL G or the F7 function key.

Snap Grid

The **snap grid** is similar to the visual grid, but it is an invisible one. You cannot see it, but you can see its effect when you are inside a command and specifying points. It is like a set of invisible magnetic points. The crosshairs jump from point to point as you move the pointing device. This allows you to lay out drawings quickly, yet you have the freedom to toggle snap off at any time.

STATUS BAR

1. Pick the **Snap Mode** button in the status bar to turn on the snap grid.

 The snap grid is different from object snap, which permits you to snap to points on objects.

You can also turn the snap grid on and off by pressing CTRL B or the F9 function key.

When you turn the snap grid on, it appears at first as though nothing has happened.

DRAW

2. Pick the **Line** button on the docked **Draw** toolbar and slowly move the pointing device and watch closely the movement of the crosshairs.

The crosshairs jump (snap) from point to point.

3. Press **ESC** to exit the LINE command.
4. Enter the **SNAP** command at the keyboard.
5. Enter **5** to specify a snap spacing of 5 mm.
6. Enter the **LINE** command once again, move the pointing device, and note the crosshairs movement.

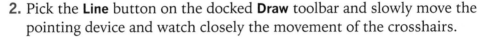

HINT — If the grid is not on, turn it on to better see the movement of the crosshairs. Notice the spacing relationship between the snap resolution and the grid. They are independent of one another.

7. Draw a line of any length and angle.

Notice that the endpoints of the line snap automatically to the invisible snap grid. As you can see, the snap grid can be useful for quickly creating fairly accurate drawings.

Changing the Aspect Ratio

The ratio of height to width of a rectangular region is called the **aspect ratio**. You can change this ratio so that the crosshairs snap one distance vertically and a different distance horizontally. This can be useful when you are laying out a drawing.

1. Enter the **SNAP** command.

Several options appear at the Command line.

2. Select the **Aspect** option or enter **A**.

AutoCAD asks for the horizontal spacing.

3. Enter **5**.

Next AutoCAD asks for the vertical spacing.

4. For now, enter **10**.

5. Enter the **LINE** command and move the crosshairs up and down, then back and forth. Note the difference between the amount of vertical movement and horizontal movement.

6. Press **ESC** to exit the LINE command.

Applying the Snap and Grid Features

The following steps demonstrate how snap and grid can be useful in creating drawings both quickly and accurately.

1. Set snap at **5** mm both horizontally and vertically.

2. Create the drawing of a plastic extrusion shown in Fig. 11-1. Use the pointing device (not the keyboard) to specify all points, and do not include the dimensions.

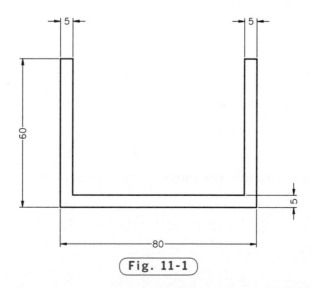

Fig. 11-1

Changing Snap and Grid Settings

The Drafting Settings dialog box permits you to review and make changes to the snap and grid settings.

1. Right-click on the **Snap Mode** or **Grid Display** button and select **Settings...** from the shortcut menu.

The Snap and Grid tab of the Drafting Settings dialog box appears.

As you can see, this dialog box enables you to change the *x* and *y* spacing of both the snap and grid. It also allows you to change the snap type to PolarSnap and use grid dots instead of grid lines. AutoCAD saves these settings with the drawing, so if you start a new drawing, the old settings return.

2. Close the dialog box.
3. Save your work in a file named **snap.dwg**, but do not close the drawing or exit AutoCAD.

Construction Lines

Construction lines, also called **xlines**, are lines of infinite length. They can be used to construct objects and lay out new drawings.

1. Select the **New** button on the Quick Access toolbar, or enter the **NEW** command to start a new drawing.

QUICK ACCESS

2. Select the **Start from Scratch** option and **Metric**.
3. Set the grid to **10** mm and the snap to **5** mm.
4. From the Draw toolbar, pick the **Construction Line** button.

This enters the XLINE command.

DRAW

5. Pick a point anywhere on the screen.
6. Move the crosshairs and notice what happens.

A line of infinite length passes through the first point and crosshairs.

7. Pick a second point anywhere on the screen.

A line freezes into place. As you move the crosshairs, notice that a second line forms.

8. Pick another point so that the intersecting lines approximately form a right angle.
9. Press **ENTER** to terminate the XLINE command.

Unlike the snap and grid construction aids, xlines are AutoCAD objects. You can select them for use with commands, and you can snap to them using object snaps.

10. Erase the xlines.

Constructing Horizontal and Vertical Xlines

The XLINE command provides several options that allow you to create specific types of construction lines.

1. Enter the **XLINE** command and select the **Hor** option on the Command line or enter **H** for Horizontal.

2. Pick a point anywhere on the screen; then move the crosshairs and pick a second point, and a third.

DRAW

Because the Horizontal option is in effect, all of the construction lines are horizontal.

3. Press **ENTER** to terminate the XLINE command and press **ENTER** or the spacebar to reenter it.

4. Select the **Ver** option on the Command line or enter **V** for Vertical and pick three points anywhere on the screen.

The Vertical option forces all the lines to be exactly vertical.

Constructing Xlines at Other Angles

To create construction lines at angles other than 0 or 90 degrees, you can use the Angle option.

1. Press **ENTER** to terminate the **XLINE** command and then reenter it.

2. Select the **Ang** option on the Command line or enter **A** for Angle and enter **45**.

3. Place three construction lines.

4. Terminate the **XLINE** command.

5. Erase all but one of the angled construction lines.

Offsetting Construction Lines

You may find many occasions to use the Offset option of the XLINE command. **Offsetting** a line means creating the line at a specific distance from another line. For example, suppose you have created a line that represents the outer edge of a sidewalk on an architectural site drawing. If the sidewalk is to be 3 feet wide, you can create a construction line at an offset distance of 3 feet to show the width of the sidewalk.

1. Enter **XLINE** and select the **Offset** option on the Command line or enter **O** for Offset.

2. Enter **7** (for 7 mm), select the construction line, and pick a point on either side of it.

You will learn more about offsetting ordinary lines in Chapter 15.

Creating Perpendicular Construction Lines

You can also create a construction line that is perpendicular to other objects on the screen.

STATUS BAR

1. Terminate **XLINE** and then reenter it.

2. Pick the Object Snap button on the status bar to turn on object snap.

3. Right-click the **Object Snap** button and select the **Perpendicular** object snap from the shortcut menu.

4. Move the crosshairs over one of the construction lines. When the Perpendicular object snap marker appears, pick a point to place the new construction line.

The new construction line is perpendicular to the one you selected.

5. Terminate **XLINE** and erase all objects from the screen.

Rays

Rays are a special type of construction line. They extend from a single point into infinity in a radial fashion. Rays are useful when you are laying out objects radially.

1. If the Menu Bar is not displayed, pick the arrow on the right side of the Quick Access toolbar and select **Show Menu Bar** from the pull-down menu.

2. From the **Draw** pull-down menu, pick **Ray**.

This enters the RAY command.

3. Pick a point near the center of the screen.

4. Move the crosshairs in any direction and then pick several points around the first point.

5. Terminate the **RAY** command.

6. Close the current drawing file and do not save your work. (**snap.dwg** should still be open.)

Orthographic Projection

Construction lines, rays, and the grid and snap features are often used to create multiview drawings. A **multiview drawing** is one that describes a three-dimensional object completely using two or more two-dimensional views. These views commonly include a front, top, and side view, although others are often necessary. The AutoCAD features discussed in this chapter, coupled with object snap tracking, help you to produce multiview drawings using a technique known as *orthographic projection.*

As mentioned in Chapter 10, *ortho* means "at right angles." **Orthographic projection**, therefore, is the projection of views at right angles to one another. Each adjacent view is projected at a right angle onto a plane, resulting in a two-dimensional view. A **plane** is an imaginary flat surface used to construct the two-dimensional view.

Figure 11-2 shows an example of a simple multiview drawing created using orthographic projection. The right-side view is located to the right of the front view. This positioning is standard. It is important to use the standard positions because doing so allows other people to understand the position of these views at a glance. The thin lines are temporary construction lines.

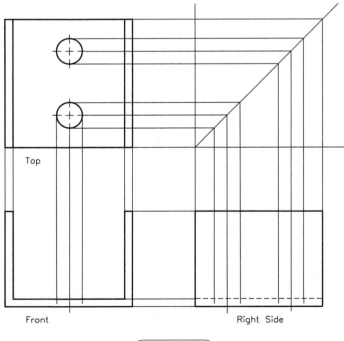

Top

Front   Right Side

Fig. 11-2

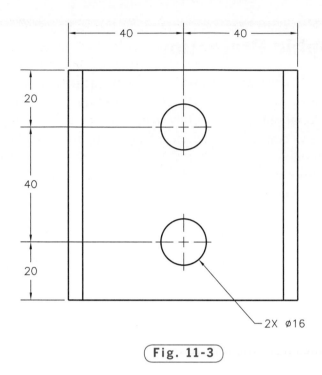

MODIFY

1. Use the **MOVE** command to move the drawing of the plastic extrusion to the lower left area of the screen.

2. Using object snap tracking, snap, and grid, create the top view as shown in Fig. 11-2 on page 151. Use the following information when constructing the top view:

   - From Fig. 11-3, obtain the length dimension of the extrusion and the location and size of the two holes.

   - The broken lines through the holes are center lines. For now, create these lines using a series of line segments. You will learn more about dashed lines, center lines, and other linetypes in Chapter 23.

   - You may need to turn off object snap when picking some of the points. However, object snap must be on for object snap tracking to work.

   - Consider using the **XLINE** command to draw a vertical construction line snapped to the midpoint of one of the horizontal lines to represent the centers of the holes. Then erase the xline.

3. Save your work.

4. After creating the top view, project the holes down to the front view. Consider the following suggestions:

   - Set **Quadrant** as a running object snap mode.

   - Use object snap tracking to acquire the diameter of the holes.

- Draw hidden lines to show the holes in the front view. (**Hidden lines** are the short dashed lines in the multiview drawing that are used to indicate features that are hidden from view.) For now, create them using two short line segments.

5. Save your work.

6. Create the right-side view by projecting lines from the top and front views. Consider the following suggestions:

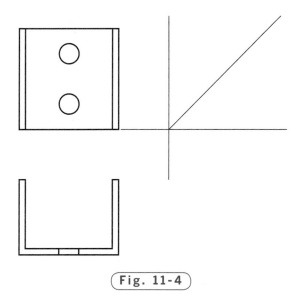

Fig. 11-4

- Create temporary lines as shown in Fig. 11-4. The vertical line is at an arbitrary distance from the top view.

- Draw the angled line at a 45° angle. (Use the Angle option of the XLINE command.) Use it to project the dimensions from the top view to the right-side view.

- Project lines from the top view to the 45° line. Then project vertical lines down from the intersections of the horizontally projected lines and the 45° lines to locate the right-side view.

- Use the height dimensions from the front view to finish the right-side view.

In practice, the views in a multiview drawing are laid out carefully to ensure that the distance between the front and top views is approximately equal to the space between the front and side views. Keep this in mind as you create the construction lines in Step 6.

7. Erase all temporary construction lines.

Your multiview drawing should now look similar to the multiview drawing shown in Fig. 11-2 on page 151, except that your drawing will not include construction lines. You will learn more about multiview drawings in Chapter 25.

8. Close any open toolbars, save your work, and exit AutoCAD.

● REVIEW QUESTIONS

Answer the following questions on a separate sheet of paper.

1. What is the purpose of the grid? How can the grid be toggled on and off quickly?

2. When would you use the snap feature? When would you toggle off the snap feature?

3. How can you set the snap feature so that the crosshairs move a different distance horizontally than vertically?

4. What is the purpose of the XLINE and RAY commands?

5. Explain the purpose of construction lines and rays. Describe a drawing situation in which you could use construction lines or rays to simplify your work.

6. What is orthographic projection? How is it used in creating engineering drawings?

7. What is the purpose of a multiview drawing?

● CHALLENGE YOUR THINKING

These questions are designed to further your knowledge of AutoCAD by encouraging you to explore the concepts presented in this chapter. Answer each question on a separate sheet of paper.

1. AutoCAD provides several different grid and snap systems. Write a short paper describing their differences and similarities.

2. It is possible to accomplish the same tasks using either construction lines and rays or AutoCAD's AutoTrack™ feature. Describe a situation in which you would prefer to use one instead of the other.

• APPLYING AUTOCAD SKILLS

Work the following problems to practice the commands and skills you learned in this chapter.

1. Draw the triangle shown in Fig. 11-5. Turn the grid off and use a snap spacing of .25.

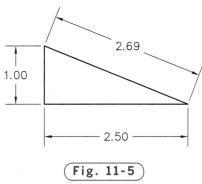

Fig. 11-5

2. Draw the rectangle shown in Fig. 11-6. Set the grid at 1 and snap at .5.

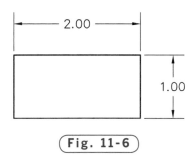

Fig. 11-6

3. Draw the front view of a pipe shown in Fig. 11-7. Use a grid of 2 and turn snap off.

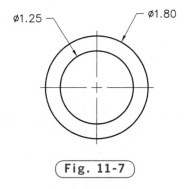

Fig. 11-7

Continued

4. Draw the two views of the rubberized hand grip used for closing the safety bar on a roller coaster car (Fig. 11-8).

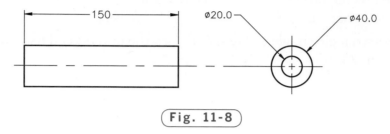

150   Ø20.0   Ø40.0

Fig. 11-8

Begin by drawing two construction lines, as shown in Fig. 11-9. Draw the front view as shown. Then draw the Ø40 circle using the tangent-tangent-radius method. Locate the Ø20 circle to be concentric with the first circle, and erase the construction lines. Do not dimension it, and do not draw the center lines.

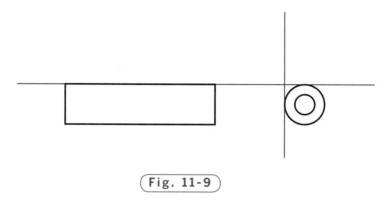

Fig. 11-9

5. Draw three views of the slotted block shown in Fig. 11-10. Start by drawing the front view using the LINE and CIRCLE commands. Do not worry about drawing the object exactly. Make the task easier by using ortho.

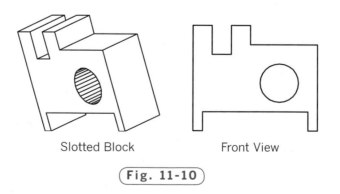

Slotted Block   Front View

Fig. 11-10

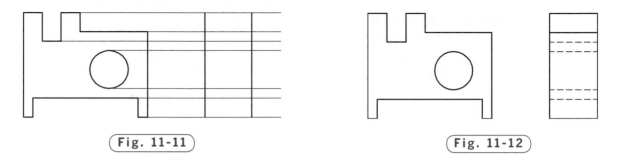

Fig. 11-11

Fig. 11-12

Use the RAY command to extend the edges in the front view into what will become the right-side view. Use the appropriate object snaps to begin the rays at the edges of the front view, as shown in Fig. 11-11.

To begin drawing the right-side view, add two vertical lines (estimate the space between them), as shown in Fig. 11-11. Use the construction lines (rays) to draw the visible lines. Draw the hidden lines as a series of very short lines.

When you are finished with the right-side view, erase the construction lines to reduce confusion in the drawing. The drawing should now look like the one in Fig. 11-12.

Repeat the process to draw the top view. The finished drawing should look like the one in Fig. 11-13.

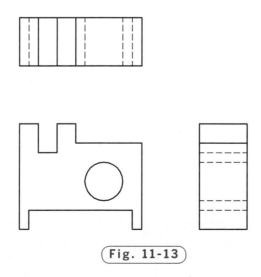

Fig. 11-13

Continued

6. Draw the top, front, and right-side views of the block support shown in Fig. 11-14. Use a grid spacing of .2 and a snap setting of .1. Consider using construction lines or rays to position the three views. For now, draw the hidden and center lines using very short lines. You may wish to change the snap spacing to .05. The grid in Fig. 11-14 is displayed as dots, instead of lines, to show the block support more clearly.

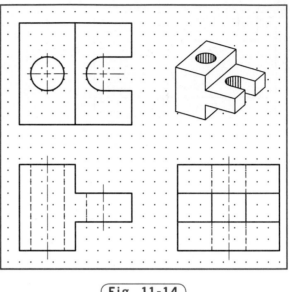

Fig. 11-14

7. Draw the three orthographic views of the bushing holder shown in Fig. 11-15. Use a grid spacing of .2 and a snap setting of .1. Consider using construction lines or rays to help you position the views. The front view is shown as a full section view; i.e., it is shown as if the object has been cut in half. The diagonal lines indicate cut material. You may want to use the Offset option of the XLINE command to create them at even intervals. The grid in Fig. 11-14 is displayed as dots, instead of lines, to show the block support more clearly.

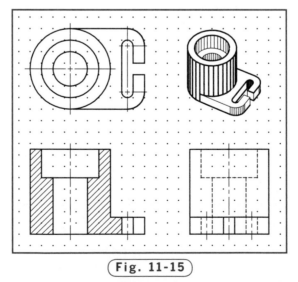

Fig. 11-15

Problems 5, 6, and 7 courtesy of Gary J. Hordemann, Gonzaga University

● USING PROBLEM-SOLVING SKILLS

Complete the following activities using problem-solving skills and your knowledge of AutoCAD.

1. Create the front and right-side views of the block shown in Fig. 11-16. Then, using the horizontal line above the front view as a basis, complete the top view of the block.

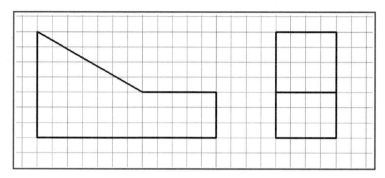

Fig. 11-16

2. Create the front and top views of the object shown in Fig. 11-17. Then, using the horizontal line to the right of the front view as a basis, complete the right-side view of the block.

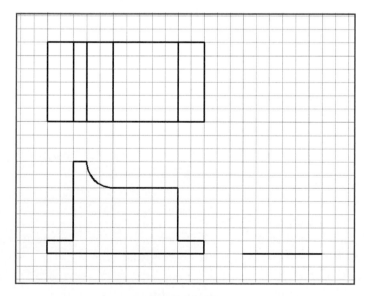

Fig. 11-17

Entrepreneur

One of the exciting things about our country is that we have the freedom to choose our own career paths. If you are interested in health care, you might decide to pursue a career in medicine or nutrition. If you wish to provide a specific type of product or service that few other companies offer, you may consider starting your own business.

Entrepreneurs—people who create and run their own businesses—are found in nearly every field. In the year 2006, about 25,000 of the 253,000 drafting jobs were held by self-employed individuals.

© Brand X Pictures/PunchStock

while others meet with clients in their spare time, in addition to working a full-time job.

Wanted: Motivated, Reliable Professionals

A successful entrepreneur must be self-motivated, patient, and persistent. Building a new business takes time, and when you're the boss, your company's fate is entirely in your hands. Good time-management and decision-making skills are helpful, too. Entrepreneurs must also be able to communicate with others; good relationships with both clients and employees are necessary in order to obtain business and to keep it. One of the most important characteristics, however, is reliability. Clients and employees must know that they can take you at your word.

Working on Your Own

Experienced, independent individuals work in many areas of drafting, but few have entered the profession on their own without first obtaining specialized training. Most have worked initially for a drafting company to learn about drafting techniques. After accumulating years of experience, these individuals decided to obtain their own clients and work for themselves. Some drafting entrepreneurs use their business as their primary means of income,

▶ Career Activities

1. Have you ever tried to run your own business? What skills did you use?

2. Interview an entrepreneur from your area. What was the most difficult problem the person faced? How was the problem solved?

Chapter 12

Zooming

Objectives

- Zoom in on portions of a drawing to view or add details
- Use the **ZOOM** command while another command is entered
- Control view resolution

Vocabulary

realtime zooming
transparent zooms
view resolution

Large drawings with detail can be difficult, and at times impossible, to view legibly on even the largest computer monitors. AutoCAD offers a zooming capability that magnifies areas of the drawing. Remarkably, this capability can magnify objects by as much as ten trillion times. While this may seem impractical for most work, scientists and astronomers who produce drawings and models of incredible scale find this capability very useful. Consider viewing a solar system, in which distances are measured in millions or billions of miles.

Taking a Closer Look

Let's apply the ZOOM command.

1. Start AutoCAD, select the **Use a Wizard** button from the Startup dialog box, pick **Quick Setup**, and pick **OK**.

If the Startup dialog box does not appear when you start AutoCAD, enter the STARTUP system variable and enter 1. Then enter the NEW command.

2. Pick the **Architectural** radio button and **Next**.

3. Enter **20′** for the Width and **15′** for the Length of the drawing area and pick **Finish**.

Be sure to include an apostrophe (') for the foot mark. If you don't, AutoCAD will assume that you mean inches, which would be much too small for the following office drawing. Note that when you work in inches, adding a double quote (") after the number is optional.

4. Select the **AutoCAD Classic** workspace from the Workspace pull-down menu on the Quick Access toolbar and close all floating tool palettes and toolbars.

5. Enter **Z** for ZOOM and **A** for All.

This zooms the drawing to an approximately 20′ × 15′ area.

6. Enter **GRID** and enter **1′** or **12** (inches).

7. Set the snap grid to **2″**.

8. Draw the bedroom shown in Fig. 12-1 using the **LINE** and **RECTANGLE** commands. Estimate the size and location of the bed, couch, coffee table, and flat-panel TV. Omit the dimension lines from your drawing.

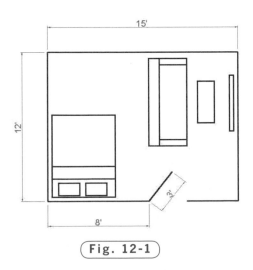

Fig. 12-1

9. Save your work in a file named **zoom.dwg**.

Methods of Zooming

AutoCAD provides several options for zooming in or out on a drawing. You can enter these options by typing Z (for ZOOM) at the keyboard and then selecting from the list of options on the Command Line or entering the first letter of the option, or you can use the many zoom buttons. These buttons are located on the Zoom toolbar and on the flyout on the Zoom Window button on the Standard toolbar. (Depress and hold the pick button to view the flyout.) You can also zoom in and out rotating the center button wheel of the pointing device.

STANDARD

1. Pick the **Zoom Window** button on the **Standard** toolbar.

This enters the ZOOM command.

2. In response to Specify first corner, place a window around the couch and coffee table, as shown in Fig. 12-2. (Pick a point for the first corner and then select a point for the opposite corner.)

The couch and coffee table should magnify to fill most of the screen.

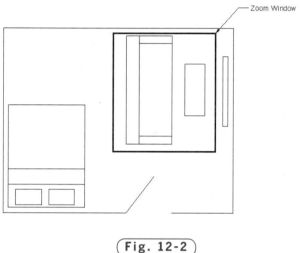

Zoom Window

3. Zoom again so that the coffee table fills most of the screen, draw the items as shown in Fig. 12-3: a water glass, napkin, a couple magazines, and a TV remote. Approximate their sizes, and omit the text. Save your work.

(Fig. 12-2)

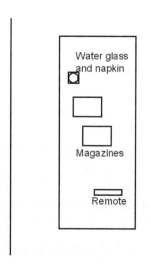

Water glass
and napkin

Magazines

Remote

(Fig. 12-3)

STANDARD

STANDARD

STANDARD

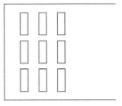

STANDARD

STANDARD

STANDARD

Now let's zoom in on the TV remote control.

4. Pick the **Zoom Window** button again, and this time place a window around the remote.

Let's zoom in further on the remote, this time using the Zoom In option.

5. From Zoom flyout on the Standard toolbar, pick the **Zoom In** button.

6. If you cannot see the left half of the remote, zoom again (using the **Zoom In** or **Zoom Window** button) so that the left half of the remote fills most of the screen.

7. On the left side of the remote, draw several small rectangles to represent buttons, as shown in Fig. 12-4. (You may need to turn snap off first.)

8. Zoom in on one button.

9. Using the **CIRCLE** and **POLYGON** commands, draw a small trademark on the button, as shown in Fig. 12-5.

10. Next, pick the **Zoom Previous** button on the Standard toolbar and watch what happens.

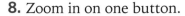

Fig. 12-4

You should now be at the previous zoom factor. The trademark should look smaller.

11. Pick the **Zoom All** button, or enter **Z** and **A** at the keyboard.

AutoCAD zooms the drawing to its original size.

12. Pick the **Zoom Extents** button or enter **Z** and **E** at the keyboard.

AutoCAD zooms the drawing as large as possible while still showing the entire drawing on the screen.

You can also zoom to the extent of a select object or set of objects.

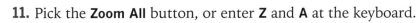

Fig. 12-5

13. Pick the **Zoom Object** button or enter **Z** and **O** at the keyboard.

14. Use a window to select the TV remote control and press **ENTER**.

AutoCAD zooms in, making the selected object the extent of the drawing screen.

15. Pick the **Zoom Extents** button or enter **Z** and **E** at the keyboard.

It is also possible to enter a specific magnification factor. AutoCAD considers the entire drawing area (the ZOOM All display) to have a magnification of 1.

By entering a different number with the ZOOM command entered, you can change the magnification relative to this. If you need to change the magnification relative to the current magnification (instead of 1), you can accomplish this by typing an x after the number.

16. Pick the **Zoom Scale** button and enter **2**.

STANDARD

This causes AutoCAD to magnify the drawing area by two times relative to the ZOOM All display (All = 1).

17. Enter **ZOOM** and **.5**.

This causes AutoCAD to shrink the drawing area by one-half relative to the ZOOM All display.

18. Pick the **Zoom In** button; pick it again.

STANDARD

19. Pick the **Zoom Out** button.

20. Pick the **Zoom Extents** button again.

STANDARD

21. Enter **ZOOM** and **.9x**.

Because you placed an x after the number, the zoom magnification decreases to nine-tenths (90%) of its current size, not nine-tenths of the ZOOM All display area.

STANDARD

22. Continue to practice using the **ZOOM** command by zooming in on the different CAD components of the drawing and including detail on each.

23. Save your work.

Realtime Zooming

AutoCAD's **realtime zooming** allows you to change the level of magnification by moving the cursor. This is one of the fastest ways to zoom.

1. From the Standard toolbar, pick the **Zoom Realtime** button.

STANDARD

This places you in realtime zoom mode. Notice that the crosshairs has changed.

2. Using the pick button, click and drag upward to zoom in on the drawing.

3. Click and drag downward to zoom out on the drawing.

4. Press **ESC** or **ENTER** to cancel realtime zooming.

5. Roll the center wheel button on the mouse to zoom in or out on the drawing.

There is also a Zoom toolbar. It may be useful to display this toolbar instead of using the button flyout on the Standard toolbar.

Zooming Transparently

With AutoCAD, you can perform **transparent zooms**. This means you can use ZOOM while another command is in progress. To do this, enter 'ZOOM (notice the apostrophe) at any prompt that is not asking for a text string.

1. Enter the **LINE** command and pick a point on the screen.

2. At the prompt, enter **'Z**.

AutoCAD lists the options preceded by >> to remind you that you are in transparent mode.

3. Zoom in on any portion of the screen.

Notice the message Resuming LINE command.

4. Pick an endpoint for the line and terminate the command.

 Picking any of the zoom buttons also permits you to perform a transparent zoom.

5. Erase the line and enter **ZOOM All**.

Controlling View Resolution

The accuracy with which curved lines appear on the screen is known as **view resolution**. The VIEWRES command allows you to set the view resolution for objects in the current drawing.

1. Draw an arc anywhere in the drawing area.

2. Enter the **VIEWRES** command.

3. In reply to Do you want fast zooms? accept the Yes default value.

Notice that the circle zoom percent (screen resolution) is 1000 by default. AutoCAD allows you to enter any number from 1 to 20,000.

4. Enter **20** for the circle zoom percent and draw a large arc.

Notice the coarse appearance of the arc.

5. Enter **VIEWRES** again; press **ENTER** and then enter **1000**.

Notice the smooth appearance of the arc on the screen.

6. Erase the arc, save your work, and exit AutoCAD.

● REVIEW QUESTIONS

Answer the following questions on a separate sheet of paper.

1. Explain why the ZOOM command is useful.

2. Cite one example of when it would be necessary to use the ZOOM command to complete an engineering drawing and explain why.

3. Describe each of the following ZOOM options.

   a. All

   b. In

   c. Extents

   d. Previous

   e. Window

4. What is a transparent zoom? When might a transparent zoom be useful?

5. Why might you often choose to use the realtime zoom option instead of other zoom methods?

6. Explain the purpose of the VIEWRES command.

● CHALLENGE YOUR THINKING

These questions are designed to further your knowledge of AutoCAD by encouraging you to explore the concepts presented in this chapter. Answer each question on a separate sheet of paper.

1. AutoCAD drawings are stored in a vector format. Many other graphics programs, such as Microsoft Paint® and Adobe Photoshop®, store their drawings in a raster format. Find out the differences between the raster and vector formats. Can you convert a drawing in a vector format to a raster format? Can you convert a raster drawing into a vector format? Explain.

2. Investigate the ZOOM Center option. What is its purpose? When might you want to use it? Explain.

Continued

● APPLYING AUTOCAD SKILLS

Work the following problems to practice the commands and skills you learned in this chapter.

1. As a general rule, the outside diameter of the washer face on a hex bolt head is the same as the distance across the hex flats. This dimension is approximately $1^1/2$ times the diameter of the bolt shank. Draw the head and washer face of the $^1/4''$ hex bolt shown in Fig. 12-6.

Fig. 12-6

Use the following guidelines:
- Create a new drawing using the Quick Setup wizard, fractional units, and a drawing area of 20 × 15.
- Create the outer edge of the washer face using a diameter of $^7/16$. Place it near the center of the screen.
- Zoom a window so that the circle almost fills the drawing area.
- Draw a hexagon (6-sided polygon) with the same center as the circle and a distance across the flats of $^7/16$. In other words, circumscribe the hexagon around a $^7/16''$ circle.
- Draw a second circle with a diameter of $^1/4''$ to represent the shank of the bolt.
- Enter ZOOM All and notice how small the bolt head is. Even if you had set the drawing area for the default of 12 × 9, it would be impossible to draw the bolt head accurately without using the ZOOM command.

2. Create a drawing such as an elevation plan of a building, a site plan of a land development, or a view of a mechanical part. Using the ZOOM command, zoom in on the drawing and include details. Zoom in and out on the drawing as necessary using the various ZOOM options and transparent zooms.

3. Draw the kitchen floor plan shown in Fig. 12-7. Then zoom in on the kitchen sink. Edit the sink to include the details shown in Fig. 12-8. Save the file as kitchen.dwg.

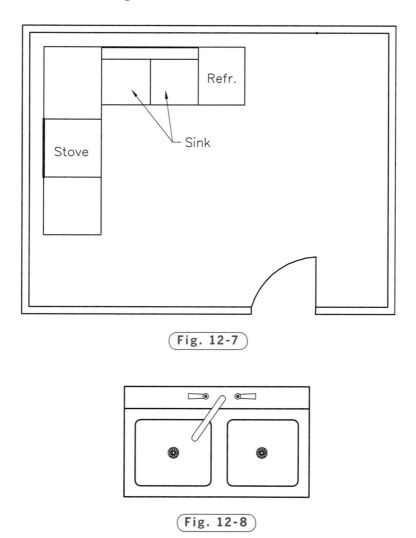

Fig. 12-7

Fig. 12-8

Courtesy Kathleen P. King, Fordham University, Lincoln Center

Continued

● USING PROBLEM-SOLVING SKILLS

Complete the following activities using problem-solving skills and your knowledge of AutoCAD.

1. Use the ZOOM command to draw the door with a fan window shown in Fig. 12-9. Approximate all dimensions.

2. Draw the gasket shown in Fig. 12-10. Do not dimension it. Use construction lines for center lines and erase them when the drawing is complete. Are you sure the 1.5 radius is in fact a radius? Use the ZOOM command to find out. One way to create this drawing is to draw two R.8 circles and one R1.50 circle and locate the four lines with the Tangent object snap. You can then erase the circles and use the ARC command to connect the lines. When you have finished the drawing, use the ZOOM command to ensure that the intersections between lines and arcs are made correctly.

Fig. 12-9

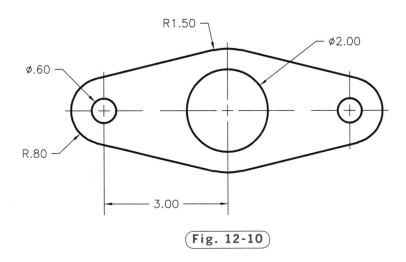

Fig. 12-10

Patent Drafter

Have you ever had a brilliant idea for an invention? Something so simple, perhaps, that you couldn't believe someone else didn't think of it first? Well, then, you have a lot of work to do to make sure that you really are the first one to come up with your idea. Conducting research on previously patented inventions is necessary to avoid "stealing" someone else's idea.

The process inventors go through to obtain a patent is complicated and time-consuming, but if their ideas are truly unique, the feeling of accomplishment is worth the effort. Inventors don't have to do all this work by themselves, however. Many people are available to help make the research and approval process easier. Attorneys and patent drafters are examples of people who play key roles in securing a patent for an invention.

© Stockbyte/Getty Images/RF

Drawing the Idea

Patent drafters produce drawings of both existing products and new product ideas. Drawings of existing products include carefully labeled components; this is necessary to identify parts that may have already been patented.

When designing drawings of new product ideas, drafters use information provided by the inventor (such as rough sketches and notes) to create a picture of what the new product will look like.

It Takes Teamwork

A patent drafter is just one of several individuals who help an idea become an actual product. The inventor's attorney is responsible for keeping the process of obtaining a patent moving. It is the inventor's attorney who typically selects and hires a patent drafter. The drafter must then work closely with the attorney and the inventor to ensure his or her work is consistent with the written description of the item.

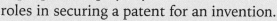

► **Career Activities**

1. Research the patent process on the Internet. How would you go about applying for a patent?

2. Think of an invention you can't imagine living without. Find out who patented it, and when, and how it has changed since its first production.

Chapter 13

Panning and Viewing

Objectives

- Pan from one zoomed area to another
- Produce and use named views

Vocabulary

panning

Zooming in on a drawing solves part of the problem of working with fine detail on a drawing. A method called **panning** solves the other part. Panning permits you to move the zoom window to another part of the drawing while keeping the same zoom magnification. This means you do not have to constantly zoom in and out to examine detail.

Panning

Note the degree of detail in the architectural floor plan shown in Fig. 13-1. The drafter who completed this CAD drawing zoomed in on portions of the floor plan in order to include detail. For example, the drafter zoomed in on the kitchen to place cabinets and appliances, as shown in Fig. 13-2. Suppose the drafter wants to include detail in an adjacent room but wants to simply "move over" to the adjacent room. This operation can be accomplished by panning.

There are several ways to pan in AutoCAD. You can use the PAN command, various panning buttons, the center wheel button on the mouse, or the scroll bars to move around in a drawing.

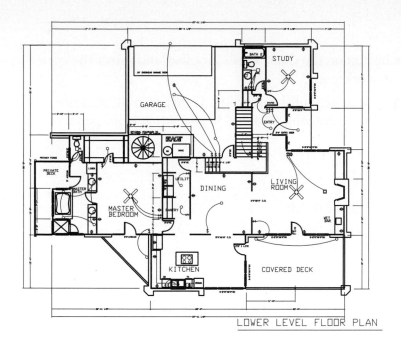

LOWER LEVEL FLOOR PLAN

Fig. 13-1

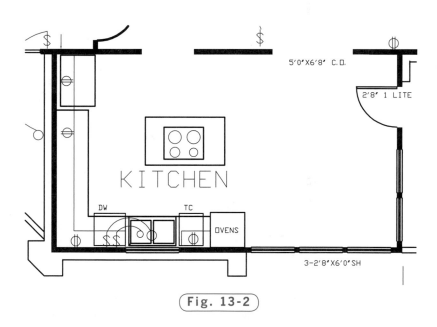

Fig. 13-2

1. Start AutoCAD, select the **Open a Drawing** button from the **Startup** dialog box, and open the drawing named **db_samp.dwg** located in AutoCAD's Sample/Database Connectivity folder.

2. Pick **Save As...** from the Application Menu or the Quick Access toolbar, open the folder with your name, and pick the **Save** button.

3. Pick the **Workspace** button on the Quick Access toolbar and select the **Drafting & Annotation** workspace.

4. Zoom in on the lower right area of the drawing. (The zoom buttons are on the Navigation Bar in the drawing area and in the Navigate 2D panel on the View tab of the Ribbon.)

Realtime Panning

Let's pan to the upper left side of the drawing.

NAVIGATE 2D

1. Pick the **Pan** button in the Navigate 2D panel of the View tab on the Ribbon or enter **P** for Pan.

The crosshairs changes to a hand.

2. In the upper left portion of the screen, click and drag down and to the right.

The drawing window moves down and to the right.

3. Experiment further with the PAN command until you feel comfortable with it. Pan in different directions and at different zoom magnifications.

4. Press **ESC** or **ENTER** to exit the realtime pan mode.

5. Click the center wheel button on the mouse and drag to pan in the drawing area.

Transparent Panning

You can use the PAN command transparently. This is particularly useful for reaching points that are not currently visible. To do this, enter 'PAN (notice the apostrophe) at any prompt that does not ask for a text string.

NAVIGATE 2D

1. Enter the **LINE** command and pick a point anywhere on the screen.

2. Pick the **Pan** button on the Ribbon or enter **'P** at the keyboard.

3. Pan to a new location and press **ESC** or **ENTER** to exit the realtime panning mode.

4. Pick a second point to create a line segment and terminate the LINE command.

5. Erase the line.

Alternating Realtime Zooms and Pans

AutoCAD provides a fast method of alternating between realtime pans and zooms.

1. Pick the **Zoom Realtime** button from the Zoom flyout on the Navigate 2D panel of the Ribbon.

2. Zoom in on the drawing.

3. Right-click the pointing device and select **Pan** from the shortcut menu.

4. Pan to a new location.

5. Right-click and pick one of the **Zoom** options on the shortcut menu.

6. When you are finished, pick **Exit** from the shortcut menu or press **ESC** or **ENTER** to exit the realtime zoom and pan modes.

NAVIGATE 2D

Capturing Views

Imagine that you are working on an architectural floor plan. You have zoomed in on the kitchen to include details such as appliances, and now you are ready to pan over to the master bedroom. Before leaving the kitchen, you foresee a need to return to the kitchen for final touches. However, by the time you're ready to do this final work on the kitchen, you may be working at a different zoom magnification or at the other end of the drawing. The VIEW command solves the problem.

1. Using realtime panning and zooming, find and zoom in on the four offices in the lower right area of the drawing.

2. On the Views panel of the View tab, pick the **View Manager** button, or enter **VIEW** at the AutoCAD prompt.

This displays the View Manager dialog box, as shown in Fig. 13-3.

3. Pick the **New...** button to open the New View/Shot Properties dialog box.

4. For View name, enter **Executives** and pick **OK**.

VIEWS

AutoCAD adds the named view to the list under Model Views.

5. Pick the **OK** button to close the View Manager dialog box.

6. Pan to a new location on the drawing.

7. On the Views panel, scroll through the named views on the menu, find **Executives**, and pick it to restore the named view.

8. Press the **ESC** key to exit the VIEW command.

9. **ZOOM All** and exit AutoCAD without saving.

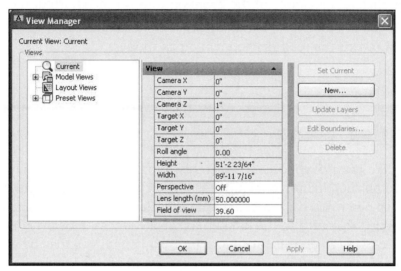

Fig. 13-3

• REVIEW QUESTIONS

Answer the following questions on a separate sheet of paper.

1. Explain why panning is useful.

2. How can you pan to another part of a drawing while another command is entered in AutoCAD?

3. Explain why named views are useful.

4. Explain the method of alternating between realtime pans and zooms.

• CHALLENGE YOUR THINKING

These questions are designed to further your knowledge of AutoCAD by encouraging you to explore the concepts presented in this chapter. Answer each question on a separate sheet of paper.

1. Many zoom and pan variations have been introduced in the last two chapters. Describe how and when you might use a combination of them, but explain also why it may be impractical to use all of them. Which methods of zoom and pan are the fastest, and which seem to be the most practical for most applications?

2. Experiment with using the Zoom and Pan buttons on the Navigation bar on the right side of the drawing area. The Navigation bar is a relatively new feature in AutoCAD. What are its advantages? Disadvantages?

• APPLYING AUTOCAD SKILLS

Work the following problems to practice the commands and skills you learned in this chapter.

1. Open the attrib.dwg file located in the Sample/ActiveX/ExtAttr folder. Use AutoCAD's zoom feature to magnify the parts list located in the upper right area of the drawing. Use realtime zoom and pan to study all the items on the list. What is the part number of the Inner Plate? How many rivets are required for the assembly?

Continued

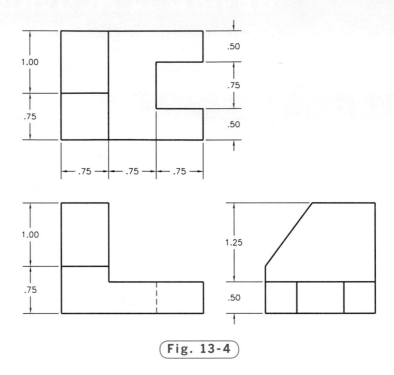

Fig. 13-4

2. Draw each of the views shown in Fig. 13-4 using the dimensions shown. However, do not include the dimensions on the drawing. Zoom in on one of the views and store it as a named view. Then pan to each of the other views and store each as a named view. Restore each named view and alter each shape. Be as creative as you wish, but be certain that all three of the views are consistent with one another.

3. Perform dynamic pans and zooms on the drawing in problem 2.

4. Open the kitchen.dwg file you created in problem 3 of Chapter 12. Add the details for the stove and refrigerator as shown in Fig. 13-5. Save the file as kitchen2.dwg.

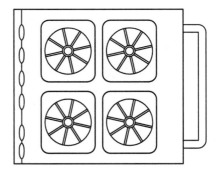

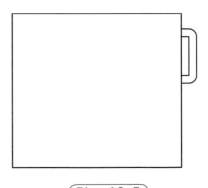

Fig. 13-5

USING PROBLEM-SOLVING SKILLS

Complete the following activities using problem-solving skills and your knowledge of AutoCAD.

1. As the person responsible for computer allocation, you need to know how many offices are without computers. Open the db_samp.dwg file in AutoCAD's Sample/Database Connectivity folder (Fig. 13-6). Using dynamic zoom and pan as necessary, make a list by office number of offices that do not have computers. Do not save any changes you may have made to the drawing.

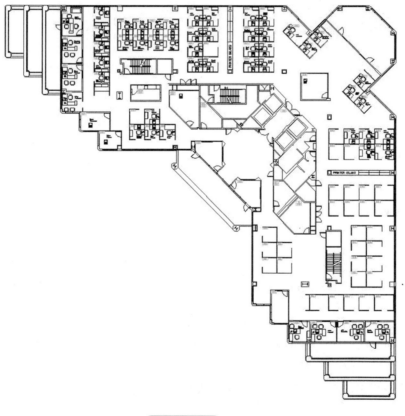

Fig. 13-6

2. Reopen db_samp.dwg and use the Zoom and Pan commands to make a list of rooms with designations of COPY, I.S., STOR., or COFFEE. Which of these rooms should have computers? Do you think office number 6001 should have a computer? Why or why not? Do not save the drawing.

Patterns and Developments

Many manufactured products begin as flat pieces of metal, plastic, or other material. Examples include cereal boxes, heating and cooling ducts, and even items of clothing. Manufacturers use flat *patterns* to cut material to the proper size and shape. The construction of these patterns is known as *development*. Fig. P2-1 shows a box constructed for a company that distributes breakfast cereals, as well as the pattern used to create the box.

▶ Description

Study the magazine holder shown in Fig. P2-2. It is to be manufactured from cardboard, and you are responsible for developing the pattern needed to cut the cardboard to size. Show the fold lines as well as the overall shape of the pattern, as demonstrated in Fig. P2-1.

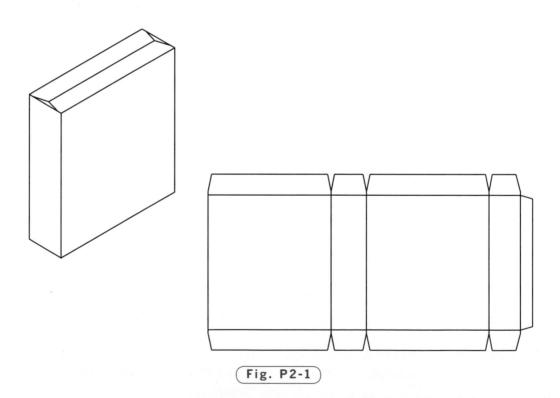

Fig. P2-1

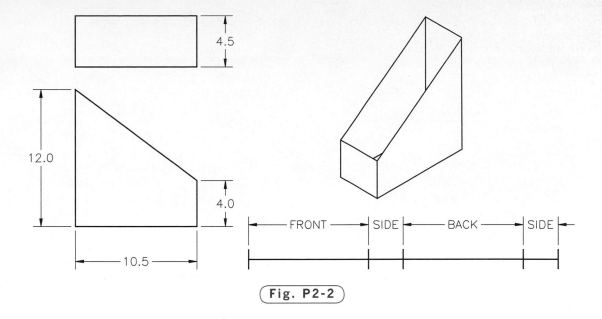

Fig. P2-2

▶ Hints and Suggestions

1. Draw the orthographic views of the magazine holder.

2. Referring to the orthographic views as necessary, create a stretchout line for the magazine holder. A *stretchout line* is a line that shows the total length of the finished pattern and the relative lengths of each part. An example stretchout line for the magazine holder is shown in Fig. P2-2.

3. Using the orthographic views and the stretchout line, create the development. Use object snaps and other construction aids as necessary, and use ZOOM and PAN to simplify your drawing task.

▶ Summary Questions

Your instructor may direct you to answer these questions orally or in writing. You may wish to compare and exchange ideas with other students to help increase efficiency and productivity.

1. Pinpoint any difficulties you encountered with this project and explain how you solved them.

2. Which construction aids did you use to create the orthographic views of the magazine holder? Which aids were used for the development? Explain how these aids helped.

3. How would you create a pattern for a soup can? In what ways would the development be different from that for the magazine holder? Explain.

Solid and Curved Objects

Objectives

- Produce solid-filled objects
- Control the appearance of solid-filled objects
- Create and edit polylines
- Create and edit splines

Vocabulary

B-splines
control points
polyline
screen regeneration
splines
traces
vertex

Many technical drawings contain thick lines and solid objects. Drawings for printed circuit boards, for instance, include relatively thick lines called **traces**. The trace lines represent the metal conductor material that is etched onto the PC board. Drawings and models that use curved lines often take advantage of polylines and spline curves. Examples are consumer products and business machines ranging from phones, pagers, and cosmetic products to printers, copiers, and fax machines. Many of the plastic injection-molded parts on these products are designed using polylines and spline curves.

Drawing Solid Shapes

Note the heavy lines in the drawing shown in Fig. 14-1. They were used to make the drawing descriptive and visually appealing. The SOLID command produces these solid-filled, two-dimensional objects.

1. Start AutoCAD, select the **Start from Scratch** button from the Create New Drawing dialog box, pick **Imperial** units, and pick **OK**.

> If the Startup dialog box does not appear when you start AutoCAD, enter the STARTUP system variable and enter 1. Then enter the NEW command.

2. Pick the **AutoCAD Classic** workspace, and close all floating toolbars and palettes.

3. Enter the **SOLID** command at the keyboard.

4. Produce a two-dimensional, solid-filled object similar to the one in Fig. 14-2. Pick the points in the exact order shown.

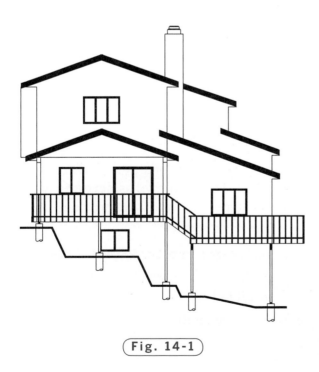

Fig. 14-1

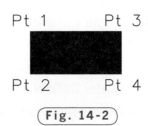

Fig. 14-2

HINT Consider using snap and grid.

5. Pick a fifth and sixth point.

AutoCAD adds a new piece to the solid 2D object.

6. Experiment with the SOLID command. (If you pick the points in a different order, AutoCAD creates an hourglass-shaped object.)

7. Press **ENTER** to terminate the SOLID command.

Keep the objects on the screen because you will need them for the following section.

Controlling Object Fill

With the FILL command, you can control the appearance of 2D solids, wide polylines, multilines, and hatches. Polylines are discussed in the next section of this chapter. You will learn more about multilines in Chapter 15 and hatches in Chapter 18.

FILL is either on or off. When FILL is off, only the outline of a 2D solid is represented on the screen. This saves time when the screen is regenerated.

1. Enter the **FILL** command at the keyboard.

2. Enter **OFF**.

To see the effect of turning FILL on or off, you must force AutoCAD to recalculate all of the vectors on the screen. This is known as **screen regeneration**. To force the screen to regenerate, AutoCAD provides the REGEN command.

3. From the **View** pull-down menu, pick **Regen**, or enter **REGEN** at the keyboard.

4. Reenter the **FILL** command and turn it **ON**.

5. Enter **REGEN** to force another screen regeneration.

6. Erase all objects on the screen.

Polylines

A **polyline** is a connected sequence of line and arc segments that is treated by AutoCAD as a single object. Polylines are often used in lieu of conventional lines and arcs because they are more versatile.

Drawing a Thick Polyline

Let's use the PLINE command to create the electronic symbol for a resistor, as shown in Fig. 14-3.

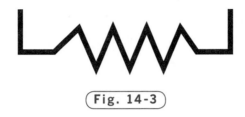

Fig. 14-3

DRAW

1. Set snap at **.5**.
2. Pick the **Polyline** button on the **Draw** toolbar, or enter **PL** (the PLINE command alias).

This enters the PLINE command.

3. Pick a point in the left portion of the screen.

The PLINE options appear.

4. Select the **Width** option or enter **W** and enter a starting and ending width of **.15** unit. (Notice that the ending width value defaults to the starting width value.)
5. Draw the object by approximating the location of the endpoints. If you make a mistake, undo the segment. Press **ENTER** when you have finished drawing the resistor.
6. Click anywhere on the polyline to display its grips.

Notice that the entire polyline is treated as a single object.

7. Save your work in a file named **poly.dwg**.

Editing a Polyline

AutoCAD supplies a special editing command called PEDIT to edit polylines. PEDIT allows you to perform many advanced functions, such as joining polylines, fitting them to curve algorithms, and changing their widths. Let's edit the polyline using the PEDIT command.

MODIFY II

1. Display the Modify II toolbar.

2. Pick the **Edit Polyline** button on the Modify II toolbar.

This enters the PEDIT command.

3. Pick the polyline.

The **PEDIT** options appear.

4. Select the **Width** option or enter **W** and specify a new width of **.2** units.

The width of the polyline changes. As you can see, using polylines can be an advantage when there is a possibility that you will need to change the width of several connected line segments at a later time.

Let's close the polyline, as shown in Fig. 14-4.

5. Select the **Close** option or enter **C**.

The PEDIT Fit option creates a smooth curve based on the locations of the vertices of the polyline. In this text, a **vertex** (plural *vertices*) is any end-point of an individual line or arc segment in a polyline. Let's do a simple curve-fitting operation.

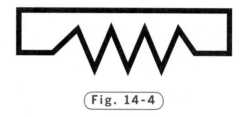

Fig. 14-4

6. Select the **Fit** option or enter **F**.

Notice how the drawing changed. The curve passes through all of the vertices of the polyline.

7. Select the **Decurve** option or enter **D** to return it to its previous form.

Next, let's move one of the vertices, as shown in Fig. 14-5.

8. Select the **Edit vertex** option or enter **E**.

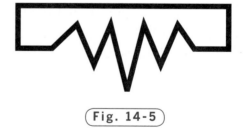

Fig. 14-5

Notice that a new set of choices becomes available. Also notice the x at one of the vertices of the polyline.

9. Move the x to the vertex you want to change by pressing **ENTER** several times.

10. Select the **Move** option or enter **M** and pick a new point for the vertex.

11. To exit the PEDIT command, enter **X** (for **eXit**) twice.

12. Save your work.

PEDIT contains many more editing features. Experiment further with them on your own.

Exploding Polylines

The EXPLODE command gives you the ability to break up a polyline into individual line and arc segments.

1. Enter the **UNDO** command and select the **Mark** option. (We will return to this point at a later time.)

2. Pick the **EXPLODE** button on the Modify toolbar.

3. Select the polyline and press **ENTER**.

MODIFY

Notice the message in the Command Line window: Exploding this polyline has lost width information. The UNDO command will restore it.

This is the result of applying EXPLODE to a polyline that contains a width other than the default. You now have an object that contains many line objects for easier editing, but you have lost the line width.

4. To illustrate that the drawing is now made up of several individual objects, pick one of them.

Drawing Curved Polylines

In some drafting applications, there is a need to draw a series of continuous arcs to represent, for example, a river on a map. If the line requires thickness, the ARC Continue option will not work, but the PLINE Arc option can handle this task.

1. Enter the **PLINE** command and pick a point anywhere on the screen.

2. Enter **A** for Arc.

DRAW

A list of arc-related options appears.

3. Move the crosshairs and notice that an arc begins to develop.

4. Select the **Width** option and enter a starting and ending width of **.1** unit.

5. Pick a point a short distance from the first point.

6. Pick a second point, and a third.

7. Press **ENTER** when you are finished.

8. Enter **UNDO** and **Back**.

Spline Curves

Splines are curves that use sampling points to approximate mathematical functions that define a complex curve. AutoCAD allows you to create splines, also referred to as **B-splines**, using two different methods. You can use the PEDIT command to transform a polyline into a spline curve, or you can use the SPLINE command to generate the spline directly.

Transforming a Polyline into a Spline

The PEDIT Spline option uses the vertices of the selected polyline as the control points of the curve. **Control points** are points that exert a "pull" on a curve, influencing its shape. The curve passes through the first and last control points and is pulled toward the other points, but it does not necessarily pass through them. The more control points you specify in an area, the more pull they exert on the curve.

MODIFY II

1. Pick the **Edit Polyline** button on the Modify II toolbar and select the polyline.
2. Select the **Spline** option or enter **S**.

Do you see the difference between the Spline and Fit options? The Spline option uses the vertices of the selected polyline as control points, whereas the Fit option produces a curve that passes through all vertices of the polyline.

3. Select the **Decurve** option or enter **D** to decurve the object.
4. Enter **X** to exit the PEDIT command.

Through the PEDIT command, AutoCAD offers two spline types: quadratic B-splines and cubic B-splines. An example of each is shown in Fig. 14-6. The type of spline created by the Spline option depends on the setting of the SPLFRAME system variable. (A system variable in AutoCAD is similar to a command, except that it holds temporary settings and values instead of performing a specific function.)

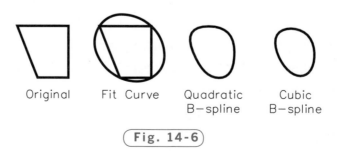

Original   Fit Curve   Quadratic   Cubic
                       B-spline   B-spline

Fig. 14-6

The system variable SPLINETYPE controls the type of spline curve to be generated. Set the value of SPLINETYPE at 5 to generate quadratic B-splines. Set its value at 6 to generate cubic B-splines.

The PEDIT Decurve option enables you to turn a spline back into its frame, as illustrated by the previous Step 3. You can view the spline curve and its frame simultaneously by setting the SPLFRAME system variable to 1.

Producing Splines Directly

With the SPLINE command, you can create spline curves using a single command. Spline curves are used extensively in industry to produce free-form shapes for automobile body panels and the exteriors of aircraft and consumer products. The spline curve itself is the building block for producing interesting and sometimes extremely complex designs.

1. Pick the **Save As...** button on the **Quick Access** toolbar and enter **splines.dwg** for the file name.
2. Erase all the objects in the drawing.
3. Pick the **Spline** button on the Draw toolbar to enter the SPLINE command.
4. Pick a point and then a second point.
5. Slowly move the end of the spline back and forth and notice how the spline behaves.
6. Continue to pick a series of points.
7. Press **ENTER** to complete the spline and terminate the command.
8. Enter **SPLINE** again and pick a series of points.
9. Select the **Close** option or enter **C** to close the spline.
10. Press **ENTER** to terminate the command.

DRAW

The Tolerance option changes the tolerance for fitting the spline curve to the control points. A setting of 0 causes the spline to pass through the control points. By changing the fit tolerance to 1, you allow the spline to pass through points within 1 unit of the actual control points.

You can easily convert a spline-fit polyline into a spline curve. Enter the SPLINE command, enter O for Object, and pick the polyline.

Editing Splines

Using grips, you can interactively edit a spline entity. The SPLINEDIT command provides many additional editing options.

1. Without any command entered, select one of the spline curves to display the grips.
2. If the floating **Quick Properties** panel appears, close it or move it out of the way.
3. Lock onto one of the grips and move it to a new location.

This is the fastest way to edit a spline.

MODIFY II

4. Press **ESC** to clear the grips.
5. From the Modify II toolbar, pick the **Edit Spline** button.

This enters the SPLINEDIT command.

6. Select your first spline.

The editing options appear. The Close option closes the spline curve. The Edit vertex option enables you to Add, Delete, and Move vertices.

7. Select the **Edit vertex** option or enter **E** to display the editing options.
8. Pick the **Add** option or enter **A** to add control points.
9. Pick several points on the spline to add new control points, and then press **ENTER**. (Turn off object snap and the snap grid before adding new control points.)
10. Press **ENTER** three times.

The Reverse option reverses the order of the control points in AutoCAD's database. The Undo option undoes the last edit operation.

Blending Splines

MODIFY

1. Pick the **Blend Curves** button on the Modify toolbar.
2. Pick a point near the end of your first spline, then a point near one end of your second spline.

AutoCAD creates a new spline to blend the two splines. The BLEND commands works with other objects as well, such as lines and arcs.

3. Experiment on your own using the BLEND command on lines, arcs, and splines.
4. Close the Modify II toolbar, save your work, and exit AutoCAD.

Chapter 14 Review & Activities

• REVIEW QUESTIONS

Answer the following questions on a separate sheet of paper.

1. How would you draw a solid-filled triangle using the SOLID command?
2. What is the purpose of FILL, and how is it used?
3. What object types does FILL affect?
4. What is a polyline?
5. In what situations might you decide to use the PLINE command instead of the LINE and ARC commands?
6. Briefly describe each of the following PLINE options.
   a. Arc
   b. Close
   c. Halfwidth
   d. Length
   e. Undo
   f. Width
7. Briefly describe each of the following PEDIT command options.
   a. Close
   b. Join
   c. Width
   d. Edit vertex
   e. Fit
   f. Spline
   g. Decurve
   h. Ltype gen
   i. Reverse
   j. Undo
8. Explain a situation in which you might need to explode a polyline.
9. Describe the SPLINETYPE system variable.
10. What is the fastest method for editing a spline?

Continued

● CHALLENGE YOUR THINKING

These questions are designed to further your knowledge of AutoCAD by encouraging you to explore the concepts presented in this chapter. Answer each question on a separate sheet of paper.

1. Discuss possible uses for the SOLID and FILL commands. Under what circumstances might you use them? Give specific examples.

2. Discuss everyday applications of spline curves.

● APPLYING AUTOCAD SKILLS

Work the following problems to practice the commands and skills you learned in this chapter.

1. Construct the building shown in Fig. 14-7 using the PLINE and SOLID commands. Specify a line width of .05 unit. Estimate the size and shape of the roof.

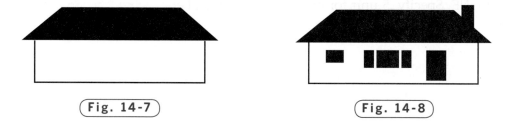

Fig. 14-7   Fig. 14-8

2. After you have completed problem 1, place the solid shapes as indicated in Fig. 14-8. Estimate their sizes and locations.

3. Use the PLINE command to draw a road sign indicating a sharp left corner, as shown in Fig. 14-9.

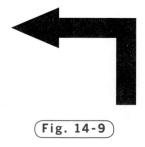

Fig. 14-9

4. The SI symbol shown in Fig. 14-10 is used on metric drawings that conform to the SI system and use what is called *third-angle projection*. This projection system, which places the top view of a multiview drawing above the front view, is the normal way of showing views in the United States. Draw the symbol, including the outlines of the letters S and I; then use the SOLID command to fill in the letters.

(Fig. 14-10)

Courtesy of Gary J. Hordemann, Gonzaga University

5. Create the approximate shape of the racetrack shown in Fig. 14-11 using PLINE. Specify .5 unit for both the starting and ending widths. Select the Arc option to draw the figure.

(Fig. 14-11)

6. Draw the symbols in Fig. 14-12 using the PLINE and PEDIT commands.

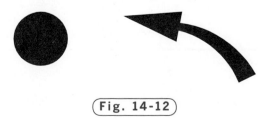

(Fig. 14-12)

Continued

7. Draw the car in Fig. 14-13 using the PLINE and PEDIT commands.

Fig. 14-13

8. The artwork for one side of a small printed circuit board is shown in Fig. 14-14. Reproduce the drawing using donuts and wide polylines. Use the grid to estimate the widths of the polylines and the sizes of the donuts.

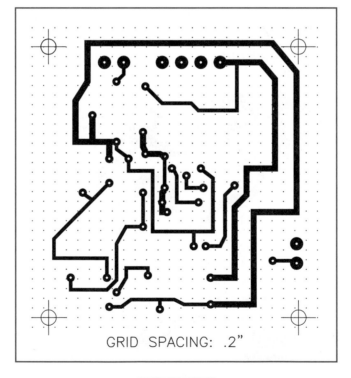

GRID SPACING: .2"

Fig. 14-14

Courtesy of Gary J. Hordemann, Gonzaga University

● USING PROBLEM-SOLVING SKILLS

Complete the following activities using problem-solving skills and your knowledge of AutoCAD.

1. Use the SPLINE command to approximate the route of Interstate 89 as it passes through Chittenden County, Vermont, as shown in Fig. 14-15. Include the Lake Champlain shoreline, but do not include the text. Save the drawing as 89.dwg.

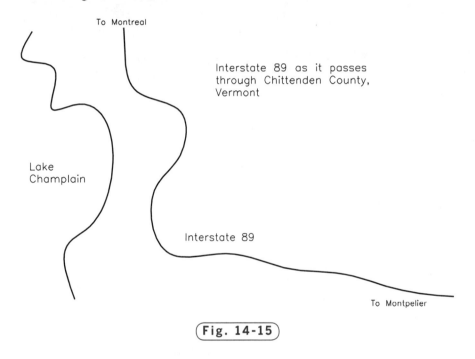

To Montreal

Interstate 89 as it passes through Chittenden County, Vermont

Lake Champlain

Interstate 89

To Montpelier

Fig. 14-15

2. Model railroads can be built to many different scales. One of the more popular scales is known as HO, in which $1/8'' = 1'$. With the PLINE command, design an HO scale railroad track layout in a figure-eight pattern. Make the center intersection at 90 degrees. At a convenient point in your layout, add a spur to terminate in a railroad yard with three parallel tracks for storing the rolling stock. Refer to books in your local library or search the Internet for any additional information you may need to complete this problem.

3. Construct the drawing shown in Fig. 14-1 on page 183.

Chapter 15

Adding and Altering Objects

Objectives

- Create chamfered corners
- Break pieces out of lines, circles, and arcs
- Produce fillets and rounds
- Offset lines and circles
- Create and edit multilines

Vocabulary

chamfers
fillets
image tiles
mline
multiline
rounds

Most consumer products contain rounded or beveled edges for improved aesthetics and functionality. AutoCAD offers quick ways to create rounded inside corners, called **fillets**, and outside corners, called **rounds**, as well as beveled edges, called **chamfers**. Other useful tools include breaking lines and offsetting them to create parallel lines. AutoCAD also gives you the flexibility to produce multiple lines at one time. This is especially useful for architectural drawing, as are the other capabilities mentioned here.

Creating Chamfers

The CHAMFER command enables you to place a chamfer at the corner formed by two lines.

1. Start AutoCAD and open the drawing named **gasket.dwg**. (This is a file you created in Chapter 7.)

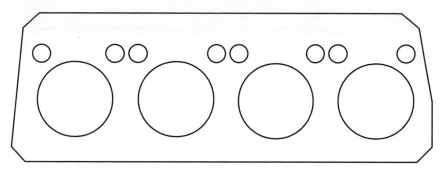

MODIFY

Fig. 15-1

2. Select the **Drafting & Annotation** workspace.

3. Pick the **Chamfer** button on the Modify panel of the Ribbon. (You may need to pick the down arrow next to the Fillet button and select the Chamfer button from the flyout.)

4. Select the **Distance** option or enter **D**.

The distance you specify is the distance from the intersection of the two lines (the corner) to the start of the bevel, or chamfer. You can set the chamfer distance for the two lines independently.

5. Specify a chamfer distance of **.25** unit for both the first and second distances.

6. Place a chamfer at each of the corners of the gasket by picking the two lines that make up each corner. Notice that AutoCAD previews the chamfer when the pointing device passes over the second line.

When you are finished, the drawing should look similar to the one in Fig. 15-1, with the possible exception of the holes.

7. Obtain AutoCAD online help to learn about the other options offered by the CHAMFER command.

HINT   Enter the CHAMFER command and press F1.

Breaking Objects

Let's remove (break out) sections of the gasket so that it looks like the drawing in Fig. 15-2 on page 198. As you know, the bottom edge of the gasket was drawn as a single, continuous line. Therefore, if you were to use the ERASE command, it would erase the entire line. The BREAK command, however, allows you to "break" certain objects such as lines, arcs, and circles.

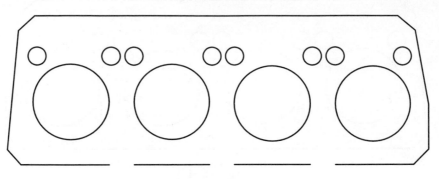

Fig. 15-2

MODIFY

1. From the expanded Modify panel on the Ribbon, pick the **Break** button.

This enters the BREAK command.

2. Turn off object snap if it is on.

3. Pick a point on the line where you would like the break to begin. Since the locations of the breaks in Fig. 15-2 are not dimensioned, you may approximate the location of each start point.

4. Pick the point where you would like the break to end.

The piece of the line between the first and second points disappears. Let's break out two more sections of approximately equal size, as shown in Fig. 15-2.

5. Repeat Steps 1 through 4 to create each break.

6. Experiment with the other BREAK options on your own. Refer to AutoCAD's online help for more information about these options.

7. Insert arcs along the broken edge of the gasket, as shown in Fig. 15-3.

8. Save your work.

Let's break out a section of one of the holes in the gasket, as shown in Fig. 15-4.

MODIFY

9. Pick the **Break** button or enter **BREAK** at the keyboard.

10. Pick a point anywhere on the circle. (You may need to turn off object snap.)

11. Instead of picking the second point, enter **F** for first point.

12. Pick the first (lowest) point on the circle.

13. Working counterclockwise, pick the second point.

A piece of the circle disappears.

QUICK ACCESS

14. Pick the **Undo** button on the Quick Access toolbar or enter **U**.

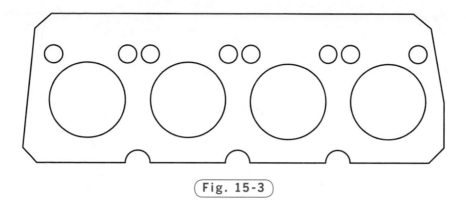

Fig. 15-3

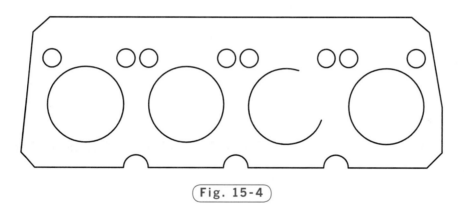

Fig. 15-4

Creating Fillets and Rounds

In AutoCAD, the FILLET command creates both fillets and rounds on any combination of two lines, arcs, or circles. Let's change the chamfered corners on the gasket to rounded corners.

1. Erase each of the four chamfered corners.

2. Pick the arrow next to the Chamfer button on the Modify panel, pick the **Fillet** button, select the **Radius** option, and enter a new radius of **.3**.

3. Produce fillets at each of the four corners of the gasket by picking each pair of lines. Notice that AutoCAD previews the fillet when the pointing device passes over the second line.

4. Save your work.

The gasket drawing should now look similar to the one in Fig. 15-5 on page 200.

MODIFY

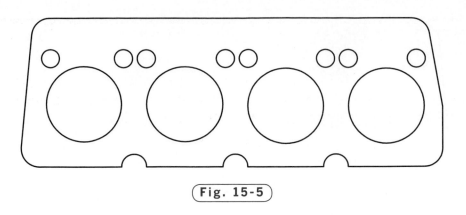

Fig. 15-5

Extending to Form a Corner

Let's move away from the gasket and try something new.

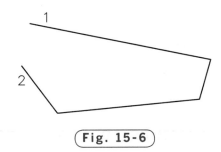

Fig. 15-6

MODIFY

1. Above the gasket, draw lines similar to the ones shown in Fig. 15-6. Omit the numbers.

2. Enter the **FILLET** command and set the fillet radius at **0**.

3. Select lines 1 and 2.

The two lines extend to meet at a corner with a 0 radius, closing the figure. This technique works with the CHAMFER command, too.

Trim Option

The Trim option is useful for making slotted holes.

MODIFY

1. Above the gasket drawing, draw two relatively short lines that are parallel.

2. Enter the **FILLET** command, select the **Trim** option or enter **T** and press **ENTER** to accept the Trim default value.

3. In response to Select first object, pick the end of one of the two lines.

4. In response to Select second object, pick the same end of the other line.

AutoCAD fits an arc to the two lines.

5. Repeat Steps 2 through 4 for the other ends of the two lines.

Offsetting Objects

Offsetting a line results in another line at a specific distance, or offset, from the original line. Offsetting a circle or arc results in another circle or arc that is concentric (shares the same center point) with the original one.

Offsetting Circles

1. From the Modify panel on the Ribbon, pick the **Offset** button.

This enters the **OFFSET** command.

MODIFY

2. For the offset distance, enter **.2**.
3. Select one of the large holes in the gasket drawing.
4. Pick a point inside the hole in reply to Specify point on side to offset.
5. Select another hole and pick a side to offset.
6. Press **ENTER** to terminate the command.
7. Delete the two offset circles.

Offsetting Through a Specified Point

1. Using the **LINE** command, draw a triangle of any size and shape.
2. Pick the **Offset** button and select the **Through** option or enter **T**.
3. Pick any one of the three lines that make up the triangle. (You may need to turn off object snap.)
4. Pick a point a short distance from the line and outside the triangle.

MODIFY

The offset line appears. Notice that the line runs through the point you picked.

5. Do the same with the remaining two lines in the triangle so that the object looks similar to the one in Fig. 15-7.
6. Enter **CHAMFER** and set the first and second chamfer distances to **0**.
7. Pick two of the new offset lines.
8. Do this again at the remaining two corners to complete the second triangle.

Fig. 15-7

9. Clean up the drawing so that only the gasket remains.
10. Save your work and close the drawing file.

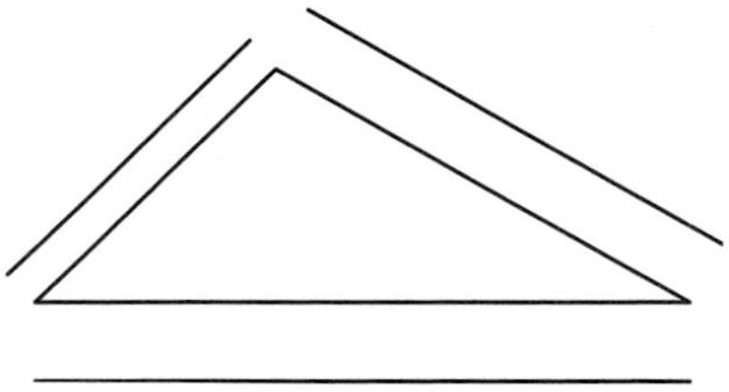

Drawing Multilines

The OFFSET command is useful for adding offset lines to existing lines, arcs, and circles. If you want to produce up to 16 parallel lines, you can do so easily—and simultaneously—with the MLINE command. This command produces parallel lines that exist as a single object in AutoCAD. The object is called a **multiline** or **mline** object.

QUICK ACCESS

1. Pick the down-facing arrow on the Quick Access toolbar and select **Show Menu Bar** from the pull-down menu.

2. On the Quick Access toolbar, pick the **New** button.

3. Select the **Start from Scratch** button, select **Imperial** units, and pick **OK**.

4. From the **Draw** menu on the **Menu Bar**, pick **Multiline**.

This enters the **MLINE** command.

5. Create a large triangle and enter **C** to close it.

Notice that the corners of the triangle meet perfectly. The triangle is a single object—an mline object.

6. Erase the triangle.

Creating Multiline Styles

The Multiline Style dialog box permits you to change the appearance of multilines and save custom multiline styles. Let's use this dialog box to produce the drawing of a room, as shown in Fig. 15-8.

1. Set snap at **.25** units and turn on either polar tracking or ortho.

2. From the Format menu on the Menu Bar, pick the **Multiline Style...** item.

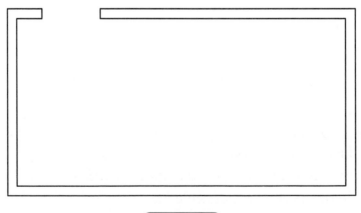

Fig. 15-8

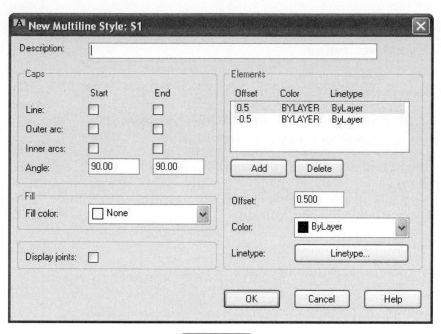

Fig. 15-9

The Multiline Style dialog box appears.

3. Pick the **New...** button.

4. Type **S1** and pick the **Continue** button.

The New Multiline Style dialog box appears, as shown in Fig. 15-9.

5. In the box located to the right of Description, type **This style has end caps**.

AutoCAD does not require you to enter a description for multiline styles you create. However, it is good practice to document your styles so that you can see at a glance which style is appropriate if you need to change or update the drawing later.

6. On the left side under Caps, check the Line Start and Line End boxes.

Many other options are available in the New Multiline Style dialog box. You can control the number of line elements and the color, linetype, and lineweight of each element. You can also set the distance between lines by changing the Offset values.

7. Pick the **OK** button.

8. Pick the **Save...** button.

9. In the Save Multiline Style dialog box, pick the **Save** button to store the S1 style in the file named acad.min.

Notice that the current style is still STANDARD. Before you can use a new style, you must make it the current style.

10. Select **S1** and pick the **Set Current** button.

Look in the Preview area in the Multiline Style dialog box. Notice that both ends of the S1 multiline are "capped" with short connecting lines.

11. Pick the **OK** button.

When you create new multilines, the S1 style will apply.

Scaling Multilines

You can set the distance between lines in a multiline using the Scale option of the MLINE command.

1. Enter **MLINE** at the keyboard.

2. Select the **Scale** option or enter **S** and enter **.25**.

3. With **MLINE** entered, create the drawing shown in Fig. 15-8 on page 202, approximating its size.

Use the Undo button if you need to back up one line segment.

4. Press **ENTER** to terminate MLINE.

5. Save your work in a file named **multi.dwg**.

Let's add the interior walls shown in Fig. 15-10.

6. Display the Multiline Style dialog box.

7. Create a new multiline style named **S2** with the following description: **This style has a start end cap.**

8. Uncheck the **Line End** check box located under Caps and pick **OK**.

9. Save S2 and make it the current style.

10. Pick **OK** to exit the Multiline Style dialog box.

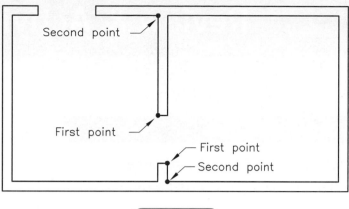

Fig. 15-10

11. Using **MLINE**, draw the interior walls to match those shown in Fig. 15-10. Press **ENTER** to complete each wall. (You may need to turn object snap off before performing this step.)

Editing Intersections

Using the Multiline Edit Tools dialog box, it is possible to edit the intersection of multilines.

1. Pick **Object** and then **Multiline...** from the **Modify** menu on the Menu Bar.

This enters the MLEDIT command and displays the Multilines Edit Tools dialog box, as shown in Fig. 15-11. The twelve items in the dialog box are called **image tiles**.

2. Pick the **Open Tee** image tile.

3. In response to Select first mline, pick one of the two interior walls.

4. In response to Select second mline, pick the adjacent exterior wall.

AutoCAD breaks the exterior wall.

5. Edit the second interior wall by repeating Steps 3 and 4.

6. Press **ENTER** to terminate the MLEDIT command.

7. On your own, explore the options in the Multiline Style and Multilines Edit Tools dialog boxes.

8. Save your work and exit AutoCAD.

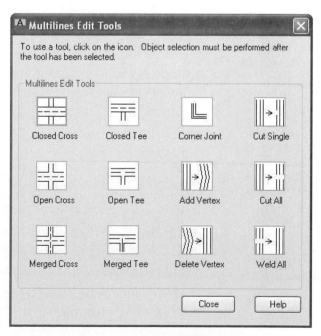

Fig. 15-11

• REVIEW QUESTIONS

Answer the following questions on a separate sheet of paper.

1. What is the function of the CHAMFER command?
2. How is using the BREAK command different from using the ERASE command?
3. If you want to break a circle or arc, in which direction do you move when specifying points: clockwise or counterclockwise?
4. On what panel on the Ribbon is the Fillet button found?
5. How do you set the fillet radius?
6. What can you accomplish by setting either FILLET or CHAMFER to 0?
7. Explain the purpose of the OFFSET command.
8. How might multilines be useful? Give at least one example.
9. Explain the purpose of the Multiline Style and Multilines Edit Tools dialog boxes.

• CHALLENGE YOUR THINKING

These questions are designed to further your knowledge of AutoCAD by encouraging you to explore the concepts presented in this chapter. Answer each question on a separate sheet of paper.

1. Explore the Angle option of the CHAMFER command. How is it different from the Distance option? Discuss situations in which each option (Angle and Distance) may be useful.
2. To help potential clients understand floor plans, the architectural firm for which you work draws its floor plans showing the walls as solid gray lines the thickness of the wall. Exterior walls are 6″ thick, and interior walls are 5″ thick. Explain how you could achieve these walls using MLINE and the Multiline Style dialog box.
3. AutoCAD allows you to save multiline styles and then reload them later. Explore this feature and write a short paragraph explaining how to save and restore multilines.

• APPLYING AUTOCAD SKILLS

Work the following problems to practice the commands and skills you learned in this chapter.

1. Create the first drawing on the left in Fig. 15-12. Don't worry about exact sizes and locations, but do use snap and ortho. Then use FILLET to change it to the second drawing. Set the fillet radius at .2 units.

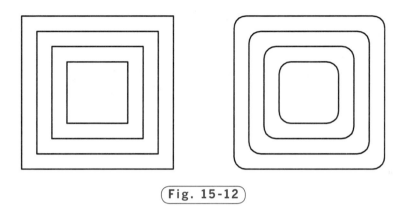

Fig. 15-12

2. Create the triangle shown in Fig. 15-13. Then use CHAMFER to change it into a hexagon. Set the chamfer distance at .66 units.

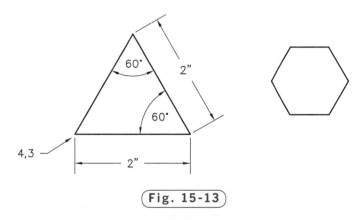

Fig. 15-13

Continued

3. Create the roller and cradle for a conveyor. Begin by drawing a circle and three lines, as shown in Fig. 15-14. Specify a fillet radius of .25 and create the fillets on the left side, as shown in Fig. 15-15. Then fillet the right side. The finished roller and cradle should look like the one in Fig. 15-16.

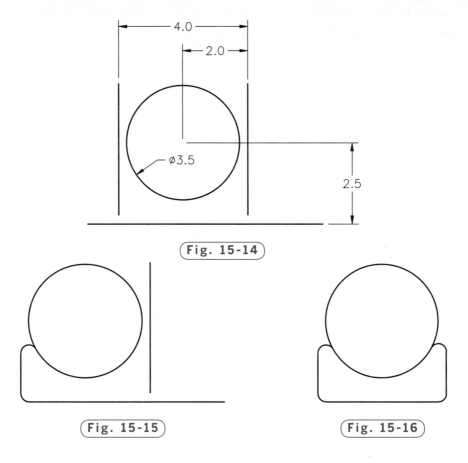

Fig. 15-14

Fig. 15-15

Fig. 15-16

4. Draw the steel base shown in Fig. 15-17. Use the FILLET command to place the fillets and rounds in the corners as indicated.

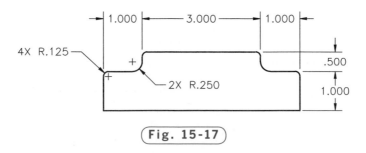

Fig. 15-17

5. Draw the shaft shown in Fig. 15-18. Make it 2 units long by 1 unit in diameter. Use the CHAMFER command to place a chamfer at each corner. Specify .125 for both the first and second chamfer distances.

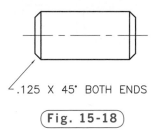

.125 X 45° BOTH ENDS

Fig. 15-18

6. Using the ARC and CHAMFER commands, construct the detail of the flat-blade screwdriver tip shown in Fig. 15-19.

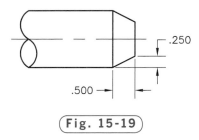

.250

.500

Fig. 15-19

7. Create the picture frame shown in Fig. 15-20. First create the drawing on the left using the LINE and OFFSET commands. Then modify it to look like the drawing on the right using the CHAMFER command.

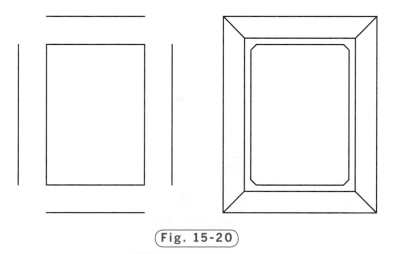

Fig. 15-20

Courtesy of Joseph K. Yabu, Ph.D., San Jose State University

Continued

8. Create the A.C. plug shown in Fig. 15-21 using the MLINE, MLSTYLE, CHAMFER, and FILLET commands.

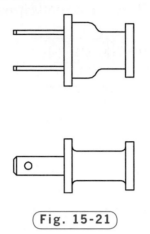

Fig. 15-21

Courtesy of Joseph K. Yabu, Ph.D., San Jose State University

9. Draw the front view of the rod guide end cap shown in Fig. 15-22. Use the OFFSET command to create the outer circular shapes and the inner horizontal lines. Use the BREAK and FILLET commands to create the inner shape with four fillets. Do not dimension the drawing.

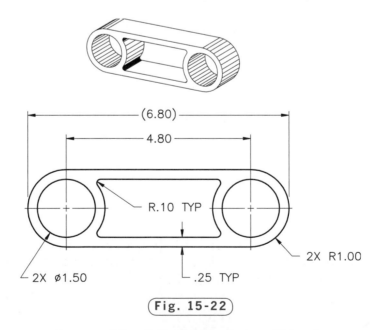

Fig. 15-22

Courtesy of Gary J. Hordemann, Gonzaga University

• USING PROBLEM-SOLVING SKILLS

Complete the following activities using problem-solving skills and your knowledge of AutoCAD.

1. Draw the front view of the ring separator shown in Fig. 15-23. Start by using the CIRCLE and BREAK commands to create a pair of arcs. Then use OFFSET to create the rest of the arcs. Do not dimension the drawing.

2. Kitchens are arranged around their three most important elements: the sink, the range, and the refrigerator. The traffic pattern is arranged in a triangle, with a vertex at each of the three items. As a result, kitchens are usually arranged in one of three patterns: U-shaped, L-shaped, or parallel. Draw an example of each type of kitchen using the MLINE command with capped ends. Do not label the appliances, but draw the appliances as shown in Fig. 15-24. Show the triangular traffic pattern for each kitchen you design. Which seems most convenient to you? Why?

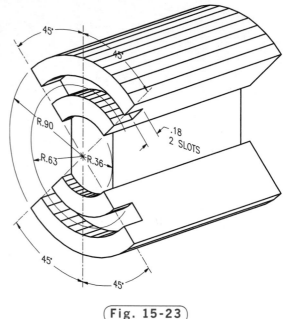

Fig. 15-23

Courtesy of Gary J. Hordemann, Gonzaga University

Symbols for Kitchen

Fig. 15-24

3. Select one of the kitchen arrangements you created in problem 2 and design a kitchen around it. Include the walls, door and window placements, and any other appliances or furniture that you think a kitchen should include.

Chapter 16

Moving and Duplicating Objects

Objectives

- Change an object's properties
- Move and copy objects
- Mirror objects and parts around an axis
- Produce rectangular and polar arrays

Vocabulary

array
associative array
circular array
mirror
mirror line
polar array
rectangular array

Earlier in this book, you learned how to move and copy objects using the grips method. Now you will learn how to perform these and other operations using commands. Why would you want to? The commands give you additional options. Also, if you choose to write scripts or programs that automate certain AutoCAD operations, you could embed these commands into them, so knowing how they work is important.

You will also learn to create arrays of objects in rectangular and polar arrangements, as well as along a path that you specify. The arraying features are useful for many purposes, including architectural drafting, facilities planning, auditorium and stadium design, and countless machine design applications. The problems at the end of this chapter provide examples.

Changing Object Properties

The Quick Properties panel provides a method for changing several of an object's properties.

1. Start AutoCAD and open the drawing named **gasket.dwg**.

2. Select the **AutoCAD Classic** workspace and close all floating toolbars and palettes.

3. Turn **Quick Properties** on.

STATUS BAR

You can turn Quick Properties on and off with the Quick Properties button on the status bar.

4. Select one of the largest holes in the gasket.

The Quick Properties panel appears, as shown in Fig. 16-1. Review the list of object properties. Note that the properties listed in the palette depend on the object selected. In this case, the options are specific to the circle you selected in Step 4.

5. Enter a new number in the Diameter box and press **ENTER**.

The diameter of the circle changes. You will use the Quick Properties panel and the larger, more thorough Properties palette many times throughout this book.

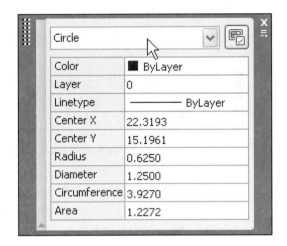

Fig. 16-1

6. Change the diameter back to its original value and press **ENTER**.

7. Move the crosshairs out of the Quick Properties panel and press the **ESC** key to remove the object selection and close the Quick Properties panel.

Moving Objects

Let's move the gasket to the top of the screen using the MOVE command.

1. From the Modify toolbar, pick the **Move** button, or enter **MOVE** at the keyboard.

MODIFY

AutoCAD displays the Select objects prompt.

2. Place a window around the entire gasket drawing and press **ENTER**.

3. In reply to Specify base point or displacement, place a base point somewhere on or near the gasket drawing, as shown in Fig. 16-2.

4. Move the pointing device in the direction of the second point (destination).

5. Pick the second point.

The gasket should move accordingly.

MODIFY

6. Practice using the MOVE command by moving the drawing to the bottom of the screen.

7. Reenter the **MOVE** command.

8. In response to the Select objects prompt, pick two of the four large holes and press **ENTER** (or right-click).

9. Specify the first point (base point). Place the point anywhere on or near either of the two holes.

10. In reply to Specify second point or displacement, move the pointing device upward. (Polar tracking should be on.)

11. Enter **3** to move the holes vertically 3 units.

12. Undo the move.

Second Point (Destination)

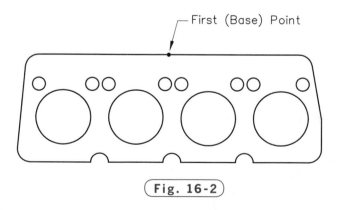
First (Base) Point

Fig. 16-2

Copying Objects

The COPY command is similar to the MOVE command. The only difference is that the COPY command keeps the selection in its original location and "moves" a copy of it to the location you specify.

1. Erase all of the large holes in the gasket except for one.
2. From the Modify toolbar, pick the **Copy** button.

MODIFY

This enters the COPY command.

3. Select the remaining large hole and press **ENTER** or right-click.
4. Select the center of the hole for the base point.

 Use the Center object snap. Also, turn on ortho or polar tracking before completing the following step.

5. Move the crosshairs and place the hole in the proper location.
6. Repeat Step 5 until all four large holes are in place; then press **ENTER**.
7. Practice using the COPY command by erasing and copying the small holes.
8. Save your work.

Mirroring Objects

There are times when it is necessary to **mirror** (produce a mirror image of) a drawing, detail, or part. A simple copy of the object is not adequate because the object being copied must be reversed, as was done with the butterfly in Fig. 16-3. One side of the butterfly was drawn and then mirrored to produce the other side. The mirroring feature can save a large amount of time when you are drawing a symmetrical object or part.

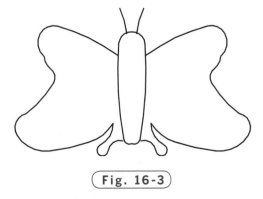

Fig. 16-3

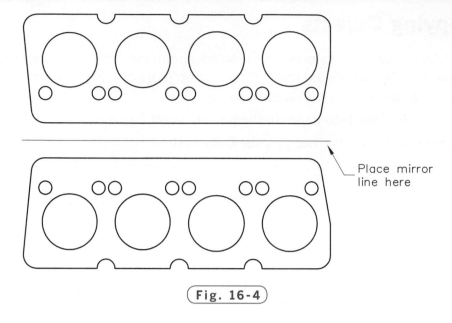

Place mirror
line here

Fig. 16-4

Mirroring Vertically

The same is true if the engine head gasket we developed is to be reproduced to represent the opposite side of an eight-cylinder engine. In this case, we will mirror the gasket vertically (around a horizontal plane).

MODIFY

1. If necessary, move the gasket to the bottom of the screen to allow space for another gasket of the same size.
2. From the Modify toolbar, pick the **Mirror** button.

This enters the MIRROR command.

3. Select the gasket by placing a window around it and press **ENTER**.
4. Create a horizontal **mirror line** (axis) near the gasket by selecting two points on a horizontal plane, as shown in Fig. 16-4.

AutoCAD asks if you want to delete the source objects.

5. Enter **N** for No, or press **ENTER** since the default is No.

AutoCAD mirrors the gasket as shown in Fig. 16-4.

Mirroring at Other Angles

You can also create mirror images around a mirror line at angles other than horizontal.

1. Draw a small triangle and mirror it with an angular (*e.g.*, 45°) mirror line.

2. Erase the triangles.

3. Save your work and close the drawing.

Producing Arrays

An **array** is an orderly grouping or arrangement of objects. AutoCAD's arraying feature can save a large amount of time when you create a drawing that includes many copies of an object arranged in a pattern.

1. Begin a new drawing by selecting the **New** button from the Quick Access toolbar.

2. Select **Use a Wizard** and **Quick Setup** and pick **OK**.

3. Pick **Architectural** for the units and pick the **Next** button.

4. Enter a width of **24′** and a length of **18′** and pick the **Finish** button.

5. Set the grid at **1′** and the snap at **1″**; **ZOOM All**.

6. Select the **Drafting & Annotation** workspace.

7. Zoom in on the lower left quarter of the drawing area and create the overstuffed armchair shown in Fig. 16-5. Do not draw the dimensions. Use the following information.

  • Create the square seating area with the **RECTANGLE** command.

  • Create the straight segments of the arms and back of the chair with the **LINE** command.

  • Use the **ARC** command to create the rounded ends of the arms and back.

  • Use the **MIRROR** command to simplify your work.

8. ZOOM All and save your work in a file named **chair.dwg**.

QUICK ACCESS

MODIFY

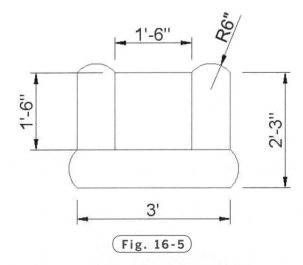

(Fig. 16-5)

Creating Rectangular Arrays

An array in which the objects are arranged in rows and columns is known as a **rectangular array**. An example is shown in Fig. 16-6.

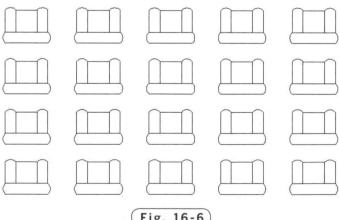

MODIFY

1. Pick the **Rectangular Array** button from the Modify panel of the Ribbon's Home tab.

2. Select the chair by placing a window around it and press **ENTER**.

AutoCAD displays a context-sensitive Ribbon tab, Array Creation, and creates a preview of the rectangular array.

3. Select the **Base point** option on the Command line, enter **B**, or pick the **Base Point** button on the Properties panel on the Ribbon and select the lower left corner of the chair as the base point.

At this point, you can specify the opposite corner of the rectangular array or the total number of chairs in the array. Let's create an array with four rows and five columns.

4. Select the **Count** option or enter **C**, **4** for the number of rows, and **5** for the number of columns.

5. Rest the crosshairs over each of the arrow-shaped grips on the array and read the information displayed in the drawing area. Note that this information also appears on the Array Creation tab of the Ribbon.

6. Click once on the horizontal spacing grip and adjust the horizontal spacing to **5′**.

7. Using the text box on the Ribbon change the vertical spacing to **3′6″**.

AutoCAD displays an array of 20 chairs, as shown in Fig. 16-6, with additional options, including the option to add a third dimension, called a level. You will learn more about drawing in three dimensions in Part 8.

8. Press **ENTER** to accept the array.

(Fig. 16-6)

Modifying Arrays

The Array command creates an **associative array,** which means that you can modify one parameter without needing to delete the array and start over.

1. Select any of the chairs in the array.

All 20 chairs become highlighted, because the array is a single object, and the Ribbon changes to a context-sensitive tab named Array. You can use the commands in the Array tab or use grips to change the number of rows, columns, and levels in the array, the spacing between elements, and the total row and column spacing of the array.

2. On the Ribbon, change the number of columns to **4** and the spacing between rows to **4′**.

3. Rest the crosshairs over each of the grips on the array and read the options that are available for each.

4. Using grips, change the number of rows to **5** and the spacing between rows to approximately **4′6″**.

5. Press **ENTER** or **ESC** to accept the modifications to the array.

The array is modified, as shown in Fig. 16-7.

6. Save your work and close the drawing.

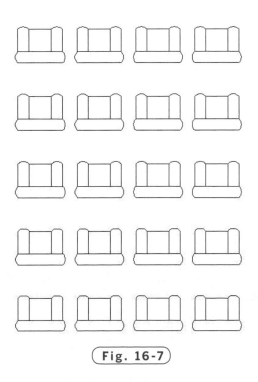

(Fig. 16-7)

Creating Polar Arrays

Next, we are going to produce the cycle wheel shown in Fig. 16-8. Notice the arrangement of the spokes. We can create them quickly by drawing only two lines and then using a circular array to create the remaining lines. A **polar array** (also called a **circular array**) is one in which the objects are arranged radially around a center point.

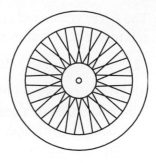

Fig. 16-8

1. Begin a new drawing from scratch.

2. Draw a wheel similar to the one shown in Fig. 16-9. Make the wheel large enough to fill most of the screen. Use the **Center** object snap to make the circles concentric.

3. Draw two crossing lines similar to the ones in Fig. 16-10. Use the **Nearest** object snap to begin and end the lines precisely on the appropriate circles.

4. Save your work in a file named **wheel.dwg**.

MODIFY

5. Pick the down-facing arrow to the right of the Rectangular Array button and select the **Polar Array** button.

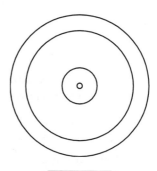

Fig. 16-9

6. Select the two lines and press **ENTER**.

7. Use the **Center** object snap to select the center of any of the circles as the center point of the array.

8. Select the **Items** option or enter **I**, then enter **18**.

9. Press **ENTER** to accept the default angle to fill, 360 degrees.

The wheel should look similar to the one in Fig. 16-8.

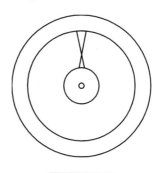

Fig. 16-10

10. Press **ENTER** to accept the array and end the command.

AutoCAD also allows you to create an associative array along a path, such as a line, arc, or spline. This type of array is called a **path array**.

11. Save your work and exit AutoCAD.

REVIEW QUESTIONS

Answer the following questions on a separate sheet of paper.

1. What is the purpose of the Quick Properties palette?
2. In what toolbar is the Move button located?
3. Explain how the MOVE command is different from the COPY command.
4. Describe a situation in which the MIRROR command would be useful.
5. During a MIRROR operation, can the mirror line be specified at any angle? Explain.
6. Name three types of arrays.
7. State one practical application for each type of array.
8. What is the effect on a polar array if you specify an angle to fill of less than 360 degrees? Explain.
9. Explain how a rectangular array can be created at any angle.

CHALLENGE YOUR THINKING

These questions are designed to further your knowledge of AutoCAD by encouraging you to explore the concepts presented in this chapter. Answer each question on a separate sheet of paper.

1. Experienced AutoCAD users take the time necessary to analyze the object to be drawn before beginning the drawing. Discuss the advantages of doing this. Keep in mind what you know about the AutoCAD commands presented in this chapter.

2. Draw the dialpad for a pushbutton telephone. Create one of the pushbuttons and use the ARRAY command to create the rest. Each pushbutton is a 13 mm × 8 mm rectangle, and the buttons are arranged as shown in Fig. 16-11. What should you specify for the space between rows and spaces between columns?

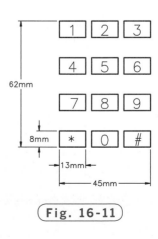

Fig. 16-11

Continued

● APPLYING AUTOCAD SKILLS

Work the following problems to practice the commands and skills you learned in this chapter.

1. Create the shim shown in Fig. 16-12. Begin by drawing the left half of the shim according to the dimensions shown, using construction lines for center lines. Then mirror the objects to create the right half of the shim. Erase any unneeded lines. Save the drawing as shim.dwg.

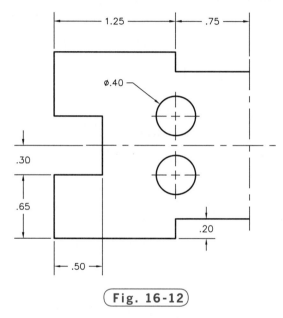

Fig. 16-12

Adapted from a drawing by Bob Weiland

2. Draw the chisel shown in Fig. 16-13 using the ARC, FILLET, LINE, COPY, and MIRROR commands.

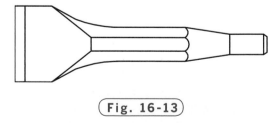

Fig. 16-13

Courtesy of Joseph K. Yabu, Ph.D., San Jose State University

3. Draw the key shown in Fig. 16-14 using the ARC, FILLET, LINE or PLINE, COPY, and MIRROR commands.

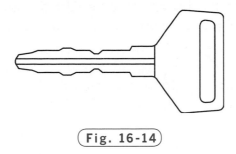

Fig. 16-14

Courtesy of Joseph K. Yabu, Ph.D., San Jose State University

4. Apply the commands introduced in this chapter to create the piping drawing shown in Fig. 16-15.

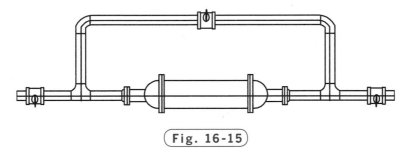

Fig. 16-15

Courtesy of Dr. Kathleen P. King, Fordham University, Lincoln Center

5. Develop a plan for an auditorium with rows and columns of seats. Design the room any way you'd like. Save your work as auditorium.dwg.

6. Create the cooling fan shown in Fig. 16-16.

Fig. 16-16

Courtesy of Joseph K. Yabu, Ph.D., San Jose State University

Continued

7. Draw the front view of the bar separator shown in Fig. 16-17. Draw one quarter of the bar separator using the LINE, ARC, and FILLET commands with a grid setting of .1 and a snap of .05. Use the MIRROR command to make half of the object; then use it again to create the other half.

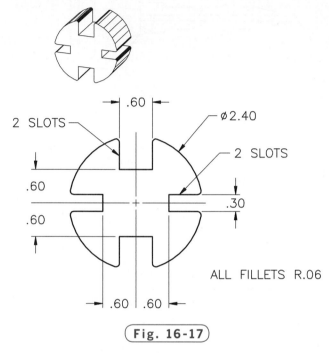

Fig. 16-17

Courtesy of Gary J. Hordemann, Gonzaga University

8. Draw the air vent shown in Fig. 16-18 using the ARRAY command. Other commands to consider using are POLYGON and COPY.

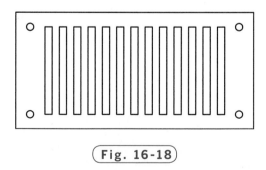

Fig. 16-18

Courtesy of Joseph K. Yabu, Ph.D., San Jose State University

9. Draw the front view of the wheel shown in Fig. 16-19. Begin by drawing one quarter of the object. Draw the three arcs and four lines, as shown in Fig. 16-20. Center the arcs on a known point. Next, use the FILLET command to create the cavity shown on the right. Erase the outside lines and the outside arc. Then use the MIRROR command to make three more copies of the cavity. Complete the drawing by adding arcs and circles. Try constructing the keyway by first drawing half of the shape, then using MIRROR to generate the other half. For more practice, try drawing one-eighth of a wheel.

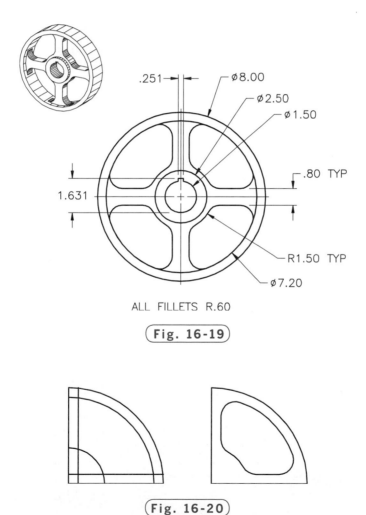

ALL FILLETS R.60

Fig. 16-19

Fig. 16-20

Courtesy of Gary J. Hordemann, Gonzaga University

Continued

10. Load the drawing named gasket.dwg. Erase each of the holes. Replace the large ones according to the locations shown in Fig. 16-21, this time using the ARRAY command. Reproduce the small holes using the ARRAY command, but don't worry about their exact locations. The diameter of the large holes is 1.25; the small holes have a diameter of .30.

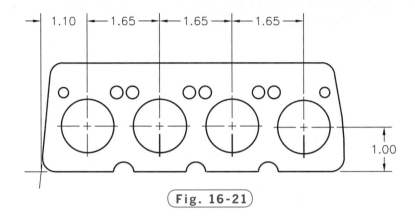

Fig. 16-21

11. Draw the snowflakes in Fig. 16-22 using the ARRAY command. Since snowflakes are six-sided, you may want to begin by drawing a set of concentric circles. Then use the ARRAY command to insert 6, 12, or more construction lines. The example in Fig. 16-23 shows the beginning of one of the snowflakes. Note that you only need to array three lines at 15° intervals for a total of 45° (two lines if you use the MIRROR command).

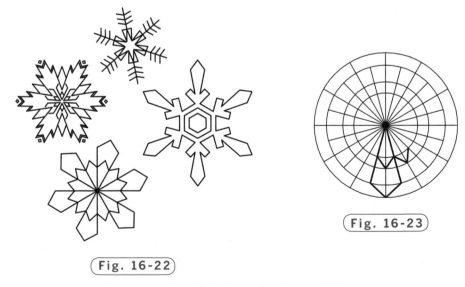

Fig. 16-22

Fig. 16-23

Courtesy of Gary J. Hordemann, Gonzaga University

12. Draw the top view of the power saw motor flywheel shown in Fig. 16-24. Use the ARRAY command to insert the 24 fins and arcs. You may find the OFFSET command useful in creating the first fin.

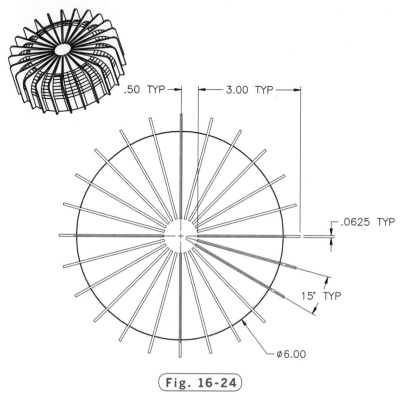

(Fig. 16-24)

Courtesy of Gary J. Hordemann, Gonzaga University

● USING PROBLEM-SOLVING SKILLS

Complete the following activities using problem-solving skills and your knowledge of AutoCAD.

1. You are creating problems for an AutoCAD textbook and need to simulate the grid displayed with a .5 spacing. With the POINT command (on the Draw toolbar), create a rectangular array within a drawing space of 12,9 so that the grid completely fills the space. If you find the point is too small to show on the screen, obtain help to find out how to make it larger. When you have completed the grid, turn AutoCAD's grid on and off to see how your grid corresponds to the real grid. Save the drawing as grid.dwg.

Continued

2. You have received a design change for the shim you created in problem 1 of this chapter's Applying AutoCAD Skills section. Open shim.dwg and use the MOVE and COPY commands to arrange the holes in a pattern approximately as shown in Fig. 16-25. Save the drawing as shimrev.dwg.

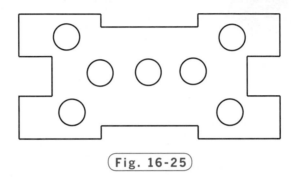

Fig. 16-25

3. Draw the wheel shown in Fig. 16-26 (you decide how).

Fig. 16-26

Adapted from a drawing by Bob Weiland

Chapter 17

Modifying and Maneuvering

Objectives

- Stretch objects to change their overall shape
- Scale objects using a scale factor or reference length
- Rotate objects to exact angles
- Trim and extend lines to specific boundaries
- Trim and extend multilines
- Join individual lines to specific boundaries

Vocabulary

rotating
scale

The more you create and edit drawings with AutoCAD, the more you will discover a need to edit objects in many ways. You will find several uses for trimming, extending, joining, and lengthening lines as you experiment with new ideas. One of the benefits of using AutoCAD is that you are not penalized for drawing something at the wrong size or in the wrong place. You can easily correct it using one of the commands covered in this textbook.

Stretching Objects

Let's start by stretching objects using the STRETCH command.

1. Start AutoCAD and choose the **Quick Setup** wizard.

2. For the units, use **Architectural**, and for the area, enter **120′** and **90′**.

Be sure to include the foot symbols (′) so that AutoCAD knows that these are feet, not inches.

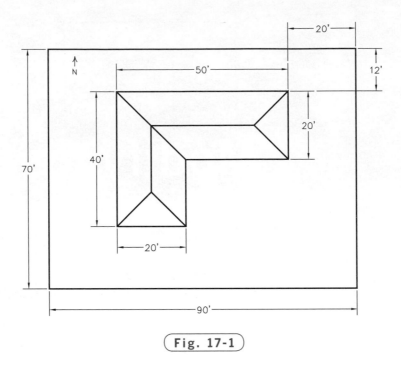

Fig. 17-1

3. Select the **Drafting & Annotation** workspace.

4. Enter **ZOOM All**.

5. Draw the site plan in Fig. 17-1 according to the dimensions shown. With snap set at **2′**, place the lower left corner of the property line at absolute point **10′,10′**. Omit dimensions. All points in the drawing should fall on the snap grid.

6. Save your work in a file named **site.dwg**.

7. Pick the **Stretch** button from the Modify panel on the Ribbon or enter **S**, the command alias for STRETCH.

MODIFY

This enters the STRETCH command.

8. Select the east end of the house, as shown in Fig. 17-2, and press **ENTER**. Use the crossing object selection procedure.

In reply to Select objects, pick a point to the right of the house and move the cursor to the left to create a green crossing window (Fig. 17-2).

9. Pick the lower right corner of the house for the base point, as shown in Fig. 17-2.

10. In reply to Second point of displacement, stretch the house **10**′ to the right and pick a point. (Snap should be on.) The house stretches dynamically.

The house should now be longer.

11. Stretch the south portion of the house **10**′.

12. Save your work.

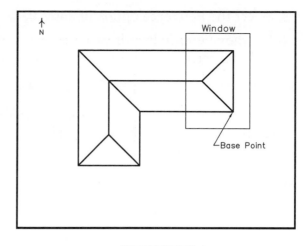

Fig. 17-2

Scaling Objects

Since the house is now slightly too large, let's scale it down to fit the lot. When you **scale** an object, you increase or decrease its overall size without changing the proportions of the parts to each other. You can use the SCALE command to scale by a specific scale factor, or you can use the existing size of the object as a reference length for the new size.

Using a Scale Factor

1. Pick the **Scale** button from the Modify panel on the Ribbon or enter **SC**, the command alias for SCALE.

2. Select the house by placing a window around it and press **ENTER**.

3. Pick the lower left corner of the house as the base point.

4. Enter **.5** in response to Specify scale factor.

Entering **.5** reduces the house to half (.5) of its previous size.

MODIFY

Scaling by Reference

Suppose the house is now too small. Let's scale it up using the SCALE Reference option.

1. Enter **SCALE**, select the house as before, and press **ENTER**.

2. Pick the lower left corner of the house again for the base point.

MODIFY

3. Select the **Reference** option or enter **R**.

4. For the reference length, pick the lower left corner of the house, and then pick the corner just to the right of it as the second point.

This establishes the length of the south wall (10′) as the reference length.

5. In response to Specify new length, move the crosshairs 4 feet to the right of the second point (snap should be on) and pick a point. You can also simply enter **14′**.

The south wall becomes 14′ long, and the rest of the house is scaled by reference to match.

Rotating Objects

Let's rotate the house on the site. Unlike scaling, **rotating** doesn't change the actual size of the house. It merely places the house at a different angle. Using the ROTATE command, you can drag objects to a new angle, or you can rotate them by a specific number of degrees if the precise angle is important.

Rotating by Dragging

MODIFY

Let's rotate the house by dragging it into place.

1. Pick the **Rotate** button from the Modify panel on the Ribbon, select the entire house, and press **ENTER**.

2. Pick the lower left corner of the house for the base point.

3. Turn off ortho and snap. As you move the crosshairs, notice that the object rotates dynamically. Drag the house in a clockwise direction a few degrees and pick a point.

4. Pick the **Undo** button to undo the rotation.

Specifying an Angle of Rotation

MODIFY

AutoCAD also allows you to specify the angle of rotation. In practice, this is a more useful option than dragging because you can control the placement of objects more precisely.

1. Enter **ROTATE**, select the house, and press **ENTER** after you have made the selection.

2. Pick the lower left corner of the house for the base point.

3. For the rotation angle, enter **25** (degrees).

The house rotates 25° counterclockwise. Your drawing should now look similar to the one in Fig. 17-3 without the additional lines.

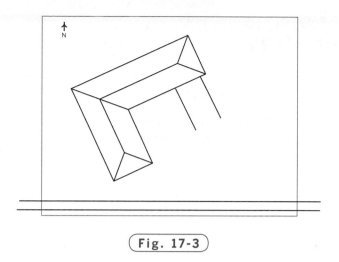

(Fig. 17-3)

Trimming Lines

The TRIM command allows you to trim lines that extend too far. You can use TRIM to clean up intersections and to convert construction lines and rays into useful parts of a drawing.

1. Draw two horizontal xlines **3′** apart to represent a sidewalk, as shown in Fig. 17-3.

2. Draw a partial driveway, as shown in Fig. 17-3. (Ortho should be off.) Make the driveway **13′** wide.

DRAW

 Copy the line that makes up the east edge of the house for the first line of the driveway. Use the OFFSET command to create the second line of the driveway exactly 13′ from the first line.

MODIFY

3. Pick the **Trim** button on the Modify panel of the Ribbon or enter **TR**, the command alias for the TRIM command.

4. Select the east property line as the cutting edge and press **ENTER**.

5. Select the sidewalk lines on the right side of the property line.

The sidewalk should now end at the property line.

6. Press **ENTER** to end the command.

7. Trim the west end of the sidewalk to meet the west property line.

8. Obtain online help to learn about the Project and Edge options.

Extending Lines

The most accurate way to extend the driveway lines so that they meet the south property line is to use the EXTEND command.

MODIFY

1. Pick the down-facing arrow to the right of the Trim button and select the **Extend** button from the pull-down menu.

2. Select the south property line as the boundary edge and press **ENTER**.

3. In reply to Select object to extend, pick the ends of the two lines that make up the driveway and press **ENTER**.

The driveway extends to the south property line.

MODIFY

4. Using the **TRIM** command, remove the short intersecting lines so that the sidewalk and driveway look like those in Fig. 17-4.

> **HINT** Use a crossing window to select all four objects, press ENTER, and then select the parts you want to remove.

5. Save your work.

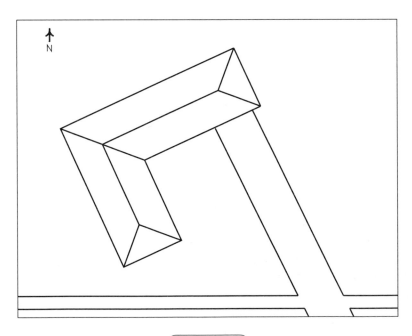

Fig. 17-4

Joining Lines

The JOIN command allows you to join similar individual objects into a single object. Let's use the JOIN command to change the roof on the house you created earlier in this chapter.

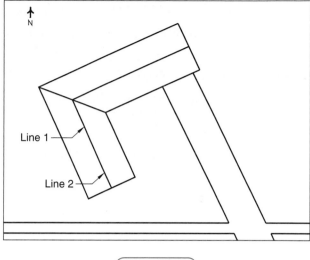

1. Change the roof line to match the one shown in Fig. 17-5. On both ends of the house, erase the diagonal lines; then draw the short line segments to connect the roof ridge to each end of the house.

Fig. 17-5

 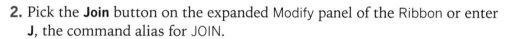 Use the Midpoint object snap.

MODIFY

2. Pick the **Join** button on the expanded Modify panel of the Ribbon or enter **J**, the command alias for JOIN.

3. Select the first line segment in reply to Select source object, select the second line segment in response to Select objects to join, and press **ENTER**.

4. Repeat Steps 2 and 3 for the other end of the house.

The roof ridge line on each side of the house is now a single line.

 The JOIN command works for a series of linear and open curved objects that share an endpoint, including lines, arcs, polylines, and splines.

5. Save your work.

Working with Multilines

The TRIM and EXTEND commands work with multilines as well.

1. Open the drawing named **multi.dwg**.

2. Open the **Multiline Style** dialog box, set the style **S1** as the current style, and pick **OK**.

3. Create a new multiline, as shown in Fig. 17-6.

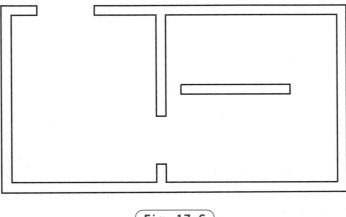

Fig. 17-6

MODIFY

4. Use the **EXTEND** command to extend the new wall to the interior wall. In reply to Enter mline junction option, select the **Open** option or type **O**.

Your drawing should look like the one shown in Fig. 17-7.

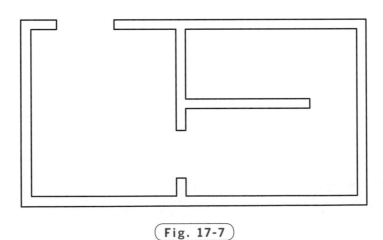

Fig. 17-7

5. Reenter the **MLINE** command and create another wall, as shown in Fig. 17-8.

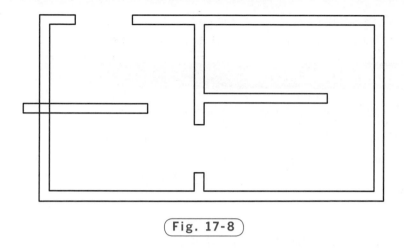

Fig. 17-8

6. Use the **TRIM** command to trim away the part of the wall that extends beyond the exterior wall of the building. This time, select the **Merged** option at the Enter mline junction option prompt.

MODIFY

The drawing should now look like the one in Fig. 17-9.

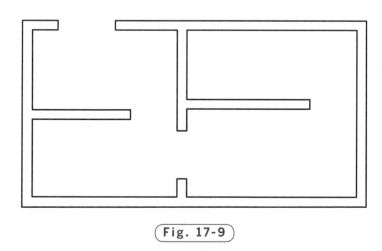

Fig. 17-9

7. Save your work and exit AutoCAD.

● REVIEW QUESTIONS

Answer the following questions on a separate sheet of paper.

1. Explain the purpose of the STRETCH command.
2. Using the SCALE command, what number would you enter to enlarge an object by 50%? ...to enlarge it to 3 times its present size? ...to reduce it to half of its present size?
3. Explain how you would dynamically scale an object up or down.
4. Can you rotate an object dynamically? Explain.
5. How would you specify a 90° clockwise rotation of an object accurately?
6. Explain the purpose of the TRIM command.
7. Describe a situation in which the EXTEND command would be useful.
8. Explain the similarities between the EXTEND and JOIN commands.

● CHALLENGE YOUR THINKING

These questions are designed to further your knowledge of AutoCAD by encouraging you to explore the concepts presented in this chapter. Answer each question on a separate sheet of paper.

1. Refer to site.dwg, which you completed in this chapter. If you were doing actual architectural work, you would need to place the features on the drawing much more precisely than you did in this drawing. Describe a way to place a 14'-wide driveway exactly 2 feet from the corner of the house. The driveway should be perpendicular to the house.
2. Both the ZOOM command and the SCALE command make objects appear larger and smaller on the screen. Write a paragraph explaining the differences between the two commands. Explain the circumstances under which each command should be used.

• APPLYING AUTOCAD SKILLS

Work the following problems to practice the commands and skills you learned in this chapter.

1. Create a new drawing and draw the kitchen range symbol shown in Fig. 17-10. The square is 5 units on each side, the large circles are R.8, and the small circles are R.6. Locate the circles approximately as shown: equidistant from the adjacent sides, with the large circles in the upper left and lower right corners.

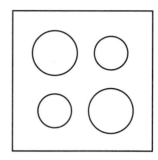

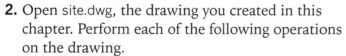

Fig. 17-10

Notice that this is an incorrect symbol for a kitchen range. The large and small circles should be reversed. You can fix this by rotating the circles 90°. To make sure the circles stay in their same relative locations, rotate them around the center of the square. To do this, first draw two diagonal lines, as shown in Fig. 17-11. Then enter the ROTATE command and select the four circles and two diagonal lines. After you have completed the rotation, erase both of the diagonal lines. Then scale the range to 25% of its current size. Save the drawing as range.dwg.

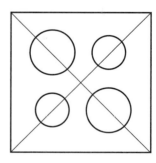

2. Open site.dwg, the drawing you created in this chapter. Perform each of the following operations on the drawing.

Fig. 17-11

- Change the roof line by adding angled (hipped) ridges on both ends of the house.

- Stretch the driveway by placing a (crossing) window around the house and across the driveway. Stretch the driveway to the north so that the house is sitting farther to the rear of the lot.

- Add a sidewalk parallel to the east property line. Use the TRIM and BREAK commands to clean up the sidewalk corner and the north end of the new sidewalk.

- Reduce the entire site plan by 20% using the SCALE command.

- Stretch the right side of the site plan to the east 10′.

- Rotate the entire site plan 10° in a counterclockwise direction.

Continued

- Place trees and shrubs to complete the site plan drawing. Use the ARRAY command to create a tree or shrub, as shown in Fig. 17-12. Duplicate and scale the tree or shrub using the COPY and SCALE commands.

When you finish, your drawing should look similar to the one in Fig. 17-12.

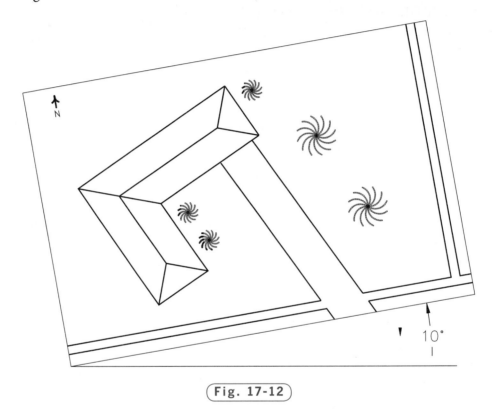

Fig. 17-12

3. Marketing wants to use the original design of your TV remote control device, but you must change it to meet competitive standards. Create the first drawing of the device (shown on the left in Fig. 17-13). Then copy the drawing. On the copy, make the following changes using the commands you learned in this chapter.

- Scale the device to 90% of its original size.
- Change the two buttons at the bottom into one large button.
- Reduce the overall length.

The final device should look similar to the drawing on the right in Fig. 17-13. Save the drawing as remote.dwg.

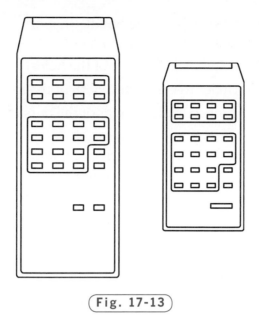

Fig. 17-13

4. Create the two armchairs shown in Fig. 17-14. Approximate their sizes, but align them as shown in the illustration so that the backs of the chairs are colinear. Use the chairs to make a matching sofa. To do this, make a copy of both chairs. Then remove the inside arms on the copies of both chairs and extend the horizontal lines to meet, as shown. Finally, use the JOIN command to join the lines that represent the back of the sofa so that the back of the sofa consists of a single line. Join the other horizontal lines as well. Save the file as furniture.dwg.

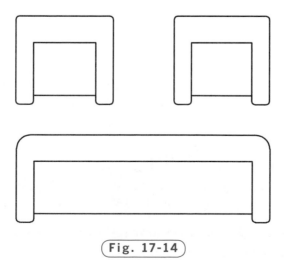

Fig. 17-14

Continued

5. Draw the front view of the tube bundle support shown in Fig. 17-15. Proceed as follows:

- Draw the outside circle and one of the holes.
- Offset both circles to create the width of the material.
- Draw a line connecting the center of the outside circle and the center of the hole.
- Offset the line to create the internal support structure that connects the hole with the center of the support.
- Draw another line from the center of the large circle at an angle of 30° to the first line.
- Use TRIM to create half of one cavity.
- Use MIRROR to create the other half of the cavity.
- Fillet the three corners.
- Use ARRAY to insert five more copies of the hole and cavity.

When you have completed the drawing, use the SCALE command to shrink it by a factor of .25 using the obvious base point. Save the drawing as tubebundle.dwg.

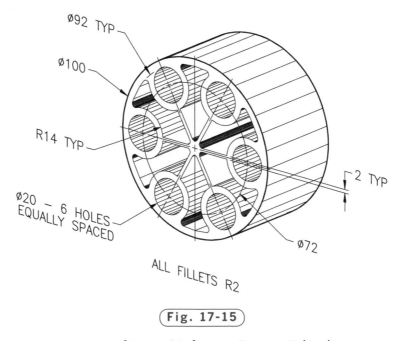

Fig. 17-15

Courtesy of Gary J. Hordemann, Gonzaga University

• USING PROBLEM-SOLVING SKILLS

Complete the following activities using problem-solving skills and your knowledge of AutoCAD.

1. You have a full single garage on your lot. The garage measures 16′ × 25′ and has a 9′ door opening. Draw a floor plan of the garage with 6″-thick walls. Show the opening, but do not include the door. Your spouse's car necessitates a larger garage. Use the EXTEND command to convert the full single garage to a full double garage measuring 25′ × 25′. Include two 9′ door openings separated by a 1′ center support. Trim where necessary. Save the drawing as garage.dwg.

2. Local zoning laws will not permit you to have a full double garage on your narrow lot. Use the TRIM command to convert the proposed full double garage to a small double garage measuring 20′ × 20′ with one large 16′ door opening. Save the drawing as garage2.dwg.

3. Create the house shown in Fig. 17-16 on a 100′ × 35′ lot.

   • Make the drawing limits the size of the lot.

   • Create a new multiline style named FRAME with two elements a total of 4″ apart to represent the walls.

   • Estimate the size of the house and of each room.

   • Use MLEDIT, TRIM, and EXTEND to create the wall intersections.

   • Save the file as frame.dwg.

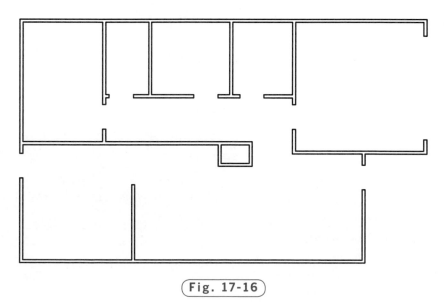

(Fig. 17-16)

Interior Designer

A variety of related careers fall under the umbrella of "drafting." While many drafters and designers create plans to manufacture or build products, others work to create attractive surroundings for daily living. Try to think of an old, run-down building in your area that has recently been restored. Although architectural designers certainly played a role in ensuring that the building was structurally secure, other design professionals were responsible for making the inside usable and aesthetically appealing. These individuals, known as *interior designers*, are responsible for designing and supervising interior construction and restoration efforts.

© Photodisc/Getty Images/RF

Details of Design

Interior designers cover a lot of territory when planning a comfortable indoor atmosphere. They create drawings and plans specifying the types and quantities of materials necessary for construction, the types of lighting systems that will best complement a room, and how to furnish an area creatively. Interior designers may be employed by design firms or may work as independent consultants. They become involved with building and restoration projects for homes, offices, and even sets for films and television shows.

Being Versatile

Interior designers wear many different hats; that is, they are responsible for various aspects of the design process. They should be able to estimate the costs involved with construction and restoration, such as specific costs of materials and labor. Leadership skills are also important, as interior designers need to be able to convey their ideas to those involved with creating or restoring an area. Interior designers must also be able to think creatively in order to design an appropriate look for various types of indoor areas.

▶ Career Activities

1. Compare the interior designer's job with that of an architectural drafter.

2. Find out about courses of study available for interior designers. What are the prerequisites for these courses?

Chapter 18

Hatching and Sketching

Objectives

- Hatch objects and parts according to industry standards to improve the readability of drawings
- Edit a hatch by changing, adding, and removing boundaries
- Change the characteristics of an existing hatch
- Use the **Tool Palettes** window to apply hatching
- Use sketching options to create objects with irregular lines

Vocabulary

associative hatch
crosshatch
full section
hatch
island
section drawings

AutoCAD's hatching and sketching features can help enhance the visual appearance and readability of a drawing. For example, on the map shown in Fig. 18-1 on page 246, hatching makes it easy to tell which parts of the area are national forest and which are urban areas. The irregular lines representing the roads were created using the SKETCH command.

In some cases, hatching and sketching are used to help make a drawing attractive. This may be true when drawings are created for use by a manufacturer's marketing or sales team. In other cases, hatching and sketching are used according to industry standards to make a technical drawing easier to read at a glance.

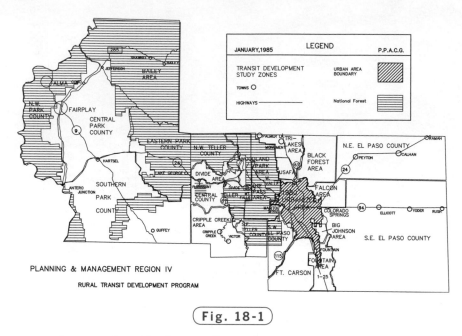

Fig. 18-1

Courtesy of David Salamon, Pikes Peak Area Council of Governments

Hatching

A **hatch** or **crosshatch** is a repetitive pattern of lines or symbols that shows a related area of a drawing. Hatching is used extensively in engineering drawing and production drafting to show cut surfaces. For example, when you draw a section of a mechanical part, you use hatching to reflect the cut, or internal, surface. **Section drawings** are used to show the interior detail of a part. Figure 18-2 shows an example of a **full section** (one that extends all the way through the part). The drawing on the left is the front view, and the drawing on the right is the section view.

Creating a Boundary Hatch

The BHATCH command creates associative hatches. An **associative hatch** is one that updates automatically when its boundaries are changed.

1. Start AutoCAD and begin a new drawing from scratch. Select **Imperial** settings.

2. Select the **AutoCAD Classic** workspace and close all floating toolbars and palettes.

3. Create a sectional view of the center support shown in Fig. 18-3. Use the dimensions shown in the figure to create it accurately. Do not include the text (Center Support), the center lines, or the dimensions.

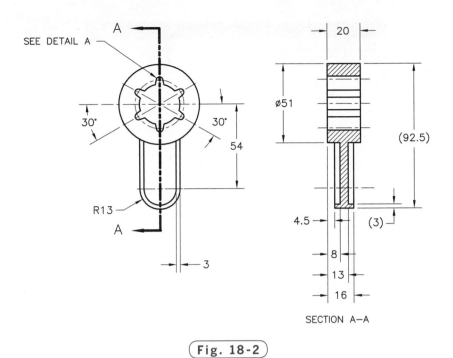

SEE DETAIL A

A

30°   30°

54

R13

A

3

20

ø51

(92.5)

4.5   (3)

8

13

16

SECTION A–A

Fig. 18-2

Courtesy of Julie H. Wickert, Austin Community College

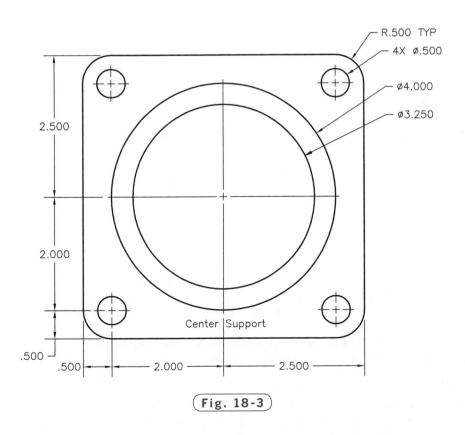

R.500 TYP

4X ø.500

ø4.000

ø3.250

2.500

2.000

.500

.500   2.000   2.500

Center Support

Fig. 18-3

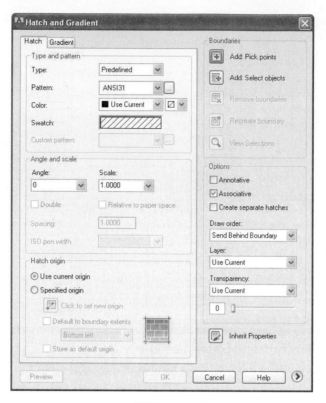

Fig. 18-4

DRAW

4. Save your work in a file named **hatch.dwg**.

5. Pick the **Hatch** button on the Draw toolbar.

This enters the BHATCH command and displays the Hatch and Gradient dialog box, as shown in Fig. 18-4.

6. Pick the **More Options** arrow (**>**) located in the lower right corner of the dialog box.

7. In the Island display style area of the dialog box, pick **Normal**, unless it is already selected.

8. Under the Hatch tab, pick the **Pattern** down arrow.

A list of hatch patterns appears. ANSI31 is the standard crosshatch pattern.

9. Pick the **...** button located to the right of the down arrow you just picked.

The Hatch Pattern Palette appears with ANSI hatch patterns. ANSI stands for the American National Standards Institute, an industry standards organization in the United States.

10. Pick the **ISO** tab.

ISO hatch patterns appear. ISO stands for the International Standards Organization, which also produces industry standards.

11. Pick the **Other Predefined** tab.

AutoCAD displays other predefined hatch patterns that are available to you.

12. Pick the **ANSI** tab, pick **ANSI31**, and pick **OK**.

Notice the Angle and scale area in the dialog box. Angle permits you to rotate the hatch pattern, while Scale enables you to scale it. The scale value should be set at the reciprocal of the plot scale so that the hatch pattern size corresponds to the drawing scale. This information will be useful when you begin to print or plot your drawings.

13. Under Boundaries, pick the **Add: Select objects** button, select the entire drawing, and press **ENTER**.

The Hatch and Gradient dialog box reappears.

14. Pick the **Preview** button located in the lower left area of the dialog box to preview the hatch.

15. As instructed by AutoCAD, right-click to accept the hatch.

Your drawing should now look similar to the one in Fig. 18-5.

16. Save your work.

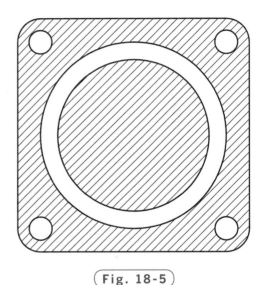

Fig. 18-5

Editing a Hatch

By default, hatches created with the BHATCH command are associative. Remember, an associative hatch is one that updates automatically when you change the size or shape of the hatched object. Associative hatches also update when you add or remove islands inside the hatched area.

Changing the Boundary

1. Select the inner cylindrical part of the center support. Be sure to select the circle and not the hatch.

2. In the status bar, pick the **Quick Properties** button to toggle Quick Properties on, unless it is already on.

STATUS BAR

3. In the Quick Properties panel, change Diameter to **2.75** and press **ENTER**.

The diameter changes, along with the associative hatch.

4. Press **ESC** to close the Quick Properties panel and de-select the circle.

5. Change the diameter of the lower right hole to **.375** and press **ESC** to close the Quick Properties panel.

Once again, AutoCAD updates the hatch pattern.

6. Undo the two changes you just made.

Adding and Removing Islands

The boundaries of a hatch can also be edited by adding and removing islands from the hatch. An **island** is an area within an object, such as a hole, that is not hatched.

The HATCHEDIT command permits you to change the boundaries of an associative hatch. To enter this command, you can enter HATCHEDIT at the keyboard, right-click and use the shortcut menu, or pick the Edit Hatch button on the Modify II toolbar.

MODIFY II

1. Display the Modify II toolbar and pick the **Edit Hatch** button.

This enters the HATCHEDIT command.

2. Pick the hatch in the drawing.

This causes the Hatch Edit dialog box to display. As you can see, it is very similar to the Hatch and Gradient dialog box.

3. Under Boundaries, pick the **Remove boundaries** button.

4. Select the inner cylindrical part of the center support and press **ENTER**.

5. Pick the **Preview** button. Then right-click to accept the edit.

AutoCAD computes the area of a hatch and updates it automatically when you edit the hatch.

STANDARD

6. Select the hatch.

7. Pick the **Properties** button on the Standard toolbar.

The Properties palette appears, which provides much more information about objects than the Quick Properties panel. Notice that the hatch area is 11.4336 square inches. (You may need to enlarge the palette and expand the Geometry table to see the Area value.)

MODIFY II

8. Reopen the **Hatch Edit** dialog box.

9. Pick the **Add: Select objects** button and add the center cylinder back into the hatched area. Pick **OK** when you are finished.

10. Select the hatch and check the area displayed in the Properties palette.

The hatch area is automatically updated to 19.7294 square inches.

11. Close the Properties palette and the Modify II toobar and save your work.

Changing Other Hatch Characteristics

The Hatch Edit dialog box also allows you to change other characteristics of a hatch. For example, you can choose a different hatch pattern, specify an origin where the hatch pattern begins, shade the hatch area, or change the angle or scale of the current hatch pattern.

1. Select the hatch, right-click, and pick **Hatch Edit...** from the shortcut menu.

2. Make the following changes:
Pattern: **ESCHER**
Angle: **30**
Scale: **1.3**

3. Pick the **OK** button.

AutoCAD applies the changes to the drawing.

4. Experiment further on your own with creating and changing associative hatch objects.

5. Using the **Edit Hatch** button, change the hatch pattern to **ANSI31**, the angle to **0**, and the scale to **1**. Pick **OK** when you are finished.

6. Save your work.

MODIFY II

Tool Palettes Window

The Tool Palettes window provides another method for inserting hatch patterns.

1. Move the center support to the right side of the drawing area, produce a copy of it on the left side of the drawing area, and erase the hatch pattern from the copy.

2. Select the **Tool Palettes Window** from the Standard toolbar.

This displays the Tool Palettes window.

STANDARD

3. Review the contents of the Tool Palettes window by clicking each of its tabs.

4. Pick the **Hatches and Fills** tab and pick the **Imperial Steel** hatch pattern.

HINT To see the name of a pattern, rest the pointer on it.

The Imperial Steel hatch pattern locks onto the crosshairs.

5. Click inside the center support's cylindrical feature.

AutoCAD hatches only the cylindrical feature.

6. Undo the hatch.

7. Select the **Steel** hatch pattern again from the **Tool Palettes** window and pick a point just inside the outer boundary of the center support. Do not select the cylindrical area.

AutoCAD hatches the center support correctly.

8. Erase the second center support and center the first one in the drawing area.

9. Close the **Tool Palettes** window and save your work.

You will learn more about the Tool Palettes window in Chapter 32.

Sketching

The SKETCH command is rarely used to create "sketches" because AutoCAD offers many other commands that make sketching (in its traditional sense) a faster, more accurate process. However, the SKETCH command is used for other purposes, such as to show irregular lines on maps like the one shown in Fig. 18-1 on page 246.

Let's try some freehand sketching.

1. Using the **Save As...** option in the Application Menu, create a new drawing named **sketch.dwg**.

2. Erase the objects that are currently in the new drawing.

3. Enter **SKPOLY** at the keyboard and specify a value of **1**.

SKPOLY is a system variable that controls whether the SKETCH command creates lines (0) or polylines (1). Setting SKPOLY to 1 allows you to smooth sketched curves using the PEDIT command.

4. Turn off snap and enter the **SKETCH** command.

5. Select the **Type** option or enter **T** to review the types of objects available in the SKETCH command.

As you can see, you can create a series of lines, a polyline, or a spline with the Sketch feature.

6. Select a **Polyline** option or enter **P**.

7. To begin sketching, pick a point where you would like the sketch to begin. The pick button toggles the pen down.

8. Move the pointing device to sketch a short line.

9. Pick a second point (to toggle the pen up) and press **ENTER**.

10. Move to an open area on the screen and sketch the golf course sand trap shown in Fig. 18-6. Set the **Increment** value at **.2**.

It's okay if your sketch doesn't look exactly like the one in Fig. 18-6.

11. Practice sketching by using the remaining SKETCH options. Draw anything you'd like.

12. When you have finished, save your work and exit AutoCAD.

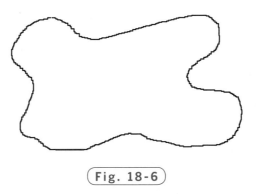

Fig. 18-6

REVIEW QUESTIONS

Answer the following questions on a separate sheet of paper.

1. Explain why hatch patterns are useful.

2. With which command can you change a hatch pattern after the hatch has been applied to an object?

3. The BHATCH command permits you to hatch drawings. Describe a second method.

4. SKETCH requires an Increment value. What does this increment determine? (If you are not sure, specify a relatively large increment such as .5 or 1 and notice the appearance of the sketch lines.)

5. What is the purpose of the SKPOLY system variable?

CHALLENGE YOUR THINKING

These questions are designed to further your knowledge of AutoCAD by encouraging you to explore the concepts presented in this chapter. Answer each question on a separate sheet of paper.

1. Find out what the Inherit Properties button on the Hatch and Gradient dialog box does. Explain how it can save you time when you create a complex drawing that has several hatched areas.

2. Investigate the Gradient tab of the Hatch and Gradient dialog box. What does this tab allow you to do? When might you want to use this feature?

• APPLYING AUTOCAD SKILLS

Work the following problems to practice the commands and skills you learned in this chapter.

1. Draw the wall shown in Fig. 18-7. Replace the angled break line on the right with an irregular line drawn with the SKETCH command to indicate a continuing edge. Hatch the wall in the Brick pattern. Select the lower left corner of the wall as the origin for the hatch. Save the drawing as brickwall.dwg.

2. Create the simplified house elevation shown in Fig. 18-8. Use the SKETCH command to define temporary boundaries for the hatch patterns on the roof.

3. Use the SKETCH command to create a map of the United States. Use Fig. 18-9 as a guide.

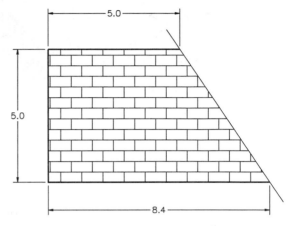

Fig. 18-7

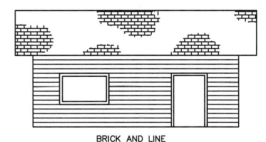

BRICK AND LINE

Fig. 18-8

Fig. 18-9

Continued

4. Use the SKETCH command to draw a logo similar to the one shown in Fig. 18-10. Before sketching the lines, be sure to set SKPOLY to 1. Use the BHATCH Solid pattern to create the filled areas.

Fig. 18-10

Courtesy of Gary J. Hordemann, Gonzaga University

5. Draw the right-side view of the slider shown in Fig. 18-11 as a full section view. Assume the material to be aluminum. See Appendix D for a table of dimensioning symbols. For additional practice, draw the top view as a full section view.

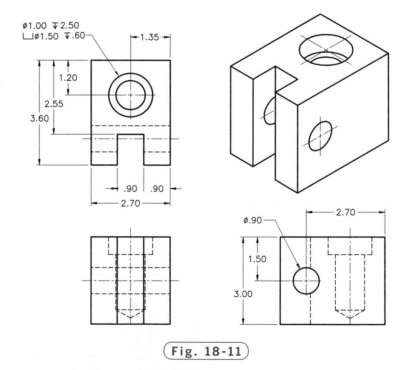

Fig. 18-11

Courtesy of Gary J. Hordemann, Gonzaga University

• USING PROBLEM-SOLVING SKILLS

Complete the following activities using problem-solving skills and your knowledge of AutoCAD.

1. Your company is trying some new bushings for the swings in the child's playground set it manufactures. Draw the front view and full section of the bushing shown in Fig. 18-12. Hatch the section as appropriate. Save the drawing as bushing.dwg.

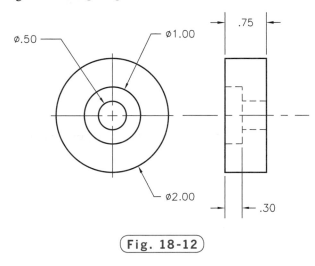

Fig. 18-12

2. An alternate design of the bushing described in problem 1 provides a hard insert to act as a wear surface within the bushing (Fig. 18-13). Make the change in the sectional view and hatch accordingly. Save the drawing as altbushing.dwg.

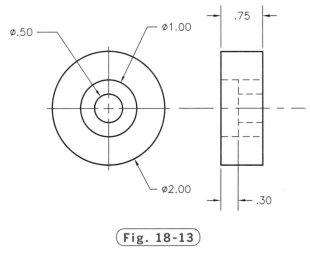

Fig. 18-13

Continued

Fig. 18-14

3. The drawing shown in Fig. 18-14 was used with a computer-driven, high-pressure water cutter to cut a shield out of a piece of $1/4''$ brass. Use the polyline option of the SKETCH command with SKPOLY set to 1 to draw a similar shield. Experiment with the PEDIT command to make the sketched drawing as smooth as possible.

Aquatic Designer

You're vacationing at a luxurious hotel in the Caribbean. You look around and see yourself surrounded by an oasis of pools and spas. Minutes later you wander outside and see fascinating aquatic structures, including a large swimming pool with many curves and bends, a waterfall, and winding water slide. Did it once cross your mind that hours of time

© Jack Hollingsworth/Getty Images/RF

and creativity went into designing and constructing such mesmerizing results?

A Swarm of Water

Commercial aquatic construction takes place worldwide varying from apartment complexes, gigantic hotel water parks, to even your neighborhood water park. The design process for these structures is crucial and in-depth. The client works with the contractor and designer to establish a design within budget. The engineers and designers of such projects have to pay very close attention to detail and effectively communicate with their client and other contractors working on the project.

AutoCAD Delivers Results

Engineers of aquatic design depend on AutoCAD software throughout their workflow. AutoCAD helps to reduce the risk of interference and construction errors, eliminate the need to redraw objects, and increase client satisfaction with easy to visualize designs. It also allows them to complete hydraulic package calculations in time to keep up with the ever-changing and very competitive leisure resort business.

▶ Career Activities

1. Research the types of firms that would employ a person skilled in aquatic design.

2. What kinds of skills would these firms expect you to have?

Signage

Throughout history, people have used visual cues in the form of signs to communicate information. Signs are typically easy to interpret, quick to read (if indeed they contain any actual text at all), and unobjectionable. Signs without text, by the way, are often considered universal in nature: they are capable of being understood in different cultures regardless of the language spoken. The purpose of a sign can be varied—advertising, public service, directions, warnings, identification—almost any message that needs to be transmitted. A very common use of signs is to regulate vehicular traffic. Traffic signs identify routes taken, geographic direction of the vehicle, speed limits, expressway exit/entrance numbers, nearby hospitals, and sources of food and fuel for the traveler, among other uses.

▶ Description

In AutoCAD, create the signs in Fig. P3-1 as assigned by your instructor, or improve upon some of the signs as they currently exist. For example, you might create a more effective graphic. Actual sizes and legal descriptions of many signs are available on the Internet.

Fig. P3-1

Alternatively, your instructor may ask you to design your own original signs. Suggestions include:

elephant crossing	no junk food
cash only	idea factory
no fish here	no fishing
no tire-changing	barbecues available
wet concrete	drive safely or don't drive

Use your imagination or seek advice from the instructor for further ideas.

▶ Hints and Suggestions

1. Consider using DRAW>SURFACE>2D SOLID to create the sign as a solid color; alternatively, you may want to use POLYGON with the Solid hatch pattern to color in the background.

2. Consider using gradients from the Gradient tab of the Hatch and Gradient dialog box. Would a gradient be effective on a highway sign?

3. Polylines allow a choice of thickness as well as color for lines, arrows, and other images that might be appropriate for your sign's communication.

4. If you use polylines, filleting closed polygons can be accomplished in a single command.

5. Utilize the various modifying commands to reuse pieces of one sign that might lend themselves to another section or sign.

▶ Summary Questions

Your instructor may direct you to answer these questions orally or in writing. You may wish to compare and exchange ideas with other students to help increase efficiency and productivity.

1. Why is technical drawing sometimes referred to as a "universal language"? Explain.

2. What might be a reason to keep all signs in a particular category, such as "informational" or "regulatory," based on a single color scheme or shape?

3. What characteristics make traffic signs easy to read as drivers go by?

4. How is it that so many countries can use similar traffic signs, yet not have a shared language or culture? Explain your reasoning.

5. Do you think there is a relationship between the size of a sign and its importance to the driver? Explain.

Chapter 19

Notes and Specifications

Objectives

- Create and edit text dynamically
- Specify the position and orientation of text
- Create and use new text styles
- Review options for fonts and adjust the quality of TrueType fonts
- Produce multiple-line text using AutoCAD's text editor
- Import text from an external word processor
- Format text to create lists

Vocabulary

font
justify
mtext
notes
specifications
title block

Notes and specifications are a critical part of most engineering drawings. Although the two terms are often used interchangeably, notes and specifications are technically two different types of text. **Notes** generally refer to the entire drawing, rather than to any one specific feature. For example, a note might describe the thickness of a part. **Specifications**, on the other hand, provide information about size, shape, and surface finishes that apply to specific portions of an object or part.

Without text, drawings would be incomplete. The drawing in Fig. 19-1 shows the amount of text that is typical in many drawings. Some drawings contain even more text. As you can see, the text is an important component in describing the drawing. Prior to computer-aided drafting, the text was carefully placed by hand, which is not only time-consuming, but tedious. With CAD, you can place the words on the screen almost as fast as you can type them.

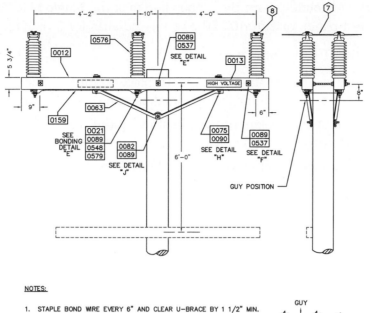

NOTES:

1. STAPLE BOND WIRE EVERY 6" AND CLEAR U-BRACE BY 1 1/2" MIN.
2. BONDING DETAILS SHOWN ON DRAWING GB1.11.
3. WASHER DETAIL SHOWN ON DRAWING GB1.10.
4. THIS CONSTRUCTION IS TO BE USED ONLY IN SPECIAL CASES WHERE OVERHEAD CLEARANCES ARE CRITICAL. GENERALLY 69B3C SHOULD BE USED.
5. OFFSET ARM ON POLE 6".
6. FOR MAXIMUM CONDUCTOR ANGLE SEE AD2.0
⑦ TABLE 1, COLUMN 3.
⑧ TABLE 1, COLUMN 9.

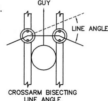

Fig. 19-1

Types of Text

AutoCAD supplies two different ways of inserting text objects. The one you use should depend on the type and amount of text you are inserting. The DTEXT command inserts one or more lines of text as single-line objects. MTEXT, on the other hand, treats one or more paragraphs of text as a single object. This chapter presents both ways of inserting text.

Dynamic Text

The DTEXT command enables you to display AutoCAD text dynamically in the drawing as you type it. It also allows you to **justify**, or align, the text in several ways, including left-aligned, right-aligned, and centered.

1. Start AutoCAD and start a new drawing from scratch.
2. Select the **Drafting & Annotation** workspace.
3. Set snap to **.25**.

TEXT

4. Select the Annotate tab on the Ribbon and pick the **Single Line Text** button on the Text panel, or enter **DT** (the command alias for DTEXT).

This enters the DTEXT command.

5. In response to Specify start point of text, place a point on the left side of the screen. The text will be left-justified beginning at this point.

6. Reply to the Specify height prompt by moving the crosshairs up **.25** units from the starting point and picking a point.

7. Enter **0** (degrees) in reply to Specify rotation angle of text.

A blinking cursor appears inside a narrow text box.

8. Type your name using both upper- and lowercase letters and press **ENTER**.

You should again see the blinking cursor, now inside a text box of two lines.

9. Type your street address or P.O. box and press **ENTER**.

Notice where the text appears in relation to the previous line.

10. Type your city, state, and zip code and press **ENTER**.

11. Press **ENTER** again to terminate the DTEXT command.

Let's enter the same information again, but this time in a different format.

12. Reenter the **DTEXT** command (press the space bar), and select the **Justify** option or enter **J**.

The justification options appear.

13. Select the **Center** option or enter **C**.

14. Place the center point near the top of the screen and set the text height by entering **.2** at the keyboard. Do not insert the text at an angle.

15. Repeat Steps 8 through 11. Be sure to press **ENTER** twice at the end.

When you are finished, your text should be centered like the example in Fig. 19-2. If it isn't, try again.

16. Save your work in a file named **text.dwg**.

Ms. Jane Doe
507 East 104th Avenue
Cadsville, CA 30160

(**Fig. 19-2**)

Multiple-Line Text

The MTEXT command creates a multiple-line text object called **mtext**. AutoCAD uses a text editor to create mtext objects.

1. Erase the text from the upper half of the screen.
2. From the Annotate tab of the Ribbon, pick the **Multiline Text** button on the Text panel or enter **T** at the keyboard.

TEXT

This enters the MTEXT command and displays the text style and height at the Command line.

3. Pick a point in the upper left area of the screen.

Notice the new list of options.

 Recall that you can access the list of options at the dynamic input cursor by pressing the down arrow key on the keyboard.

4. Select the **Width** option or enter **W**.
5. Enter **4.5** in response to Specify width or, with ortho and snap on, pick a point **4.5** units to the right of the first point.

AutoCAD displays a new Ribbon tab, Text Editor, with text formatting panels and a ruler above the selection point. Notice that the font is txt and the default text height is 0.2500 units.

 A **font** is a set of characters, including letters, numbers, punctuation marks, and symbols, in a particular style. Arial and Roman Simplex are examples of fonts used in AutoCAD.

6. From the font drop-down list on the Formatting panel, change the font from txt to **Arial**.
7. Type the following sentence:

   I'm using AutoCAD's text editor to write these words.

8. Highlight the word **using** and pick the **B** (bold) button.
9. Highlight the word **words** and pick the **U** (underline) button.

The sentence should now look like the one in Fig. 19-3 on page 266.

CLOSE

10. Pick the **Close Text Editor** button on the **Ribbon** and save your work.

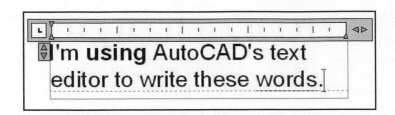

Fig. 19-3

Importing Text

Instead of typing text within AutoCAD, you can create it using a standard word processing program and then import it. Another option is to copy text from a word processing program and paste it into AutoCAD. Many people find it easier to create lengthy text passages using word processors with which they are familiar.

1. Minimize AutoCAD.
2. From the Windows Start menu, pick Programs (or All Programs), Accessories, and **Notepad**.
3. Enter the following text:

   AutoCAD's text editor permits you to import text from a file. This text will appear in AutoCAD shortly.

4. From Notepad's File pull-down menu, pick **Save As...**, select the folder with your name, and name the file **import.txt**.
5. Exit Notepad and maximize AutoCAD.

TEXT

6. From the Annotate tab of the Ribbon, pick the **Multiline Text** button on the Text panel.
7. Near the top center of the screen, pick the first corner.
8. Produce a rectangle that measures **1** unit tall by **3** units wide.

The multiline text editor appears.

9. Right-click inside the text box to display the shortcut menu.
10. Select **Import Text...** from the menu.
11. Locate and select the new file named **import.txt**.

This imports the text into the multiline text editor.

12. Highlight the text and pick the **Arial** font from the font drop-down list on the Ribbon.

The text changes to the Arial font.

CLOSE

13. Pick the **Close Text Editor** button on the Ribbon and save your work.

Formatting Text

The multiline text editor in AutoCAD offers many of the same formatting capabilities as standard word processing programs. For example, you can format text into bulleted, numbered, or lettered lists. You can also create paragraph indents and tabs.

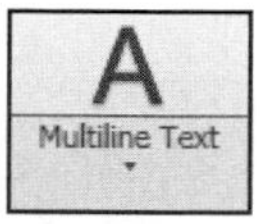

TEXT

1. From the Text panel on the Ribbon, pick the **Multiline Text** button, or enter **T** at the keyboard.

2. Pick a point in an open area of the screen, select the **Width** option, and enter **6** for the width.

3. Type the following lines. Be sure to press **ENTER** at the end of each line.
   Notes:
   All dimensions in inches.
   Material: Aluminum 6061-T6 bar stock.
   All fillets and rounds R.125 unless otherwise specified.
   Break all sharp corners.

4. Highlight all of the typed text, change the font to **Arial**, and make the text height **0.2000** unit.

5. Highlight the first line and underline it.

6. Highlight every line except the first line, pick the **Bullets and Numbering** button on the Paragraph panel, and select **Numbered** from the menu.

7. If there is too much space between the numbers and text, pick the small tab marker and triangle in the ruler bar and slide them to the left to align the text closer to the numbers.

The list should now look like the one in Fig. 19-4.

8. Pick the **Close Text Editor** button.

9. Save your work.

Notes:
1. All dimensions in inches.
2. Material: Aluminum 6061-T6 bar stock.
3. All fillets and rounds R.125 unless otherwise specified.
4. Break all sharp corners.

Fig. 19-4

Text Styles and Fonts

It is possible to create new text styles using the STYLE command. During their creation, you can expand, condense, slant, and even draw the characters upside-down and backward.

Creating a New Text Style

1. Select the **Text Style** pull-down menu on the top of the Text panel on the Ribbon and select the **Manage Text Styles...** option, or enter **ST** at the keyboard for the STYLE command.

The Text Style dialog box appears, as shown in Fig. 19-5.

 You can also open the Text Style dialog box from the Format menu in the Menu Bar. In the AutoCAD Classic workspace, the Text Style... button is on the docked Styles toolbar. It is located at the right of the Standard toolbar.

2. Pick the **New...** button, enter **comp1** for the new text style name, and pick the **OK** button.
3. Under Font Name, display the list of fonts by picking the down arrow.
4. Using the scroll bar, find the font file named **complex.shx** and select it.

Notice that the text sample in the Preview box changes to show the appearance of the complex.shx font.

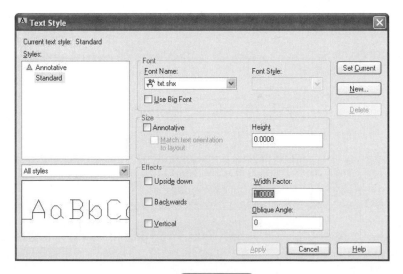

Fig. 19-5

5. Study the other parts of the dialog box and then pick the **Apply** and **Close** buttons.

You are ready to use the new comp1 text style. Notice that the comp1 name appears in the Text Style box on the Text panel.

6. Select the **Single Line Text** button on the Text panel to enter the DTEXT command and select the **Justify** option.

7. Right-justify the text by selecting the **Right** option.

8. Place the endpoint near the right side of the screen.

9. Set the height at **.3** unit.

10. Set the rotation angle at **0**.

11. For the text, type the three lines shown in Fig. 19-6. Be sure to press **ENTER** twice after typing the third line.

TEXT

Assemble and ream for No. 0 taper pin.

Fig. 19-6

With the STYLE command, you can develop an infinite number of text styles. Try creating other styles of your own design. The romans.shx font is commonly used in engineering drawings.

As you create more text styles within a drawing file, you may occasionally want to check their names.

12. Select the down arrow to the right of the Text Style text box on the Text panel to display the defined text styles.

If you want to learn more about the text styles, you will need to display the Text Style dialog box.

13. Select the **Manage Text Styles...** option from the pull-down menu.

14. In the Text Style dialog box, under Styles, select **Standard**.

The font name (txt.shx) and other characteristics of the style appear. This is the default text style.

15. Pick the **Cancel** button.

Text Fonts

AutoCAD supports TrueType fonts and AutoCAD compiled shape (SHX) fonts.

1. Reopen the **Text Style** dialog box and pick the **New...** button.
2. Type **tt** for the new text style name and pick the **OK** button or press **ENTER**.
3. Under Font Name, find and select **Swis721 BT**.

The Swis721 BT font displays in the preview area. Notice the overlapping T's located at the left of the font name. This indicates that it is a TrueType font. Notice also that Roman displays under Font Style.

4. Under Font Style, display the list of options, pick **Italic**, and notice how the font changes in the Preview area.

Note the 0.0000 value in the text box under Height. This 0 value indicates that the text is not fixed at a specific height, giving you the option of setting the text height when you enter the DTEXT command.

5. Pick the **Apply** and **Close** buttons.
6. Enter the **DTEXT** command and pick a point anywhere in the drawing area.
7. Enter **.2** for the height and **0** for the rotation angle.
8. Type **This is TrueType.** and press **ENTER** twice.

The TrueType text should look like the text in Fig. 19-7.

This is TrueType.

Fig. 19-7

Text Quality

The TEXTQLTY system variable sets the resolution of text created with TrueType fonts. A value of 0 represents no effort to refine the smoothness of the text. A value of 100 represents a maximum effort to smooth the text. Lower values decrease resolution and increase plotting speed. Higher values increase resolution and decrease plotting speed. You will learn more about plotting in Chapter 24.

1. Enter the **TEXTQLTY** system variable at the keyboard.
2. Press the **ESC** key.
3. Experiment with other TrueType fonts on your own.
4. Save your work.

Setting the Current Style

Let's set a new current text style.

1. Select the down arrow to the right of the **Text Style** text box on the **Text** panel and pick the **comp1** text style.

The comp1 style is now the current text style.

2. Place some new text on the screen and then terminate the **DTEXT** command.

3. Change back to the **tt** text style.

4. Save your work.

Applying Text in Drawings

Now that you have learned many ways of creating text, let's apply that knowledge to the creation of a title block. A **title block** is a portion of a drawing that is set aside to give important information about the drawing, the drafter's name, the company, and so on.

QUICK ACCESS

1. Pick the **New** button on the Quick Access toolbar, or type **NEW** at the keyboard and begin a new drawing from scratch. Save it as **title.dwg**.

2. Set the grid at **.25** and snap at **.0625**.

3. Create the title block shown in Fig. 19-8 using the following information:

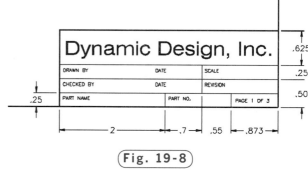

(Fig. 19-8)

- For Dynamic Design, Inc., create a new style named **Swiss** using the **Swis721 Ex BT** font. Make the text **.25** tall.

- Create a new text style named **Roms** using the **romans.shx** font and use it to produce the small text, which is **.06** tall.

- Use the **Justify** options of the **DTEXT** command to position the text accurately.

4. Save your work and exit AutoCAD.

Many companies use their own customized title blocks. AutoCAD also supplies several ready-to-use title blocks in the Templates folder.

● REVIEW QUESTIONS

Answer the following questions on a separate sheet of paper.

1. What is the most efficient way to enter a single line of text? What command should you use?
2. What might be a benefit of using the MTEXT command?
3. What is the purpose of the DTEXT Style option?
4. Name at least six fonts provided by AutoCAD.
5. What command do you enter to create a new text style?
6. Briefly describe how you would create a tall, thin text style.
7. Explain how to import text from a word processor into AutoCAD. Why might you choose to do this?
8. What is the purpose of a title block?
9. Explain the differences between the DTEXT command and the MTEXT command. Describe a situation in which each command would clearly be a better choice.

● CHALLENGE YOUR THINKING

These questions are designed to further your knowledge of AutoCAD by encouraging you to explore the concepts presented in this chapter. Answer each question on a separate sheet of paper.

1. Obtain on-screen help and then experiment with the COMPILE command. What are the advantages and disadvantages of compiling fonts? When might you want to use this option?
2. In addition to standard AutoCAD and TrueType fonts, AutoCAD can also work with PostScript fonts. Write a paragraph comparing TrueType and PostScript fonts. Are there any situations in which one type might be preferable to the other? Explain.
3. Which word processor do you use on your computer? Make a list of all the word processing programs you have used. Most likely, any of these can be used to import text into AutoCAD. Try importing text from several word processing programs.

● APPLYING AUTOCAD SKILLS

Work the following problems to practice the commands and skills you learned in this chapter.

1. Type your name and today's date in Windows Notepad. Import this text into the drawings you created in Chapter 17. Save your work.

2. Create a new text style using the following information:

   Style name: cityblueprint
   Font file: CityBlueprint
   Height: .25 (fixed)
   Width factor: 1
   Oblique angle: 15

3. Create a new text style using the following information:

   Style name: ital
   Font file: italic.shx
   Height: 0 (not fixed)
   Width factor: .75
   Oblique angle: 0

4. Use the DTEXT command to place the text shown in Fig. 19-9. Use the cityblueprint text style you created in problem 2. Right-justify the text. Do not rotate the text.

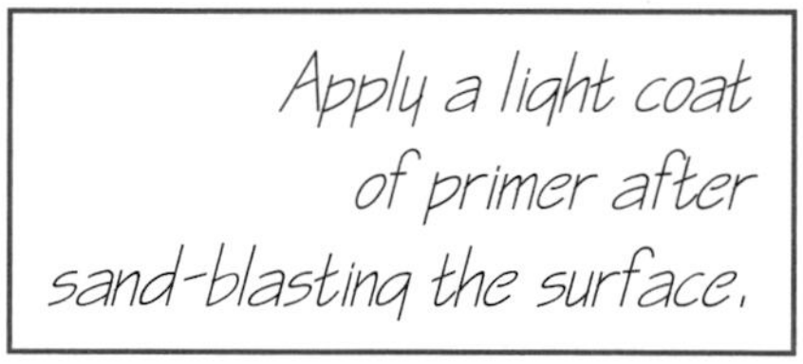

Apply a light coat
of primer after
sand-blasting the surface.

Fig. 19-9

5. Use the MTEXT command and the ital text style you created in problem 3 to create the text shown in Fig. 19-10. Set the text height at .3 unit. Rotate the text 90 degrees. Set the width of the text line to 4 units.

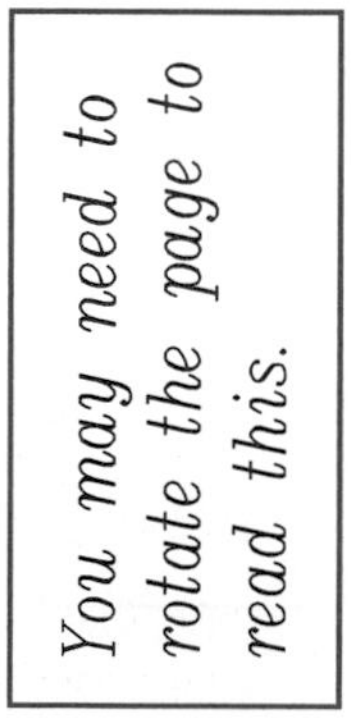

You may need to rotate the page to read this.

Fig. 19-10

Continued

6. Open the remote.dwg file you created in Chapter 18, "Applying AutoCAD Skills" problem 3. If you have not yet created the drawing, do so now. Then add the text shown in Fig. 19-11.

7. Shown in Fig. 19-12 is a block drawing algorithm for a program that sorts numbers into ascending order by a method known as *sorting by pointers*. Use the LINE and OFFSET commands to draw the boxes. Then insert the text using the romans.shx font. For the words MAIN, SORT, and SWAPINT, use the romanc.shx font and make the text larger. For the symbol >, use the symath.shx font, character N.

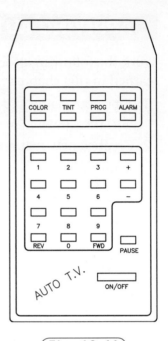

Fig. 19-11

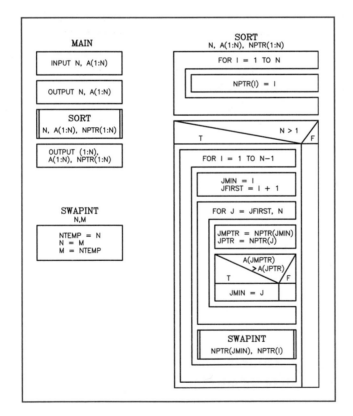

Fig. 19-12

Courtesy of Gary J. Hordemann, Gonzaga University

● USING PROBLEM-SOLVING SKILLS

Complete the following activities using problem-solving skills and your
knowledge of AutoCAD.

1. Most companies have standard operating procedures for creating engineering drawings. Your company is new, and the procedures have not
   yet been formalized. However, it has become standard practice to place a
   rectangular border around drawings and enclose the drawing identification in a title block. Open shaftbearing.dwg, which you created in
   Chapter 9 ("Applying AutoCAD Skills" problem 2). Draw a border
   around the periphery of the drawing. Across the bottom, draw a title
   block .75 unit high and divide it into three equal parts. In the first part,
   place your name; in the second part, place the name of the drawing; and
   in the third part, place today's date. Save the drawing.

2. Open the shim.dwg or shimrev.dwg drawing from Chapter 16. Make
   a border and title block and complete the information as in problem 1.
   Add the following information to the drawing in the form of drawing
   notes. Use the text style and font of your choice, or as specified by your
   company's procedures. Place the notes in the lower left corner of the
   drawing, just inside the border. Save the drawing.

   Notes:
   1. Material: SAE 1010 Carbon Steel
   2. Break sharp edges
   3. All dimensions +/−.01 unless otherwise specified

Chapter 20

Text Editing and Spell Checking

Objectives

- Edit text using AutoCAD's text editor
- Create special text characters
- Find and replace text
- Check the spelling of text in a drawing using AutoCAD's spell checker

Vocabulary

in-place text editor
special characters

AutoCAD offers several commands and features for editing and enhancing both types of text objects. Its text editor and spell checker help you produce error-free notes and specifications. With AutoCAD, it is possible to find and replace text using a feature similar to that in word processing programs.

Editing Text

AutoCAD provides two commands to change the contents of text. DDEDIT edits text created with the DTEXT command. MTEDIT edits multiple-line text created with the MTEXT command. In each case, AutoCAD displays an edit box, or **in-place text editor**, that allows you to edit the selected text.

1. Start AutoCAD and open the file named **text.dwg**.
2. Select the **AutoCAD Classic** workspace and close all floating toolbars and palettes.
3. Create a new text style named **roms** using the romans.shx font. Apply all of the default values and close the dialog box.

4. Using the **DTEXT** command, enter the following text at a height of **.15**.

NOTE: ALL ROUNDS ARE .125

Editing Dynamic Text

You can edit text that was created with the DTEXT command right on the screen.

1. Double-click the text you just created.

The text becomes highlighted against a blue background.

2. Move the pointer to the area containing the text and pick a point between the words ALL and ROUNDS.

3. Enter **FILLETS AND** so that the line now reads:

NOTE: ALL FILLETS AND ROUNDS ARE .125

4. Press **ENTER** twice to terminate the command.

Editing Mtext Objects

You can use a similar method to edit text created with the MTEXT command.

1. Double-click the mtext object located in the upper left area of the screen.

This displays an in-place text editor and the **Text Formatting** toolbar.

2. With the pointing device, position the cursor in the text and change a couple of words in the same way as you would with a word processor or text editor.

3. Change the style to **roms** and the height to **.125** and pick **OK**.

 To change the text height, you must first highlight the text.

The mtext changes according to the instructions you placed in the dialog box.

Justifying Mtext

1. Double-click the mtext object located in the upper left area of the screen.

2. Change the text style to **tt**.

3. Highlight the text.

4. In the Text Formatting toolbar, select the **MText Justification** button (the second button on the bottom row).

AutoCAD offers nine different ways to justify mtext objects. Top Left is the default setting. Text can be center-, left-, or right-justified with respect to the left and right text boundaries. Also, text can be middle-, top-, or bottom-aligned with respect to the top and bottom text boundaries.

5. Select **Top Right** and pick **OK** on the Text Formatting toolbar.

AutoCAD right-justifies the text along the top and right boundaries.

Creating Special Characters

The in-place text editor permits you to insert special characters such as the degree (°) and plus/minus (±) symbols. **Special characters** are those that are not normally available on the keyboard or in a basic text editor.

DRAW

1. Enter the **MTEXT** command.
2. In an empty area of the screen, create a rectangle measuring about **1** unit tall by **2** units wide.
3. Enter the following specification, but omit the special characters.

   Drill a Ø.5 hole at a 5° angle using a tolerance of ±.010.

4. Pick a point prior to **.5** and right-click.
5. From the shortcut menu, select **Symbol** and **Diameter**.

AutoCAD inserts the diameter symbol.

6. Pick a point after **5**, right-click, and select **Symbol** and **Degrees**.

The degree symbol appears.

7. Pick a point before **.010**, right-click, and select **Symbol** and **Plus/Minus**.

The plus/minus symbol appears.

8. Pick the **OK** button on the Text Formatting toolbar.
9. Double-click the text to redisplay the in-place text editor.
10. Move the pointer inside the text box, right-click, and rest the pointer on **Symbol**.

The menu includes several commonly used symbols such as Angle, Center Line, Delta, Ohm, and Not Equal.

11. Select **Other...** from the menu.

The character map displayed provides access to many additional characters.

12. Close the character map and pick the **OK** button to close the Text Formatting toolbar.

Finding and Replacing Text

Suppose you have lengthy paragraphs of text and you want to find and replace certain words.

1. Rest the crosshairs on top of the mtext object located in the upper left area of the screen.

2. Right-click and pick the **Find...** item.

3. Pick the **More Options** arrow from the lower left corner of the dialog box.

This displays the expanded Find and Replace dialog box, as shown in Fig. 20-1.

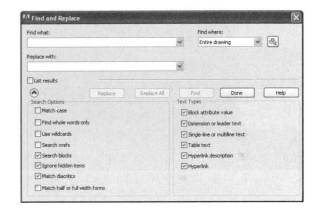

4. In the Find what: text box, enter **write**.

5. In the Replace with: text box, enter **draft**.

6. Pick the **Find** button.

AutoCAD displays the text in the dialog box and finds and highlights the word write.

Fig. 20-1

7. Pick the **Replace** button.

AutoCAD replaces the word write with the word draft.

8. Pick the **OK** button to close the information dialog box that reports the results of the find and replace.

The Find and Replace dialog box offers additional options. If you check the Match case check box, AutoCAD finds the text only if the case of all the characters in the text matches the text in the Find what: text box. If you check the Find whole words only check box, AutoCAD matches the text in the Find what: text box only if it is a single word. If the text is part of another word, AutoCAD ignores it. The Text Types section allows you to select the types of text you want to search. By default, all options are selected.

9. In the upper right area of the dialog box, pick the down arrow under the Find where: heading.

As you can see, you can search the entire drawing, the current space/layout, or the selected objects.

10. Pick the **Close** button.

Using AutoCAD's Spell Checker

AutoCAD's spell checker examines text for misspelled words.

1. Enter **SP** at the keyboard to enter the SPELL command.

The Check Spelling dialog box appears.

2. Pick the down arrow under Where to check, and pick the **Selected objects** option.

3. Pick the **Select objects** button, select the text NOTE: ALL FILLETS AND ROUNDS ARE .125, and press **ENTER**.

4. Select the **Start** button.

If you spelled the words correctly, AutoCAD displays the message box shown in Fig. 20-2.

(Fig. 20-2)

5. Pick **OK** and close the message box and **Close** to close the Check Spelling dialog box.

6. Using the **DTEXT** command, add the term **TrueType** to the screen.

7. Enter the **SPELL** command. With Entire drawing selected in the Where to check text box, pick the **Start** button.

AutoCAD displays the words in the document that are not in the spell checker's dictionary. These include words you might have misspelled and words like TrueType (and perhaps your last name). For each word that AutoCAD displays, you can choose to Ignore the spelling or select one of AutoCAD's suggestions and Change the word's spelling. You can also choose to Ignore All instances of the word, Change All instances of the word, or add the word to AutoCAD's dictionary by picking the Add to Dictionary button.

8. Pick the **Change** button to correct words that are misspelled and the **Ignore** button to skip words that are not misspelled.

9. When the spelling check is complete, pick **OK** to close the message box and **Close** to close the Check Spelling dialog box.

Other Options

AutoCAD provides other options for changing the appearance of text created with the MTEXT command.

1. Double-click the mtext in the top right corner of the drawing area.

2. Move the pointer inside the text box and right-click.

3. Select the **Paragraph...** item.

This displays the Paragraph dialog box, which permits you to adjust the indentation and tabs for mtext objects.

 You can also display the Paragraph dialog box by right-clicking the ruler above the mtext and selecting the Paragraph... item.

4. Pick the **Cancel** button.

5. Highlight all of the mtext.

6. Right-click to display the shortcut menu and rest the pointer on **Change Case**.

This allows you to change the mtext to all upper- or lowercase letters. Below Change Case is AutoCAPS. This converts all newly typed and imported text to uppercase, but it does not affect existing text.

7. Select **Cancel** to close the shortcut menu.

8. Save your work and exit AutoCAD.

● REVIEW QUESTIONS

Answer the following questions on a separate sheet of paper.

1. How does selecting and right-clicking text permit you to edit dynamic text? How does it permit editing of mtext?

2. Why might the character map be useful?

3. How would you add a diameter symbol to mtext?

4. Explain a situation in which AutoCAD's spell checker may find a word that is spelled correctly. What might cause this?

5. Suppose you have paragraphs of text that contain several words or phrases that need to be replaced with new text. What is the fastest way to replace the text?

● CHALLENGE YOUR THINKING

These questions are designed to further your knowledge of AutoCAD by encouraging you to explore the concepts presented in this chapter. Answer each question on a separate sheet of paper.

1. Look again at the mtext object you edited in this chapter. If you changed the word text to test and the word write to rite, what misspellings do you think the spell checker would find? Try it and see. Then write a short paragraph explaining the proper use of a spell checker.

2. As you read in this chapter, AutoCAD allows you to change the dictionary it uses to check the spelling of words. AutoCAD also allows you to create one or more custom dictionaries. Under what circumstances might you want to use a custom dictionary? Might you ever need more than one? Explain.

● APPLYING AUTOCAD SKILLS

Work the following problems to practice the commands and skills you learned in this chapter.

1. Open shaftbearing.dwg, which you updated in the previous chapter to include text. Change the date to be the date you actually created the original drawing. Add your middle initial to your name. If you already have your middle initial, remove it. Spell-check the title block. Did your name appear for correction? If it did, add it to the dictionary. Save the drawing as shaft-rev1.dwg.

2. Edit the drawing notes in shimrev.dwg. Remove note number 2 (Break sharp edges). Renumber note 3, and add periods at the end of the notes. Spell-check the notes and the title block. Your name should not appear for correction. Save the drawing.

3. Refer again to shimrev.dwg. Your method of entering +/– for a plus/minus value in the original note 3 is not acceptable to your supervisor. You have been asked to change it to the standard format of ±. Erase your notes, reenter the text using the MTEXT command, and insert the proper symbol. Save the drawing.

● USING PROBLEM-SOLVING SKILLS

Complete the following activities using problem-solving skills and your knowledge of AutoCAD.

1. Open Mechanical - Text and Tables.dwg located in AutoCAD's Sample/ Mechanical Sample folder and save it in your named folder as Text and Tables 2.dwg. Zoom in so that you can read the individual lines. Search and replace each occurrence of PIPE SUPPORTS with the term PIPE BRACKETS. Create a new text style called Notes using the Myriad Web Pro font. Select all the text in the General Notes and change the text to Notes, make the text 4 units in height, and pick OK. Save your work.

Continued

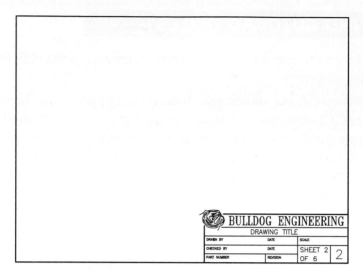

Fig. 20-3

2. The border and title block shown in Fig. 20-3 are suitable for a paper size of 8.5″ × 11″. The border allows .75″ white space on all four sides. Draw the border and title block using the dimensions in the detail drawing shown in Fig. 20-4. The three sizes of lettering in the title block are .0625″, .125″, and .25″. Use the romans.shx font. Replace "Bulldog Engineering" with your own name and logo. Position the text precisely and neatly. Use the snap and grid as necessary. Omit words, such as BY in DRAWN BY, and misspell a few words. Then, using AutoCAD's editing and spell-checking capabilities, fix the text.

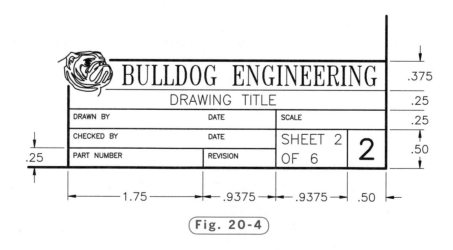

Fig. 20-4

Tables

Objectives

- Create a table and populate it with text
- Apply styles to tables
- Edit tables
- Use formulas in tables to perform calculations

Vocabulary

cells
formula
table

Engineering drawings often contain a parts list or bill of materials in tabular form. These tables list the requirements for the raw materials used to manufacture parts, all the individual detail parts that make up an assembly, and other critical information. AutoCAD allows you to create tables in your drawings and provides easy editing capabilities. You can also use tables to perform calculations as you would in a spreadsheet program.

Creating Tables

A **table** is a set of rows and columns that includes text. The basic units of a table—the individual boxes within the table's grid—are called **cells**. Creating a table in AutoCAD is similar to creating one in Microsoft Word®. After creating a table, you are permitted to populate it with text.

1. Start AutoCAD and open **text.dwg**.
2. Select the **Drafting & Annotation** workspace.
3. Pick the **Table** button on the Tables panel under the Annotate tab of the Ribbon, or enter the **TABLE** command.

TABLES

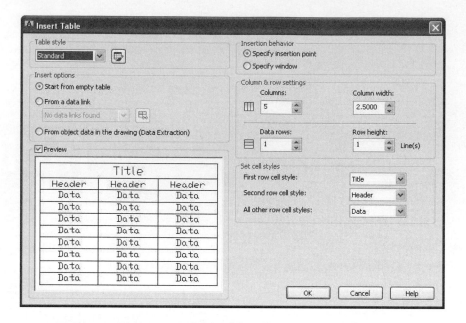

Fig. 21-1

This displays the Insert Table dialog box, as shown in Fig. 21-1.

4. In the Insertion behavior area, pick **Specify insertion point**.

5. In the Columns & rows settings area, enter **3** for the number of columns and **5** for the number of rows, and pick the **OK** button.

A table appears that you can drag on the screen.

6. Insert the table in an open area in the lower part of the screen. Pan and zoom as needed.

7. On the Style panel of the Text Editor tab of the Ribbon, select **tt** for the text style and enter **.18** for the text height.

8. Type **Fasteners** and then press the tab key.

9. Select **tt** for the text style, enter **BOLT**, and press the tab key.

10. Select **tt** for the text style, enter **.3125 x 16 x .75**, and press the tab key.

11. Select **tt** for the text style, enter **10**, and press the tab key.

12. Pick the **Close Text Editor** button to terminate the process.

CLOSE

Table Styles

In most cases, the same or similar text styles are used in all of a table's cells. AutoCAD allows you to create a table style to streamline the process of entering text.

1. Undo the insertion of the table.

2. Pick the **Table** button on the Ribbon.

3. In the Table style area of the Insert Table dialog box, pick the **Table Style** button located to the right of the down arrow.

TABLES

This displays the Table Style dialog box.

4. Pick the **New...** button, enter **Fasteners** for the new style name, and pick the **Continue** button.

The **New Table Style** dialog box appears, as shown in Fig. 21-2.

5. Select **Data** in the Cell styles text box and pick the **Text** tab.

6. Select **tt** for the text style and enter **.125** for the text height.

The Borders tab controls the appearance of the border lines. The General tab allows you to format the cells in the table, specify the alignment of the data in the cells, and control the spacing of the horizontal and vertical margins between the border of the cell and the cell contents. Under General, the Table direction text box allows you to change the table so that it reads from bottom to top instead of from top to bottom.

7. Select **Header** in the Cell styles text box, pick the **Text** tab, enter **tt** for the text style, and enter **.15** for the text height.

8. Select **Title** in the **Cell styles** text box, select **tt** for the text style, and enter **.18** for the text height.

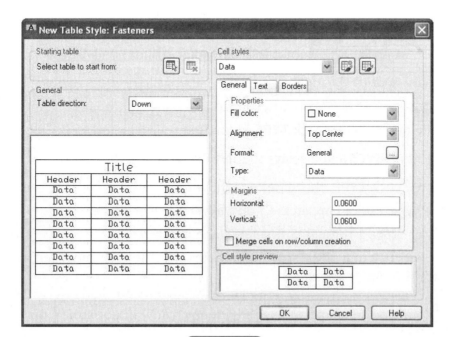

Fig. 21-2

FASTENERS		
TYPE	SIZE	QUANTITY
HEX BOLT	.375 x 16 x 1	10
HEX NUT	.375 - 16	10
PLAIN WASHER	.5 x 1 x .0625	20
CAP SCREW	.5 x 13 x 2	5

Fig. 21-3

9. Pick the **OK** button; pick the **Close** button in the Table Style dialog box.

10. In the Insert Table dialog box, select **Fasteners** for the table style and pick the **OK** button.

11. Insert the table in an open area in the lower part of the screen.

12. Enter **FASTENERS** and press the tab key.

13. Enter the text shown in Fig. 21-3 into the cells. Be sure to press the tab key to advance from one cell to the next. Click outside the table when you are finished.

14. Save your work.

Editing Tables

It is possible to edit the contents of a table. In fact, you can also change the properties of the table itself.

Editing Text in Tables

Making changes to the text within a table is very similar to editing mtext.

1. In the table, double-click **TYPE** in the first cell.

This selects the cell and displays the Text Editor tab on the Ribbon.

2. Double-click **TYPE**, this time to highlight it.

3. Type **Type** with an uppercase **T** and lowercase **y, p,** and **e**.

4. Repeat Steps 1 through 3 with **SIZE** and **QUANTITY**.

5. Double-click to activate (place a text cursor in) the left-most cell in the bottom (empty) row of cells.

6. Type **CAP SCREW** and press the tab key.

7. For the size, type **.5 x 13 x 1.5**; for the quantity, type **15**.

8. Pick the **Close Text Editor** button on the Ribbon or click outside the table to finish the editing session.

9. Pick the arrow in the lower right corner of the Tables panel on the Ribbon.

This displays the Table Style dialog box.

10. Select the **Fasteners** table style and pick the **Modify...** button.

11. Under Cell styles, pick **Data**.

12. In the General tab, pick **Middle Left** from the drop-down menu located to the right of Alignment.

13. In the Text tab, select **red** from the drop-down menu to the right of Text color.

14. Pick **OK** and **Close** to close the two dialog boxes.

The data text is now red and left-justified.

15. Undo your changes.

Changing the Table Border Lines

AutoCAD allows you to change the table itself, which is represented by border lines.

1. Select any one of the border lines.

This selects the entire table.

2. On the table's left-most vertical line, click the middle grip box and move it to the right to reduce the width of the column.

3. With the table still highlighted, right-click to display the shortcut menu and select **Size Columns Equally**.

AutoCAD adjusts the columns so they are of equal width.

4. Pick the arrow in the lower right corner of the Tables panel to display the Table Style dialog box and pick the **Modify...** button.

5. Under Cell styles, pick **Title**.

6. In the Borders tab, select **0.30 mm** for Lineweight and **Yellow** for Color.

7. Pick the **All Borders** button (the first button at the bottom) to make the title border lines yellow, pick the **OK** button, and pick the **Close** button.

 If the line weight did not change, pick the Show/Hide Lineweight button on the status bar.

8. Reopen the **Modify Table Style** dialog box and change the title border lineweight and color to **ByBlock**.

9. Save your work.

Inserting Formulas in Tables

A **formula** performs an automatic mathematical equation. You can insert simple formulas into cells in tables to calculate sums, counts, and averages.

1. In a blank area of the screen, create a table with two columns and five rows. Use the **Fasteners** table style. (It should be the current table style.)
2. Populate the cells with the text shown in Fig. 21-4.

Part Count	
Item	Quantity
Bolt	8
Washer	16
Nut	8
Spacer	4
TOTAL	

Fig. 21-4

3. Make the word **TOTAL** in the bottom left cell bold.
4. Select the empty cell in the lower right corner of the table.
5. Right-click and pick **Insert** from the shortcut menu, then pick **Formula** and **Sum** from the cascading menus.
6. In response to **Select first corner of table cell range**, click in the first cell below **Quantity**.
7. In response to **Select second corner of table cell range**, click in the cell in the number column next to **Spacer**.

The formula Sum(B3:B6) appears in the Field dialog box. This formula will add the contents of cells B3 through B6.

8. Pick **OK** to close the dialog box and close the **Text Editor**.

The cell now contains the number 36, which is the sum of the part quantities listed in the table. Notice that the number in the cell is shaded, indicating that the cell contains a formula.

 You can also type formulas directly into cells. In this example, typing Sum(B3:B6) directly into the cell would have yielded the same result.

9. Save your work and exit AutoCAD.

Chapter 21 Review & Activities

● REVIEW QUESTIONS

Answer the following questions on a separate sheet of paper.

1. What is a table in AutoCAD?
2. Describe the quickest method of editing text in a cell of a table.
3. Suppose that you created a table, but later decided to change the weight of the table's border lines. How would you go about it?
4. How many cells are in a table that contains five rows and seven columns?
5. Describe how to create a table style. Explain why this method is faster than formatting cells individually.
6. What is the purpose of a formula in a table?
7. How can you tell whether or not a number in a cell is the result of a formula?
8. Explain how you could make sure that all of the columns in a table are exactly the same width.

● CHALLENGE YOUR THINKING

These questions are designed to further your knowledge of AutoCAD by encouraging you to explore the concepts presented in this chapter. Answer each question on a separate sheet of paper.

1. Other than for parts lists and bills of materials, when might you use a table in AutoCAD?
2. AutoCAD allows you to build a table with the title and headers on the bottom and the list on top. Why do you think this option exists? When would it be useful?

Continued

• APPLYING AUTOCAD SKILLS

Work the following problems to practice the commands and skills you learned in this chapter.

1. Create the door schedule shown in Fig. 21-5 using a table style that includes cell data that is .125 in height, column heads that are .18 in height, and a title that is .22 in height. All text should use the Romans font. Name the file schedule.dwg.

DOOR SCHEDULE			
QUAN.	TYPE	SIZE	REMARKS
2	FLUSH	3−0 x 6−8	HOLLOW CORE
3	FLUSH	2−8 x 6−8	HOLLOW CORE
1	FLUSH	2−6 x 6−8	SOLID CORE, OAK
3	SLIDING	4−0 x 6−8	HOLLOW CORE

Fig. 21-5

2. Using Save As..., create a new file from schedule.dwg and name the file schedule2.dwg. In this new file, change the text in the door schedule so that it matches the schedule in Fig. 21-6. Use the tt text style for the column heads and title and make all of the border lines green.

Door Schedule			
Quantity	Type	Size	Comments
2	FLUSH	2−10 x 6−8	HOLLOW CORE
4	FLUSH	2−8 x 6−8	HOLLOW CORE
1	FLUSH	3−0 x 6−8	SOLID CORE, OAK
2	BI−FOLD	5−0 x 6−8	HOLLOW CORE

Fig. 21-6

• USING PROBLEM-SOLVING SKILLS

Complete the following activity using problem-solving skills and your knowledge of AutoCAD.

1. The frame.dwg file, which you created in "Using Problem-Solving Skills" problem 3 in Chapter 17, is shown in Fig. 21-7. (If you have not yet created the drawing, create it now.) Use the drawing and editing commands to insert windows and doors into the floor plan. Do research if necessary to find standard window and door widths so that you can create them accurately. Then create a table to show the sizes, kinds, and prices for the windows and doors. In the last column, show the number of each type of window or door. In the last row, use one or more formulas to find the total number of windows and doors.

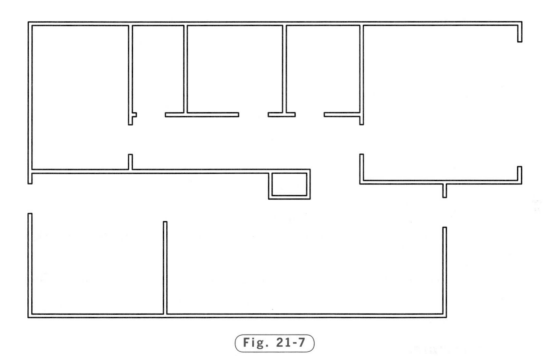

Fig. 21-7

Manufacturer's Product Table

Even in the modern electronic age, written communication remains a significant form of exchanging information. Even in a technical drawing, dimensions alone often do not convey sufficient information to enable the drawing to be used as intended. A few well-chosen words can provide enough information to minimize the number of views or necessary dimensions, making the drawing cleaner in appearance.

▶ Description

In this project, you will use written language to convey the information needed for the following scenario: A manufacturer of automotive wheels wishes to organize its product line in various ways to make it easier for consumers seeking to find a product that suits them. A decision has been made to organize the wheels in four different ways: by type of finish/color, by rim size (in diameter only—it is assumed each size is available in the most commonly needed widths), by material, and by price range.

Organization

Organize the following data into tables using AutoCAD's TABLE command.

- Finish/colors available include brushed aluminum and shiny aluminum; shiny and matte black; chromed steel and gunmetal gray steel (6 categories total)
- Rim sizes include diameters from 16″ to 24″ in 1″ increments (9 categories)
- Materials include aluminum, magnesium-aluminum alloy, and steel alloy (3 categories)

Price Structure

The price structure, per wheel, is as follows: aluminum 16″ wheels start at $299, magnesium-aluminum alloy wheels at $349, and steel alloy wheels at $229. For each line of wheel, the price increases by 5% (of the base price) for each incremental increase in diameter. Thus an aluminum wheel costs $299 plus 5% of $299 for a 17″ rim; an 18″ aluminum wheel goes for $299 plus 10% of $299, and so on. There is no distinction in price for color or finish.

Fig. P4-1

Accompanying Drawings

Create drawings of one or more styles of wheel that might be available. If you would like to see some of the many wheel designs being sold, simply do an Internet search for "automobile wheels." See Fig. P4-1 for other ideas.

▶ Hints and Suggestions

1. Consider including instructions for consumers on how to use the information in the table to find and calculate the cost of their order.

2. AutoCAD can insert formulas directly into tables. Your instructor may therefore ask you to include, for example, the potential to calculate taxes for the purchase of one or more wheels.

3. Make sure each table and column heading is clearly labeled.

4. Take advantage of text editing features such as bulleted and numbered lists.

5. When illustrating the wheels, consider using polar arrays.

▶ Summary Questions

Your instructor may direct you to answer these questions orally or in writing. You may wish to compare and exchange ideas with other students to help increase efficiency and productivity.

1. Why might it be preferable to use a table instead of a list or graph to convey this information?

2. How would a company benefit by facilitating the ability of consumers to initiate and price out their own orders?

Chapter 22

Drawing Setup

Objectives

- Explain the purpose of a template file and list settings that are commonly included
- Choose the appropriate unit of measurement for a drawing
- Determine appropriate sheet size and drawing scale for a drawing
- Check the current status of a drawing file

Vocabulary

drawing area
limits
template file

The first 21 chapters of this book have provided a foundation for using AutoCAD. Now it is time to apply many of the pieces. As part of this effort, you will identify the scale and sheet size for a particular drawing. In doing this, you will determine the drawing units and drawing area—key elements of the drawing setup process. You will also consider settings of features such as the grid and snap that you can store in a template file and use over and over again.

Template Files

In AutoCAD, a **template file** is a file that contains drawing settings that can be imported into new drawing files. This feature is helpful to people who need to create the same types of drawings (drawings with similar settings) frequently. The purpose of a drawing template file is to minimize the need to change settings each time you begin a new drawing. Some users and

Table 22-1. Template Development

Step	Description
1.	Determine what you are going to draw (mechanical detail, house elevation, etc.).
2.	Determine the drawing scale.
3.	Determine the sheet size. (Steps 2 and 3 are normally done simultaneously.)
4.	Set the drawing limits.
5.	Set the drawing area.
6.	Set the grid.
7.	ZOOM All. (This will zoom to the new drawing area.)
8.	Set the snap resolution.
9.	Enter STATUS to review the settings.
10.	Determine how many layers you will need and what information will be placed on each layer; establish the layers with appropriate colors, linetypes, lineweights, etc.
11.	Set the linetype scale.
12.	Create new text styles.
13.	Set the dimension scale, dimension text size, arrow size, etc.
14.	Store as a drawing template file.

companies include a standard border and title block in their template files. This further shortens the time they spend preparing for a new drawing.

Once you have set up a template file, subsequent drawing setup for similar drawings is quick and easy. When you use a template, its contents are automatically loaded into the new drawing. The template's settings thus become the settings for the new drawing. Template development includes the steps shown in Table 22-1. This and the next several chapters will provide an opportunity to practice these steps in detail.

Using a Predefined Template File

AutoCAD provides many standard templates that can be used as is or modified to fit individual needs.

1. Start AutoCAD and pick the **Use a Template** button on the **Startup** dialog box.

AutoCAD displays a list of template files available to you. Notice that all template files have a DWT file extension.

2. Scroll down the list and single-click several of the files, noting the files with borders and blocks.

Notice that the names of a few of the templates contain "3d." These templates are for 3D modeling. You will learn more about 3D modeling and these templates in Parts 8 and 9 of this book.

3. Find and double-click **Tutorial-imfg.dwt**.

The contents of the template file display on the screen in the default Drafting & Annotation workspace and the D-Size Layout tab. This template is set up to fit a D-size drawing sheet according to standards established by the American National Standards Institute (ANSI).

4. Pick the **Model** tab at the bottom of the drawing area.

As you can see, the Model space does not display the border and title block. You will learn more about Model (model space) and Layout (paper space) in Chapter 25.

Initial Template Setup

Many AutoCAD users choose to create custom template files. We will create one also. The first step is to identify the type of object (mechanical part, building, etc.) for which the new template will be used. For this exercise, let's create a template that can be used for stair details in drawings for homes and commercial buildings.

Next, we will determine the drawing scale for the stair detail. In this sense, a scale is a means of reducing or enlarging a representation of an actual object or part that is too small or too large to be shown on a drawing sheet. This information will give us a basis for establishing the drawing area, the linetype scale, and the scale for the dimensions. Let's use a $1/2'' = 1'$ scale, and let's base the template on a sheet size of 8.5" × 11".

1. Close the current drawing (do not save changes), and pick the **New** button from the Quick Access toolbar.

2. Pick the **Use a Wizard** button.

This begins our setup of the template file. We will create and save it as a regular drawing file. Then, when we have finished defining the settings, we will save it as a template file.

3. Read the description of the Advanced Setup wizard.

Advanced Setup uses the template file acad.dwt. So does Quick Setup.

4. Pick the **OK** button to proceed with the Advanced Setup.

Advanced Setup consists of five steps: Units, Angle, Angle Measure, Angle Direction, and Area. Completing these steps as part of the process of creating a new file can simplify part of the template setup process.

Specifying the Unit of Measurement

The unit of measurement you choose for the template file depends on the types of drawings for which you will use the template. AutoCAD provides several different types of units from which to choose.

1. Pick the **Architectural** radio button and review the sample drawing located at the right.

2. Under Precision, pick the down arrow and review the list.

3. Choose the **0'-0 1/16″** default setting.

This means that $1/16″$ is the smallest fraction that AutoCAD will display.

4. Pick the **Next** button to go to the next step.

If you choose not to use a wizard to set up your drawing files and templates, you can set up the drawing units using AutoCAD's UNITS command. This command displays the Drawing Units dialog box, from which you can set drawing units and precisions for lengths and angular units.

Setting Angle Measurements

Decimal Degrees is the default setting for angle measurements. This is what we want to use for our template, so no selection is required. When Decimal Degrees is selected, AutoCAD displays angle measurements using decimals.

1. Under Precision, pick the down arrow to adjust the angle precision, and pick **0.0**.

A setting of 0.0 means that AutoCAD will display angle measurements to one decimal place, as shown in the sample.

2. Pick the **Next** button to proceed to the next step.

The next dialog box permits you to control the direction for angle measurements. As discussed earlier in the book, AutoCAD assumes by default that 0 degrees is to the right (east). Let's not change it.

3. Pick the **Next** button.

We also established that angles increase in the counterclockwise direction by default. Do not change this, either.

4. Pick the **Next** button to proceed.

Setting the Drawing Area

The next step is to set the **drawing area**, or **limits**, of the drawing. The drawing area defines the boundaries for constructing the drawing, and it should correspond to both the drawing scale and the sheet size. The default drawing area is 12 × 9 units. Using architectural units, the drawing area is 1′ × 9″. (See Appendix C for a chart showing the relationships among sheet size, drawing scale, and drawing area.)

Actual scaling does not occur until you plot the drawing, but you should set the drawing area to correspond to the scale and sheet size. The drawing area and sheet size can be increased or decreased at any time using the LIMITS command. The plot scale can also be adjusted prior to plotting. For example, if your drawing will not fit on the sheet at ¹/4″ = 1′, you can enter a new drawing area to reflect a scale of ¹/8″ = 1′. Likewise, you can enter ¹/8″ = 1′ instead of ¹/4″ = 1′ when you plot.

As mentioned before, the drawing area should reflect the drawing scale and sheet size. Let's look at an example.

If the sheet size is 11″ × 8.5″ and the drawing scale is $^{1}/_{4}″ = 1'$, what is the drawing area? Since each plotted inch on the sheet would occupy 4 scaled feet, it's a simple multiplication problem: 11 × 4 = 44 and 8.5 × 4 = 34. The drawing area, therefore, would be 44′ × 34′ because each plotted inch represents 4′.

Since our scale is $^{1}/_{2}″ = 1'$, what should the drawing area be?

 How many $^{1}/_{2}″$ units would 11″ occupy, and how many would 8.5″ occupy?

1. Enter **22**′ for the width and **17**′ for the length. Be sure to include the apostrophe for the foot mark.

2. Pick the **Finish** button.

You have completed AutoCAD's Advanced Setup process.

3. Save your work in a file named **tmp1.dwg**.

Establishing Other Settings

Let's continue setting up the template file by setting the grid and snap and zooming out to display the new drawing area.

1. Enter the **GRID** command and set the grid at **1**′. (Be sure to enter the apostrophe.)

The purpose of setting the grid is to give you a visual sense of the size of the objects and the drawing area. After you enter ZOOM All in the next step, the grid will fill the drawing area, with a distance of 1′ between grid lines.

2. Enter **ZOOM All**.

The grid reflects the size of the sheet.

3. Enter **SNAP** and set it at **6**″.

4. Position the crosshairs in the upper right corner of the grid and review the coordinate display.

It should read approximately 22′-0″, 17′-0″, 0′-0″. The 0′-0″ is the z coordinate.

Status of the Template File

Let's take a moment to review the settings we have created up to this point.

1. Enter **STATUS** at the keyboard to enter the STATUS command.

The AutoCAD Text Window appears with the status of the drawing tmp.1dwg. Note each of the components found in STATUS.

2. Close the AutoCAD Text Window.

3. Press **ENTER** to exit the STATUS command.

Tmp1 now contains several settings specific to creating architectural drawings at a scale of $1/2'' = 1'$. It will work well for the stairway detail mentioned earlier.

We have developed the basis for a template file. Technically, we have not yet created the template, because we have not stored the contents of our work in a template (DWT) file. We will do this later when we continue its development.

The next steps in creating a template file deal with establishing layers. You will create layers in the next chapter and complete the final steps for creating a template file in a subsequent chapter.

4. Save your work and exit AutoCAD.

5. Produce a backup copy of the file named **tmp1.dwg**.

Backup copies are important because if you accidentally lose the original (and you may, sooner or later), you will have a backup. You can produce a backup copy in seconds, and it can save you hours of lost work. If you don't remember how to make a backup copy, see Chapter 8 for details on producing copies of files.

• REVIEW QUESTIONS

Answer the following questions on a separate sheet of paper.

1. Explain the purpose and value of template files.
2. What settings are commonly included in a template file?
3. If you select architectural units, what does a precision of 0'-0 1/4" mean?
4. What information is used to determine the size of the drawing area?
5. If the sheet measures 22" × 16" and the scale is 1" = 10', what should you enter for the drawing area?
6. Describe the information displayed as a result of entering the STATUS command.

• CHALLENGE YOUR THINKING

These questions are designed to further your knowledge of AutoCAD by encouraging you to explore the concepts presented in this chapter. Answer each question on a separate sheet of paper.

1. Discuss unit precision in your drawings. If necessary, review the precision options listed in the Drawing Units dialog box for units and angles. (*Hint:* You can display this dialog box by entering the UNITS command.) Then describe applications for which you might need at least three of the different settings.
2. For a drawing to be scaled at $1/8" = 1'$ and plotted on a 17" × 11" sheet, what drawing area should you establish?

Continued

● APPLYING AUTOCAD SKILLS

Work the following problems to practice the commands and skills you learned in this chapter.

1. Create a new drawing using the Acad -named plot styles.dwt template file. Use the STATUS command to review its settings. Create another new drawing using the Acadiso -named plot styles.dwt template file and review its settings also. What differences do you see between AutoCAD's templates for ANSI and ISO standards?

2. Establish the settings for a new drawing based on the information below, setting each of the values as indicated. Save the file as prb22-1.dwg.

   Scale: 1″ = 2″
   Sheet size: 17″ × 11″
   Units: Decimal
   Drawing area: (You determine it.)
   Grid value: .5″
   Snap resolution: .25″

   Be sure to ZOOM All.

   Review settings with the STATUS command.

3. Establish the settings for a new drawing based on the information below, setting each of the values as indicated. Save the file as prb22-2.dwg.

   Scale: $^1/8″$ = 1″
   Sheet size: 24″ × 18″
   Units: Architectural (You choose the appropriate options.)
   Drawing area: (You determine it.)
   Grid value: 4″
   Snap resolution: 2′

   Reminder: Be sure to ZOOM All.

   Review settings with the STATUS command.

4. Consider the following drawing requirements:

   Drawing type: Architectural drawing of a detached garage
   Dimensions of garage: 32′ × 20′

   Other considerations: Space around the garage for dimensions, notes, specifications, border, and title block

   Based on this information, write the missing data for drawing setup on a separate sheet of paper. Suggest values for scale, paper size, units, drawing area, grid, and snap resolution.

5. Consider the following drawing requirements:

Drawing type: Mechanical drawing of the wheel and bearings for an in-line skate

Approximate diameter of wheel: 2.5″

Other considerations: Space for dimensions, notes, specifications, border, and title block

Based on this information, write the missing data for drawing setup on a separate sheet of paper. Suggest values for scale, paper size, units, drawing area, grid, and snap resolution.

USING PROBLEM-SOLVING SKILLS

Complete the following activities using problem-solving skills and your knowledge of AutoCAD.

1. Your architectural office needs a drawing template setup created that will accommodate the floor plans of houses measuring 16 m × 8.4 m. Create the template. In addition to the units, angles, and area, specify reasonable settings for both snap and grid. Save the template as arch1.dwt.

2. Electronic engineering requires large drawings of small parts. Set up a drawing for electronic circuits measuring $1/4″ \times 3/32″$. Use decimal settings, and set the appropriate snap and grid. Save the template as elec.dwt.

Landscape Designer

Parks and outdoor recreation centers are perfect locations for picnics, Frisbee games, and other social activities. The next time you visit your local park, try to take notice of the surroundings: healthy green grass, immaculate flower beds, a ball diamond or two, a playground, and perhaps even a water fountain for kids to splash in. The design and placement of such items, all of which make up a visually appealing outdoor atmosphere, require strategic planning.

© Royalty-Free/CORBIS

Working Together

Landscape designers typically work closely with a landscape architect. While architects supply practical and artistic views of how they want to arrange outdoor items, landscape designers possess the knowledge to create the actual plans for the design. Before construction begins, the landscape designer must prepare detailed drawings for plans that result in landscapes that are both attractive and economical.

Employment Opportunities

Large development corporations, private sectors, and government agencies need people with landscape design skills. Landscape designers create plans for golf courses and college campuses. They also design outdoor lighting systems and irrigation systems.

All landscape designers must be able to produce accurate drawings based on conceptualizations and rough sketches. It is also important for landscape designers to understand mathematical and measurement concepts, since the plans must be drawn to scale.

▶ Career Activities

1. Interview someone who can tell you about landscape designers in your area. Are apprenticeships available?

2. Do you think landscape designers should have a background in ecology and science? Why?

Layers and Linetypes

Objectives

- Create layers with appropriate characteristics for the current drawing or template
- Use layers to control the visibility and appearance of objects in a drawing
- Change an object's properties
- Use object properties to filter object selections
- Apply a custom template file

Vocabulary

center lines
color palette
filtering
freeze
layers
phantom lines

AutoCAD gives you the option of separating classes of objects into layers. It may be helpful to think of AutoCAD's **layers** as separate sheets of transparency film that helps you organize your drawing.

One of the benefits of using layers is the ability to make them visible and invisible. For example, you can place construction lines and reference notes on a layer and then turn the layer off when you're not using it. A house floor plan could be drawn on a layer called Floor and displayed in red. The dimensions of the floor plan could be drawn on a layer called Dimension and displayed in yellow. A layer called Center could contain blue center lines.

Table 23-1 on page 308 shows an example set of layers. Note the layer names, colors, linetypes, and lineweights. We will create these layers in this chapter.

Table 23-1. Typical Layers

Layer Name	Color	Linetype	Lineweight
0	white	Continuous	Default
Border	cyan	Continuous	0.50 mm
Center	magenta	Center	0.20 mm
Dimensions	blue	Continuous	0.20 mm
Hidden	green	Hidden	0.30 mm
Notes	magenta	Continuous	0.30 mm
Objects	red	Continuous	0.40 mm
Phantom	yellow	Phantom	0.50 mm

Be sure to make a backup copy of tmp1.dwg before you begin working with it in this chapter if you have not already done so.

Creating New Layers

LAYERS

1. Start AutoCAD and open the drawing named **tmp1.dwg**.
2. Select the **Drafting & Annotation** workspace.
3. Select the **Home** tab of the Ribbon and pick the **Layer Properties** button on the Layers panel, or enter **LA**, the alias for the LAYER command.

This enters the LAYER command and displays the Layer Properties Manager palette. First, let's create layer Objects.

4. Pick the **New Layer** button near the top center of the palette.

This creates a new layer with a default name of Layer1. Notice that the name Layer1 is highlighted and a cursor appears in the edit box.

5. Using upper- or lowercase letters, type **Objects** and press **ENTER**.

The name Objects replaces Layer1. If you make a mistake, you can single-click the layer name and edit it.

6. Create the layers **Border**, **Center**, **Dimensions**, **Hidden**, **Notes**, and **Phantom** on your own, as listed in Table 23-1. (You will set the color, linetype, and lineweight of each layer later in this chapter.)

Changing the Current Layer

Let's change the current layer to Objects.

1. Select **Objects** and pick the **Set Current** button.

 The Set Current button is the green check mark at the top of the palette. You can also make a layer current by double-clicking the icon in the Status column next to the layer name.

Objects is now the current layer.

2. Close the **Layer Properties Manager** palette.

Notice that Objects now appears in the Layer text box on the Layers panel.

3. Save your work.

Assigning Colors

The Layer Properties Manager permits you to assign screen colors to the layers.

1. Pick the **Layer Properties** button or enter **LA**.

2. Find the Color heading, which is located at the left of the Linetype column heading. If the Color column heading and the color names are visible, skip to Step 5. If not, proceed to Steps 3 and 4.

3. Position the pointer between the two headings until the pointer changes to a double arrow.

4. Click and drag to the right until the names of the colors appear.

5. At the right of Objects, pick the black box under the **Color** heading.

LAYERS

The Select Color dialog box appears. It contains a color palette. A **color palette** is a selection of colors, similar to an artist's palette of colors.

6. Select the **True Color** tab.

This offers a choice of more than 16 million colors.

7. Select the **Color Books** tab.

This allows you to specify colors using third-party color books, such as Pantone, or user-defined color books.

8. Move the vertical slider bar up and down and select the down arrow in the **Color book** area to review options.

9. Select the **Index Color** tab.

10. In the bottom half of the dialog box, pick the color red (next to yellow) and pick the **OK** button.

AutoCAD assigns the color red to Objects.

11. Assign colors to the other layers as indicated in the layer listing in Table 23-1 on page 308. Cyan is light blue and magenta is purple.

12. When you have finished assigning the colors, auto-hide the Layer Properties Manager palette and save your work.

Notice that the color red appears beside Objects in the Layers panel.

AutoCAD's COLOR command allows you to set the color for subsequently drawn objects, regardless of the current layer. This allows you to set the color of each object individually, giving you a great deal of flexibility, but it can become confusing. It is recommended that you avoid use of the COLOR command and that its setting remain at ByLayer. The ByLayer setting means that the color is specified by the layer on which you draw the object.

The Color Control drop-down box on the Properties panel also allows you to set the current color. The recommended setting is ByLayer.

13. Draw a circle of any size on the current Objects layer.

It should appear in the color red.

14. Select **Hidden** from the Layer drop-down box in the Layers panel to make Hidden the current layer.

15. Draw a concentric circle inside the first circle.

It should appear in the color green.

Assigning Linetypes

Various types of lines are used in drafting to show different elements of a drawing. By convention, for example, hidden lines (those that would not be visible if you were looking at the actual object) are shown as dashed or broken lines. **Center lines** (imaginary lines that mark the exact center of an object or feature) are shown by a series of long and short line segments. Let's look at and load some of the different linetypes AutoCAD makes available to you.

1. Display the Layer Properties Manager palette.

2. At the right of layer Hidden, and under the Linetype heading, pick **Continuous**.

This displays the Select Linetype dialog box.

3. Pick the **Load...** button.

This displays the Load or Reload Linetypes dialog box, as shown in Fig. 23-1.

AutoCAD stores the list of linetypes in a file named acad.lin, as listed in the box at the right of the File... button. These linetypes conform to International Standards Organization (ISO) standards. This is important because ISO is the leading organization for the establishment of international drafting standards.

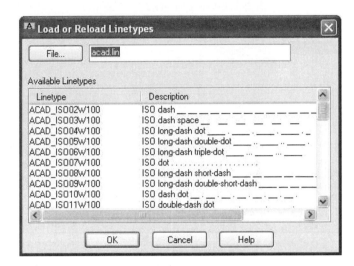

Fig. 23-1

At this point, you can select the individual linetypes that you want to load and use in the current drawing.

4. Using the scroll bar, review the list of linetypes.

5. Find and select the **CENTER**, **HIDDEN**, and **PHANTOM** linetypes.

HINT Press the CTRL key when selecting them. This allows you to make multiple selections.

Hidden lines are used to show invisible edges on drawings. This is why they are called "hidden." Center lines are used to show the centers of holes, cylinders, rounded corners, and fillets. In AutoCAD, lineweights are applied to **phantom lines** based on their use. When used for cutting planes in sectional views, a thick lineweight is applied.

6. Pick the **OK** button in the Load and Reload Linetypes dialog box and notice the new list of linetypes in the Select Linetype dialog box.

7. Pick **HIDDEN** for the linetype and pick **OK**.

8. Auto-hide the Layer Properties Manager palette.

The inner circle on layer Hidden changes from a continuous line to a hidden line. However, it's unlikely that you will be able to see the hidden line until after we scale the linetypes.

9. Save your work.

 You can also load linetypes using the LINETYPE command. Pick the Linetype... item from the Format menu in the Menu Browser or enter LT (for LINETYPE) and pick the Load button.

Scaling Linetypes

The LTSCALE command permits you to scale the linetypes so that they are correct for the drawing scale.

1. Enter **LTS** for the LTSCALE command.

Let's scale the linetypes to correspond to the scale of the template drawing. This is done by setting the linetype scale at $1/2$ the reciprocal of the plot scale. When you do this, broken lines, such as hidden and center lines, are plotted to ISO standards.

 As you may recall, we are creating a template file based on a scale of $1/2'' = 1'$. Another way to express this is $1'' = 2'$ or $1'' = 24''$. This can be written as $1/24$. The reciprocal of $1/24$ is 24, and half of 24 is 12. Therefore, in this particular case, you should set LTSCALE at 12.

2. In response to Enter new linetype scale factor, enter **12**.

The green circle should now appear in the hidden linetype.

3. If you are not sure, zoom in on it.

4. Display the Layer Properties Manager palette.

5. Assign the **CENTER** linetype to layer Center and the **PHANTOM** linetype to layer Phantom.

6. Save your work.

> The Linetype drop-down box, located on the Properties panel, permits you to change the current linetype regardless of the current layer. This gives you a great deal of flexibility, but it can become confusing. It is recommended that you leave the linetype at its default setting of ByLayer.

Setting Lineweights

Lineweights are important in technical drawing. For example, object lines should be thicker so they stand out more than dimension lines. Normally, object lines are thick, hidden lines and text are of medium thickness, and center lines, dimensions, and hatch lines are thin. The exact lineweights vary from industry to industry and from company to company. In AutoCAD, the default lineweight is .25 mm.

1. Display the Layer Properties Manager palette.
2. At the right of layer Objects, and under the Lineweight heading, pick the word **Default**.

This displays the Lineweight dialog box, as shown in Fig. 23-2.

3. Scroll down the list to review the lineweight options.
4. Select **0.40 mm** and pick **OK**.

This assigns a lineweight of 0.40 mm to layer Objects. This lineweight replaces the word Default in the Layer Properties Manager palette.

5. Pick the **Show/Hide Lineweight** button on the status bar.

This turns on the display of lineweights in the drawing area.

6. Pick the **Show/Hide Lineweight** button to toggle it off; toggle it back on.

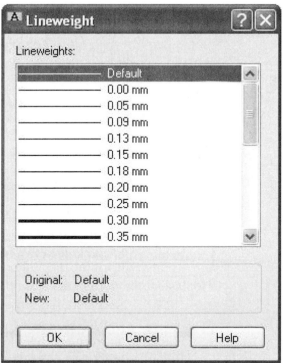

Fig. 23-2

7. Assign lineweights to the other layers, as shown in Table 23-1 on page 308.

8. Auto-hide the Layer Properties Manager palette and save your work.

 The Lineweight drop-down box on the Properties panel allows you to set the lineweight for objects regardless of layer. However, it is recommended that you leave the lineweights at their default setting of ByLayer to avoid confusion.

Working with Layers

Setting up the appropriate layers for a drawing gives you much more control over the drawing. It also gives you more flexibility. For example, suppose you are working on the plans for a new residence. You can set up separate layers for plumbing, electrical, and other work that is often subcontracted. By placing the plans for these elements on separate layers, you can use the same drawing to generate plans for each of the subcontractors. For the electrical plans, you can turn off or **freeze** the plumbing layer so that the plumbing specifications don't clutter up the electrical drawing, and so on.

Turning Layers On and Off

1. Display the Layer Properties Manager palette.

2. At the right of Objects, click the light bulb.

This toggles off layer Objects.

3. Auto-hide the palette and notice that the red circle has disappeared.

4. Display the Layer Properties Manager palette.

5. Toggle on layer Objects by clicking the darkened light bulb.

The red circle reappears.

Freezing and Thawing Layers

1. Display the Layer Properties Manager palette and make Hidden the current layer.

2. In the Freeze column, and at the right of Objects, pick the symbol representing the sun.

This changes the symbol to a snowflake and freezes the layer.

3. Move the pointing device to auto-hide the palette and notice that the red circle has disappeared.

4. Display the Layer Properties Manager palette again, and change the snowflake into a sun by picking it.

The circle reappears. As you can see, freezing and thawing layers is similar to turning them off and on. The difference is that AutoCAD regenerates a drawing faster if the unneeded layers are frozen rather than turned off. Therefore, in most cases, Freeze is recommended over the Off option. Note that you cannot freeze the current layer.

Locking Layers

AutoCAD permits you to lock layers as a safety mechanism. This prevents you from editing objects accidentally in complex drawings.

1. Display the Layer Properties Manager palette.

2. At the right of Objects, click the symbol that looks like a padlock.

The lock closes, and the red circle appears less bright, indicating that the layer is now locked.

3. Try to edit the circle.

A locked padlock appears next to the crosshairs. AutoCAD will not permit you to edit the circle because it now resides on a locked layer.

 AutoCAD does allow you to create new objects on a locked layer. However, once you have created them, you cannot edit them.

4. Redisplay the Layer Properties Manager palette, and unlock layer Objects by clicking the locked padlock.

Managing Layers

There are several ways to control layers, layer properties, and the current layer in AutoCAD. The Make Object's Layer Current button allows you to change the current layer based on a selected object. The Layer Properties Manager allows you to change layer properties and delete unused layers.

1. Pick the **Make Object's Layer Current** button on the Layers panel of the Ribbon.

LAYERS

Notice that the AutoCAD prompt reads Select object whose layer will become current.

2. Pick the red circle.

Objects is now the current layer.

3. Display the Layer Properties Manager palette and pick the **New Property Filter** button located in the upper left area.

This displays the Layer Filter Properties dialog box. This dialog box is available to filter layers based on name, state, color, and linetype. **Filtering** means selectively including or excluding items using specific criteria. Filtering layers is especially useful in complex drawings that contain a large number of layers.

4. Pick **Cancel** to close the Layer Filter Properties dialog box.

The New Group Filter button creates a layer filter containing layers that you select and add to the filter. The Layer States Manager button displays the Layer States Manager. This dialog box enables you to save the current property settings for layers in a named layer state that you can later restore.

5. Pick the **Objects** layer and then right-click to display the shortcut menu.

6. Select **Change Description** from the shortcut menu.

This produces a text box under the Description column.

7. Enter **Object lines** for the description and press **ENTER**.

The description can be a helpful reference when you use these layers in the future.

Focus your attention on the Plot Style and Plot headings. Colors appear under Plot Style. These are the colors you see in the drawing area. Color_1 is red, Color_2 is yellow, and so on, as shown. They are grayed out and unavailable because this drawing does not use plot styles.

Under the Plot heading, you can pick the printer icon to turn off plotting for a layer. When plotting is turned off, AutoCAD still displays the objects on that layer, but they do not plot.

8. Pick the printer icon located at the right of layer Center.

A red symbol appears on top of the printer, indicating that plotting has been turned off for this layer. You will learn more about plotting in the following chapter.

9. Turn on plotting for layer Center and auto-hide the palette.

Changing and Controlling Layers

AutoCAD offers a convenient way of changing and controlling certain aspects of layers.

1. In the Layers panel of the Ribbon, select the down arrow to the right side of the **Layer** text box.

This displays the list of layers in the current drawing. By picking the symbols, you can quickly change the current status of the layers.

2. Pick **Center**.

Layer Center becomes the current layer.

3. Display the list again.
4. Beside Objects, click the padlock.

The padlock changes to its locked position, and layer Objects becomes locked.

5. Pick it again to unlock layer Objects.
6. Beside Hidden, pick the sun.

The sun changes to a snowflake. Layer Hidden is now frozen.

7. Display the list and pick the snowflake.

This thaws layer Hidden. Notice that you can toggle layers on and off and freeze and thaw layers using buttons on the Layers panel.

8. Make Objects the current layer and save your work.

Working with Objects

Now that you have created a number of layers and assigned properties to them, you have much more flexibility in working with objects in the drawing. You can control their appearance and style by placing them on appropriate layers. You can also change an object's properties and even filter objects according to their characteristics.

Changing an Object's Properties

The Quick Properties palette provides a quick method for reviewing and changing the properties of an object.

1. Double-click the green circle.

This displays the Quick Properties palette. The Quick Properties palette provides a lot of information about the circle, including its color, linetype, lineweight, and layer name. Suppose you created the object on the wrong layer—something you will certainly do in the future—and want to move it to a different layer.

2. At the right of Layer, pick **Hidden**.

This produces a down arrow.

3. Pick the down arrow to display the list of layers and pick **Center**.

4. With the cursor anywhere in the drawing area, press **ESC** to remove the selection.

The circle now resides on layer Center.

5. Using the same method, change the same circle back to layer Hidden.

6. Close the Quick Properties palette.

The Quick Properties palette is a handy way of quickly making changes to the objects in the drawing.

Matching Layer Properties

CLIPBOARD

The Match Properties button provides a quick way to copy the properties of one object to another.

1. Pick the **Match Properties** button on the Clipboard panel of the Ribbon.

This enters the MATCHPROP command.

2. In reply to Select source object, pick the red circle.

Notice the change to the pointer.

3. In reply to Select destination object(s), pick the green circle.

This copies the properties of the first object to the second object.

4. Press **ENTER** to terminate the command.

PROPERTIES

5. Undo the last operation.

6. Pick the **Match Properties** button again and pick the green circle.

7. Select the **Settings** option or type **S**.

This displays the Property Settings dialog box. The checked items are copied from the source object to the destination object(s), allowing you to control the properties that are copied.

Table 23-2. Layers for Stair Detail

Layer Name	Color	Linetype	Lineweight
Risers	yellow	Continuous	0.30 mm
Treads	green	Continuous	0.60 mm

8. Pick the **Cancel** button in the dialog box and press **ENTER** to terminate the command.

9. Close the Layer Properties Manager palette.

10. Erase both circles and save your work, but do not close the drawing.

Filtering by Object Property

The Quick Select dialog box allows you to filter a selection set based on an object's properties. You can use the Quick Select dialog box with the Properties palette to make changes quickly to a large number of objects.

1. Start a new drawing by picking the **New** button on the Quick Access toolbar or by typing **NEW** at the keyboard. Use the **Quick Setup** wizard to specify **Architectural** units and a drawing area of **17′ × 11′**.

2. Create two new layers as shown in Table 23-2 and **ZOOM All**.

3. Create the stair detail shown in Fig. 23-3. Place the risers on layer Risers and the treads on layer Treads. (The *risers* are the vertical lines, and the *treads* are the horizontal lines.)

4. Zoom in so that the stairs fill most of the screen.

5. Save your work in a file named **stairs.dwg**.

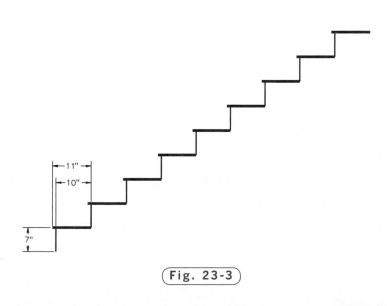

Fig. 23-3

Suppose your supervisor reviews the drawing and asks you to place both the treads and the risers on the Risers layer. However, you must keep the differences in lineweight and color for display purposes.

UTILITIES

6. Pick the **Quick Select** button on the Utilities panel of the Ribbon.

This displays the Quick Select dialog box. The Properties box lists several properties by which you can select objects.

7. Pick **Layer** in the Properties selection box.

8. Pick a point in the Value box to see a list of layers, and pick **Treads**.

We will leave the rest of the values at their default values. The operator is an equal sign, and the radio button for Include in new selection set is selected. This means that all the lines in the drawing that are on layer Treads will be included in the selection set.

PALETTES

9. Pick **OK**.

All of the stair treads are selected in a single operation.

10. In the View tab of the Ribbon, pick the **Properties** button on the Palettes panel.

11. Pick **Layer** in the Properties panel, pick the down arrow to see a list of layers, and choose **Risers**.

12. Auto-hide the Properties palette and press **ESC** to remove the grips.

All of the stair treads are now located on layer Risers, but they are now all the same lineweight and color. Because they are all on the same layer, you can't select according to layer this time. However, you know that all of the treads are exactly 11 inches long. The Quick Select dialog box allows you to select them based on their length.

UTILITIES

13. Pick the **Quick Select** button on the Utilities panel of the Home tab again.

14. Scroll down the Properties selection box and pick **Length**.

15. In the Value box, enter **11** and pick **OK**.

All of the 11″ lines in the drawing—all of the treads—are selected.

16. Pick **Lineweight** in the Properties palette.

17. Pick the down arrow and select **0.60 mm** from the drop-down list.

18. Pick **Color**, pick the down arrow, and choose green.

The treads are now on the layer Risers, but they are still 0.60 mm thick and green.

19. Close all open palettes, save your work, and close the file.

Applying the Custom Template

The tmp1.dwg file that you created in the last chapter should still be open. This template file is nearly complete. The last few steps (12, 13, and 14 in Table 22-1 on page 297) typically involve creating new text styles and setting the dimensioning variables and DIMSCALE. All of this is covered in Chapters 26 and 27.

The template file concept may lack meaning to you until you have actually applied it. Therefore, let's convert tmp1.dwg to a template file and use it to begin a new drawing. First we will create the template file.

1. Be sure to save your work if you haven't already.
2. Select the **Application** button and rest the pointing device over the **Save As** option.
3. To the right of Save As, select **Drawing Template**.
4. Find and select the folder with your name, enter **tmp1** for the file name, and pick the **Save** button.

The Template Options box appears.

5. Under Description, enter **Created for stair details** and pick **OK**.
6. Close **tmp1.dwt**.
7. Select the **New** button from the Quick Access toolbar and pick the **Use a Template** button.
8. Pick the **Browse...** button.
9. Find and open the **tmp1.dwt** template file. Be sure to look in your named folder.

AutoCAD loads the contents of tmp1.dwt into the new drawing file.

10. Review the list of layers.

Look familiar? All the layers from the template file are already loaded into your new drawing file. You can see why template files are valuable time-saving tools.

11. Save the new drawing file as **staird.dwg** and exit AutoCAD.

Chapter 23 Review & Activities

• REVIEW QUESTIONS

Answer the following questions on a separate sheet of paper.

1. Name at least two purposes of layers.
2. How do you change the current layer using the Layer Properties Manager?
3. Describe the purpose of the LTSCALE command and explain how to set it.
4. Describe a situation in which you would want to freeze a layer.
5. Name five of the linetypes AutoCAD makes available.
6. What is the purpose of locking layers?
7. If you accidentally draw on the wrong layer, how can you correct your mistake without erasing and redrawing?
8. Explain how to freeze a layer using the Layer drop-down box.
9. Describe a way to select several similar objects using a single operation.

• CHALLENGE YOUR THINKING

These questions are designed to further your knowledge of AutoCAD by encouraging you to explore the concepts presented in this chapter. Answer each question on a separate sheet of paper.

1. When might you use the Layer drop-down box in the Layers panel instead of using the Layer Properties Manager palette? Explain.
2. AutoCAD allows you to use linetypes that correspond to ISO standards. Find out more about ISO standards. When and where are they used?

• APPLYING AUTOCAD SKILLS

Work the following problems to practice the commands and skills you learned in this chapter.

1-2. Create two new drawings using the tmp1.dwt template file. Change the layers to match those shown in Tables 23-3 and 23-4. Name the drawings prb23-1.dwg and prb23-2.dwg. Make Objects the current layer in prb23-1.dwg, and make 0 the current layer in prb23-2.dwg.

Table 23-3. Layers for Problem 1

Layer Name	State	Color	Linetype	Lineweight
0	Frozen	white	Continuous	Default
Border	On	cyan	Continuous	0.50 mm
Center	On	yellow	CENTER	0.20 mm
Dimensions	On	green	Continuous	0.20 mm
Hidden	On	yellow	HIDDEN	0.30 mm
Objects	On	red	Continuous	0.40 mm
Phantom	On	blue	PHANTOM	0.50 mm
Text	Frozen	magenta	Continuous	0.30 mm

Table 23-4. Layers for Problem 2

Layer Name	State	Color	Linetype	Lineweight
0	On	white	Continuous	Default
Center	Frozen	blue	CENTER	0.20 mm
Dimensions	On	yellow	Continuous	0.20 mm
Electrical	On	cyan	Continuous	0.30 mm
Found	On	magenta	DASHED	0.40 mm
Hidden	On	blue	HIDDEN	0.30 mm
Notes	On	yellow	Continuous	0.30 mm
Plumbing	Frozen	white	Continuous	0.30 mm
Title	Frozen	magenta	Continuous	0.30 mm
Walls	On	red	Continuous	0.40 mm

3. Create a drawing template for use with an A-size drawing sheet. Use a scale of $1/4'' = 1'$. Use architectural units with a precision of $1/16''$. Set up the following layers: Floor, Dimensions, Electrical, Plumbing, and Furniture. Name the template ch23tmp.dwt.

Continued

4. Create a new drawing from scratch using Quick Setup. Set up the layers shown in Table 23-5. Then create the slide shown in Fig. 23-4 on layer Object. Place the center lines on layer Center, positioning them as shown in the illustration. Change back to the Object layer and create the hole (circle) for the slide so that its center point is at the intersection of the two center lines. To finish the drawing, offset the hole by .2 units to the outside and trim the center lines to the outside circle. Erase the temporary trim circle, and save the drawing as slide.dwg.

Table 23-5. Layers for Problem 4

Layer Name	Color	Linetype	Lineweight
Object	white	Continuous	0.40 mm
Center	blue	CENTER2	0.20 mm
Dims	red	Continuous	Default

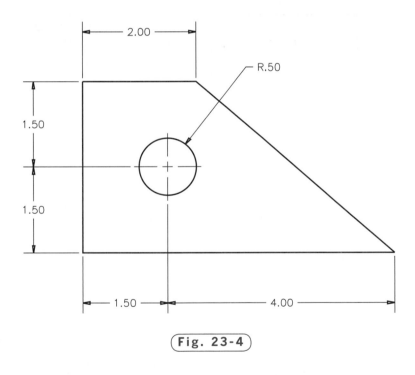

Fig. 23-4

5. Using the ch23tmp.dwt template file you created in problem 3 on page 323, create a simple floor plan for a house. Make the floor plan as creative and as detailed as you wish. Place all the objects on the correct layers.

6. The graph in Fig. 23-5 shows the indicated and brake efficiencies as functions of horsepower for a small engine. Reproduce the graph as follows: Using a suitable scale, draw the grid and plot the given points as shown; then draw a spline through each set of points. Place the border, title block, and curves on a layer named Visible; the grid and point symbols on layer Grid; and the text on layer Text. Trim the grid around the text and arrows, and trim the curves and grid out of the symbols. Use the appropriate text justification to align the axis numbers and titles properly.

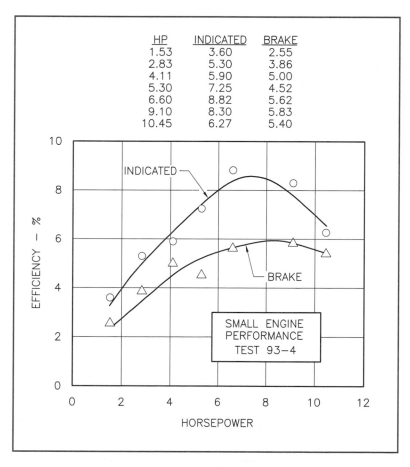

HP	INDICATED	BRAKE
1.53	3.60	2.55
2.83	5.30	3.86
4.11	5.90	5.00
5.30	7.25	4.52
6.60	8.82	5.62
9.10	8.30	5.83
10.45	6.27	5.40

SMALL ENGINE
PERFORMANCE
TEST 93–4

Fig. 23-5

Courtesy of Gary J. Hordemann, Gonzaga University

Continued

● USING PROBLEM-SOLVING SKILLS

Complete the following activities using problem-solving skills and your knowledge of AutoCAD.

1. Create the drawing of the washing machine spacer from the graph paper sketch in Fig. 23-6. Each square represents .25 in. Add a title and include a note saying that the material is SAE 1010 carbon steel. Make the appropriate drawing setup and create the necessary layers. Add center lines for all holes. Save the drawing as ch23spacer.dwg.

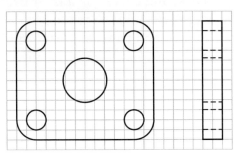

(**Fig. 23-6**)

2. Figure 23-7 shows a sketch of the step block for a bowling alley pinsetter. Create the drawing using appropriate layers. Each square represents 5 cm. Add a title and these notes: 1) Break all sharp edges, 2) fillets and rounds 5 cm, and 3) material is UNS S30451 stainless steel. Save the drawing as ch23pinsetter.dwg.

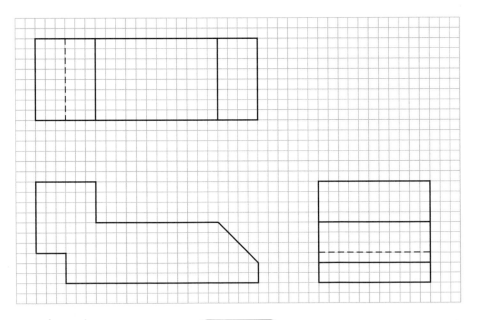

(**Fig. 23-7**)

Chapter 24

Plotting and Printing

Objectives

- Preview a plot
- Adjust plotter settings
- Plot an AutoCAD drawing to scale
- Create DWF and PDF files

Vocabulary

landscape
plot scale
plot style
plotters
portrait
printers
rendered
viewports
visual style
wireframe

CAD users plot and print drawings using plotters and printers. Many years ago, there was a distinct difference between a plotter and a printer, but today, the distinction has blurred to the extent that there is now little difference between the two. Both use the same basic technology—usually inkjet or laser—and both print in black, grayscale, or color onto paper and other sheet materials. Some people still refer to devices that handle large sheets, such as 36″ × 48″ and even larger, as **plotters**. They think of **printers** as smaller, tabletop devices handling sheets up to 11″ × 17″. Note that this is not a universally accepted difference, however. The terms *plotter* and *printer* and *plotting* and *printing* are used interchangeably in this book.

Let's create and plot a set of stairs.

1. Start AutoCAD and open the file named **staird.dwg**.

This file, which is based on the tmp1.dwt template file, was created in the previous chapter.

2. Select the **Drafting & Annotation** workspace.

3. On layer Objects, draw the stair step shown in Fig. 24-1. Use polar tracking and enter the lengths at the keyboard. (The Show/Hide Lineweight button on the status bar should *not* be depressed.)

4. Copy the objects five times to produce the stairs as shown in Fig. 24-2. Use points 1 and 2 in Fig. 24-1 as the base point and second point, respectively. Be sure to add the lines that make it complete.

5. **ZOOM All** and save your work.

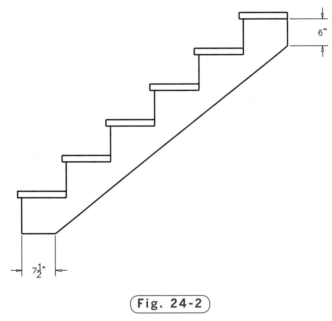

Fig. 24-1

Fig. 24-2

Previewing a Plot

AutoCAD allows you to preview a plot before you send it to the plotter. This feature allows you to catch mistakes before you spend the time and supplies to create the actual plot.

QUICK ACCESS

1. From the Quick Access toolbar, pick the **Plot** button or enter the **PLOT** or **PRINT** command at the keyboard.

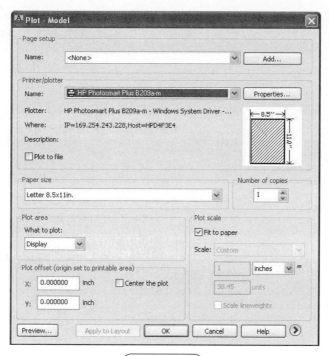

Fig. 24-3

The Plot dialog box appears, as shown in Fig. 24-3. In the illustration, notice that HP Photosmart Plus B209a-m is the current printer/plotter. The device you see in this area will probably be different. AutoCAD stores plotter settings for each configured plotter, so other settings may also differ.

If no plotter names appear in the upper left area named Printer/plotter, you must configure one before you proceed with the following steps. Consult your instructor or the head of your company's IT department for details.

Note the preview button located in the lower left area of the dialog box.

2. Pick the **Preview...** button.

This feature gives you a preview of how the drawing will appear on the sheet, as shown in Fig. 24-4 on page 330.

3. Use the pick button to zoom in on the drawing.

4. Right-click to display a shortcut menu, and pick **Pan** to move around in the drawing.

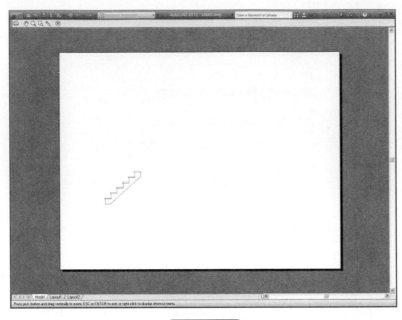

Fig. 24-4

Zooming and panning enable you to examine the drawing closely before you plot it. Plotting large drawings can take considerable time and supplies, so spotting errors before you plot can save time and money.

5. Press **ESC** or **ENTER**.

6. If your computer is connected to the configured plotter or printer and it is turned on and ready, continue with Step 7. If it is not, prepare the device. If you do not have access to an output device, pick the **Cancel** button and skip to the next section, titled "Tailoring the Plot."

7. Pick **OK**.

Since we did not make changes to the plot settings before plotting, the plot may be incorrect.

Tailoring the Plot

QUICK ACCESS

The following steps consider sheet size, drawing scale, and other plotter settings and parameters.

1. Pick the **Plot** button.

2. In the bottom right corner, pick the arrow (**>**) to expand the dialog box to include other options.

Table 24-1. Plotting Options

Option	Function
Display	Plots the current view
Extents	Similar to ZOOM Extents; plots the portion of the drawing that contains objects
Limits	Plots the entire drawing area
Window	Plots a window whose corners you specify

Specifying Sheet Size

1. Under Paper size, display the list of options.

2. Review the list of sizes available for the configured printer.

3. Pick **Letter** if it is available. If not, pick the option that is closest to $11'' \times 8.5''$.

Defining the Plot Area

1. In the Plot area, select the **Limits** option.

We selected Limits because we want to plot the entire drawing area as defined by the $22'' \times 17''$ drawing limits. In the future, you may choose to select one of the other plotting options. Table 24-1 explains what each option means.

Setting the Drawing Scale

1. Focus your attention on the **Plot scale** area of the dialog box.

The **plot scale** is the scale at which the drawing is plotted to fit on a drawing sheet.

2. Uncheck the Fit to paper check box.

The scale options become available.

3. Pick the down arrow next to **Scale:** and review the list of options.

As you may recall, the scale for the stairs drawing is $1/2'' = 1'$.

4. Select **1/2″ = 1′-0″**.

The numbers 1 and 24 (or 0.5 and 12) are entered automatically in the inches and units boxes because 1 inch = 24 units is another way to express our plot scale.

You can manage the list of scales you use for viewports, page layouts, and plotting in the Edit Scale List dialog box. To open the list, pick Scale List... from the Format pull-down menu or type the command SCALELISTEDIT. Scales can be added, deleted, and modified, and you can rearrange the scale list to display the most commonly used scales at the top.

Setting the Drawing Orientation

Notice the illustration to the right of the Portrait and Landscape radio buttons in the Drawing orientation area. In **portrait** orientation, the drawing is positioned so that "north" or "up" falls on the narrower edge of the paper. In **landscape**, the wide edge of the paper is north.

1. Pick each radio button while watching the illustration to see the difference between portrait and landscape.

2. In the Drawing orientation area, pick **Landscape**.

Reviewing Other Settings

Let's look briefly at the other settings available from the Plot dialog box.

1. Focus on the Page setup area and pick the **Add** button.

This button allows you to save the current settings in the Plot dialog box to a named page setup. You can modify the page setup using the Page Setup Manager.

2. Pick **Cancel** in the Add Page Setup dialog box.

3. Notice the Plot to file check box in the Printer/plotter area.

When the Plot to file check box is checked, AutoCAD sends plot output to a file rather than to a device and creates a PLT file type.

4. Pick the **Properties...** button located in the Printer/plotter area.

This displays the Plotter Configuration Editor, which presents information relevant to the printing device.

5. Pick the **Cancel** button to close the Plotter Configuration Editor.

6. Now focus your attention on the Plot offset area of the dialog box.

It specifies an offset of the plotting area from the left corner of the sheet.

7. Focus your attention on the Plot style table (pen assignments) area.

The Plot style table (pen assignments) area allows you to create and edit plot style tables. A **plot style** in AutoCAD is a collection of property settings that are saved in a plot style table.

8. Focus your attention on the Shaded viewport options area.

This area specifies how shaded and rendered **viewports** are plotted and determines their resolution. A viewport is a bounded area that displays some portion of a drawing. Viewports are discussed further in the next chapter.

9. Select the down arrow located to the right of Shade plot.

The Shade plot area allows you to select a visual style for the plot. The As displayed option plots objects the way you see them on the screen. The other options are different visual styles for plotting your drawing. A **visual style** is the visual representation of objects in the drawing area or on a plot. It includes the way the edges are displayed and the way the object is shaded. A **wireframe** is the representation of a 3D object using lines and curves to show its boundaries. Hidden plots objects with the hidden lines removed. In this context, hidden lines are those lines and curves of an object that are hidden from view. A **rendered** 3D object is one that has shading applied. The other options—Conceptual, Realistic, Shaded, Shaded with edges, Shades of Gray, Sketchy, and X-Ray—all plot rendered objects. You will learn more about visual styles and rendering in Parts 8 and 9.

10. Select the down arrow located to the right of Quality.

The Quality area allows you to select the resolution of the plot, from Preview quality—150 dots per inch (dpi)—up to Presentation quality (600 dpi). You can also specify the maximum quality or specify the dpi resolution yourself by selecting Custom in the Quality box.

11. Focus on the Plot options area.

This area specifies options for lineweights, plot styles, and the order in which objects are plotted.

Plotting the Drawing

Let's preview and plot the drawing.

1. Pick the **Preview...** button.

2. Review the preview and then press **ESC**.

3. Prepare the device for plotting.

4. Pick the **OK** button to initiate plotting.

After plotting is complete, examine the output carefully. The dimensions on the drawing should measure correctly using a $1/2'' = 1'$ scale.

Creating DWF and PDF Files

AutoCAD users often need to share a design or drawing with business associates who do not use AutoCAD. These associates may be managers, sales representatives, customers, or suppliers. AutoCAD offers other output options in addition to printing that allow users to share drawings with non-AutoCAD users. Note that these output options allow non-AutoCAD users to view a file, but do not offer them any of the commands or features of AutoCAD. We will discuss two such output options: the DWF format and the PDF format.

Autodesk (the maker of AutoCAD) offers a free program called DWF Viewer. An AutoCAD user can save a drawing in the DWF format, then send the DWF file to or share it with a non-AutoCAD user who has installed the DWF Viewer software.

QUICK ACCESS

1. Pick the **Plot** button.

2. In the Printer/plotter area, pick the down arrow to the right of Name.

The drop-down menu displays a list of output options, including printing options and file output options.

3. Select **DWF6 ePlot.pc3** from the drop-down menu and pick **OK**.

4. Save the new DWF file as **staird-Model.dwf** in the folder with your name.

Another common file sharing format is the Adobe® Portable Document format (PDF). PDF files can be created in most popular programs, and can be read by Abode Reader, another free file-viewing program.

QUICK ACCESS

5. Pick the **Plot** button.

6. In the Printer/plotter area, pick **DWG To PDF.pc3** from the drop-down menu, then pick **OK**.

7. Save the new PDF file as **staird-Model.pdf** in the folder with your name and exit AutoCAD.

After saving a DWF or PDF file, you would typically share it with others by sending the file as an e-mail attachment or copying it to a secure server. Keep in mind that the recipient of the file must have the viewing software installed to be able to view the file.

• REVIEW QUESTIONS

Answer the following questions on a separate sheet of paper.

1. Explain why a plot preview is useful.

2. In the Plot dialog box, what is the purpose of the Plot to file check box?

3. Briefly describe each of the following plot options.

   a. Limits

   b. Extents

   c. Display

   d. View

4. The drawing plot scale for a particular drawing is 1 = 4″. What does the 1 represent and what does the 4″ represent?

5. Describe how you would create a version of an AutoCAD file that could be viewed by someone who does not have the AutoCAD software.

• CHALLENGE YOUR THINKING

These questions are designed to further your knowledge of AutoCAD by encouraging you to explore the concepts presented in this chapter. Answer each question on a separate sheet of paper.

1. Explore ways of sending PLT files to a printer or plotter.

2. Suppose you wanted to plot all of your drawings to PLT files and store them automatically in the folder of your choice. How can you set the default folder for PLT files?

Continued

• APPLYING AUTOCAD SKILLS

Work the following problems to practice the commands and skills you learned in this chapter. In problems 1 and 2, prepare to plot a drawing using the information provided. Choose any drawing to plot.

1. Paper size: Letter

   Units: Inches
   Drawing orientation: Landscape
   Plot scale: 1 inch × 2 units
   Plot with object lineweights
   Do not plot with plot styles
   Shade Plot: As Displayed
   Quality: Normal
   Number of copies: 1
   Do not plot to file
   Perform a preview

2. When starting a drawing from scratch, select Metric.

   Paper size: A4
   Units: Millimeters
   Drawing orientation: Landscape
   Plot area: Limits
   Plot scale: 1 mm × 10 units
   Plot with lineweights
   Number of copies: 1
   Perform a preview
   Plot to file

3. Open slide.dwg (from problem 4 in Chapter 23). Zoom in so that the slide fills most of the screen. Enter the PLOT command and choose the Limits option in the Plot area portion of the dialog box. Plot the drawing. Then create two more plots, one with the Extents radio button selected, and one with the Display radio button selected. Compare the drawings.

4. Choose and plot a drawing that you created in an earlier chapter. Consider the scale and drawing area so that dimensions measure correctly on the plotted sheet. Text and linetypes should also measure correctly on the sheet. For example, 1/8″ text should measure 1/8″ in height.

● USING PROBLEM-SOLVING SKILLS

Complete the following activities using problem-solving skills and your knowledge of AutoCAD.

1. The R&D department has requested a full-size plot of the step block for the bowling alley pin-setter you created in Chapter 23. Adjust the plot area, scale accordingly, and plot the drawing.

2. An outside sales representative in your company has requested a PDF of the washing machine spacer you created in Chapter 23. Open the file and plot the spacer as a PDF file. Open the PDF file of the spacer in Adobe Reader® to verify that it is accurate. (Adobe Reader is free software that you can download from the Adobe Website if it is not already installed on your computer.)

Technical Illustrator

You have just purchased a new stereo and you are eager to hear how your favorite CD sounds through the system's powerful speakers. To figure out how to connect each component properly, you consult the instruction sheet:

Connect wire A from component 1 to component 2. Connect wire B from component 1 to component 3. Connect wires C and D from component 1 to external speakers. How do you know which is wire A? How is it different from wire D? Fortunately, the instructions also include an illustrated example of the setup procedure.

© Photodisc Collection/Getty Images/RF

Assisting the User

Technical illustrators produce drawings for instruction manuals and appliance or equipment operating manuals. These illustrations allow users to identify specific parts and verify that they have followed the directions properly. Technical illustrators consult with the manufacturer of the product to ensure that their understanding of the item, its parts, and the intended procedure is accurate.

Attention to Detail

As with many other types of drafting careers, technical illustrators must pay close attention to detail. If an unusual piece of equipment is mentioned in writing but a visual example is not provided, the person following the directions may become confused. A combination of illustrations and text offers multiple opportunities to understand exactly how to build or operate an item.

Many drafters are drawn to the field of technical illustration. Even though technical illustrators rarely create dimensioned technical drawings, their work must be accurate and complete. In fact, many technical illustrators have a strong drafting background.

Career Activities

1. Find and compare instructions that are completely text-based and instructions that consist only of illustrations. Which do you find easier to follow? Why?

2. Investigate technical illustrator careers using the Internet. What are forecasts for future employment?

Chapter 25

Multiple Viewports

Objectives

- Create and use multiple viewports in model space
- Create and use viewports in paper space
- Position objects in paper space viewports
- Edit, position, and plot paper space layouts

Vocabulary

layout
model space
paper space

In AutoCAD, viewports are portions of the drawing area that you define to show a specific view of a drawing. By default, AutoCAD begins a new drawing using a single viewport. By defining and using multiple viewports, you can see more than one view of a drawing at once.

AutoCAD includes two different environments—**model space** and **paper space**—for working with objects. Paper space is used to arrange views of an object or 3D model and embellish them for plotting. Model space is used to construct and modify the objects that make up the drawing. Because model space is where you will perform most of your drafting and design work, AutoCAD opens a single viewport in model space when you create a new drawing without using a template.

Each new drawing also has two default layouts. A **layout** is an arrangement of one or more views of an object on a single sheet. When you pick a layout, AutoCAD automatically switches to paper space. After you have set up the paper space parameters, AutoCAD positions model space objects against a white "paper" background to show you how they will appear on the printed sheet.

Viewports in Model Space

You can apply viewports to both model space and paper space. In model space, you can use viewports to draw and edit in more than one view at a time. The magnification of each viewport can be set individually, so viewports provide capabilities that would otherwise be impossible.

After creating a drawing in model space, you can create floating viewports in paper space to display different views of the drawing. Because they float, you can easily position them for plotting. Depending on your needs, you can set options that determine what is plotted and how the viewports fit on the sheet.

Fig. 25-1 gives an example of applying multiple viewports in model space to zoom.dwg. Notice that each viewport is different both in its window extents and in magnification.

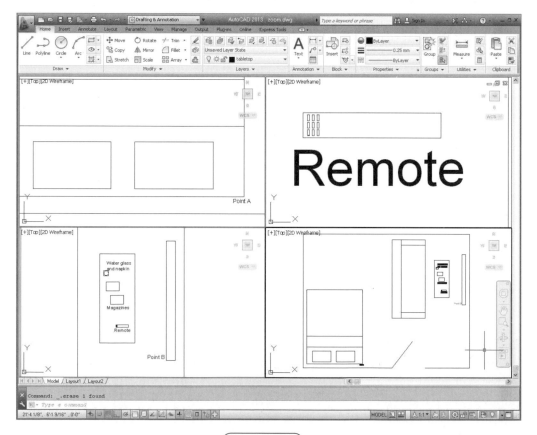

Fig. 25-1

Creating Viewports in Model Space

Viewports are controlled with the VPORTS (or VIEWPORTS) command.

1. Open the drawing file named **zoom.dwg**.
2. Select the **Drafting & Annotation** workspace.
3. Enter **ZOOM All** to make the drawing fill the screen.
4. In the upper left corner of the drawing area, pick **[-]**, then **Viewport Configuration List**, then **Configure...** from the cascading menu, or enter the **VPORTS** command at the keyboard.

The Viewports dialog box appears.

5. Select the **New Viewports** tab unless it is already selected.
6. Under Standard viewports, click on each of the options.

Each viewport option appears in the Preview area.

7. Pick **Four: Equal** and pick the **OK** button.

AutoCAD produces four viewports of equal size, each with an identical view of the drawing.

8. Move the pointer to each of the four viewports.

The crosshairs appears only in the viewport with the bold border. This is the current (active) viewport.

9. Move to one of the three nonactive viewports.

An arrow appears in place of the crosshairs.

10. Press the pick button on the pointing device.

This viewport becomes the current one.

Using Viewports in Model Space

Let's modify zoom.dwg using the viewports.

1. Refer to Fig. 25-1 and create four similar viewports using AutoCAD's zoom and pan features.

Make one of the viewports current and then zoom and pan. Repeat this process for the other three viewports.

2. Make the upper left viewport current.
3. Enter the **LINE** command and pick point **A**. (Refer to Fig. 25-1 for point **A**.)
4. Move to the lower left viewport and make it current.

Notice that the LINE command is now active in this viewport.

5. Pick point **B** and press **ENTER**.

View all four viewports. The line appears in other views because each viewport is just a different view of the same drawing area.

So you see, you can easily begin an operation in one viewport and continue it in another. Any change you make is reflected in all viewports. This is especially useful when you are working on large drawings with lots of detail.

Let's move the remote from one viewport to another.

6. Make the upper right viewport current and select the entire remote.

7. Right-click and pick **Move** from the shortcut menu.

8. Pick a corner of the remote as the base point.

9. Move to the upper left viewport and make it current. (Make sure ortho is off.)

10. Place the remote on one of the pillows by picking a point at the appropriate location.

Notice that the remote location changed in the other viewports.

Viewport Options

Let's combine two viewports into one.

1. Select the **View** tab on the Ribbon and pick the **Join Viewports** button on the Model Viewports panel.

**MODEL
VIEWPORTS**

2. Choose the upper left viewport in reply to Select dominant viewport.

3. Now choose the lower left viewport in reply to Select viewport to join.

As you can see, the Join option enables you to expand—in this case, double—the size of a viewport.

4. Pick **[+]** in the **Viewport Controls** in the upper left corner of the drawing area and select the **Single** option.

The screen changes to single viewport viewing. This single viewport is inherited from the current viewport at the time you selected Single.

AutoCAD also allows you to subdivide current viewports into two or more additional viewports.

5. Pick **[-]** in the Viewport Controls and select the **Three: Right** option from the Viewport Configuration List.

6. Make the upper left viewport the current one.

7. Redisplay the **Viewports** dialog box, select the **New Viewports** tab, and pick **Two: Vertical**.

8. Under **Apply to**, located in the lower left corner, pick the down arrow, select **Current Viewport**, and pick **OK**.

As you can see, AutoCAD applies Two: Vertical to the current viewport.

9. Try the remaining viewport options on your own. Practice drawing and editing using the different viewport configurations.

10. Save your work and close the drawing file, but do not exit AutoCAD.

 The REGEN command affects only the current viewport. If you are using multiple viewports and you want to regenerate all of them, you can use the REGENALL command.

Viewports in Paper Space

When you are creating a layout in AutoCAD, you can consider viewports as objects with a view into model space that you can move and resize. By default, AutoCAD presents a single viewport in the paper space layout.

1. Open the file named **staird.dwg**.

2. Pick the **Quick View Layouts** button on the status bar.

STATUS BAR

This produces three thumbnails of the views in the drawing: Model, Layout1 and Layout2. Notice the four buttons below the thumbnails. You can pin the quick views open, create a new layout, publish the sheet set, or close the Quick View Layouts.

3. With the pick button, pick the **Layout1** thumbnail.

When you pick a layout view, a sheet with margins displays, reflecting the paper size of the currently configured plotter and printable area of the sheet.

In paper space, you can view the exact size of the sheet and see how the drawing will appear on the sheet when you plot. The dashed line represents the plotting boundary, and the solid line represents the single viewport. (AutoCAD's paper space defaults to a single viewport if you do not specify more than one.)

The paper space icon, located in the lower left corner of the drawing area, replaces the standard coordinate system icon. The paper space icon is present whenever paper space is the current space. The coordinate system icon is present whenever model space is the current space.

4. Right-click the Layout1 thumbnail, select **Rename** from the menu, enter **My Layout,** and press **ENTER.**

AutoCAD renames the layout tab.

5. Pick the **Close** button to close Quick View Layouts.

In the status bar, notice that PAPER is displayed on the Model or Paper space toggle button, indicating that you are now in paper space.

6. Pick the viewport (the solid red border).

As you can see, a viewport in paper space is an object.

7. Move the viewport a short distance.

The stair detail moves too, because it belongs to the viewport.

8. Undo the last step.

Plotting a Single Viewport in Paper Space

Plotting a viewport containing an object in paper space is different than plotting the same object in model space. Before plotting to scale, you must fit the objects in the viewport using the ZOOM command.

1. Double-click inside the viewport.

The outline of the viewport becomes bold and the coordinate system icon returns, indicating that you are now in model space within the layout.

2. Enter the **ZOOM** command and select the **Scale** option.

3. For the scale factor, enter **1/24xp.**

The "1 to 24" reflects the scale of the drawing. As you may recall, we established a scale of $^1/_2'' = 1'$ for tmp1.dwt, which we used for the stair detail. You can also express the scale as $1'' = 24''$. When you enter a value such as 1/24 followed by xp, AutoCAD specifies the scale relative to the paper space scale. (The term xp means "times paper space scale.") If you enter .5xp, AutoCAD displays model space at half the scale of the paper space scale.

QUICK ACCESS

4. Double-click outside the viewport to return to paper space.

5. Pick the **Plot** button from the **Quick Access** toolbar.

Notice under Plot area that Limits is not available; Layout appears in its place.

6. Produce the following settings:

   Paper size: Letter or 11″ × 8.5″
   Drawing orientation: Landscape
   Plot area: Layout
   Scale: 1:1
   1 inch = 1 unit

The plot scale is 1=1 because we do not want to scale the layout up or down. We did that in Step 3. This is one distinct difference between plotting from model space and plotting from paper space.

7. Perform a full preview and then press **ESC**.

8. Make sure that your plotter is ready, and pick the **OK** button to initiate plotting.

AutoCAD plots the stairs to scale.

9. Save your work and close the drawing file.

Adding Viewports in Paper Space

You can add viewports in paper space using a method similar to the one you used in model space, but you must be in paper space. Let's set up a drawing with multiple paper space viewports.

1. Create a new drawing and pick the **Quick Setup** wizard.

2. Pick **Next** to accept **Decimal** units.

3. Enter **11** for the width and **8.5** for the length, pick the **Finish** button, and **ZOOM All**.

4. Create the layers shown in Table 25-1, making **Vports** the current layer.

5. Set snap at **.5″** and create a new text style named **romans** using the **romans.shx** font. Use the default settings.

6. Save your work in a file named **pspace.dwg**.

7. Pick the **Layout1** tab at the bottom of the drawing area to enter paper space.

8. Erase the viewport that appears on layer Vports.

Table 25-1. Layers

Layer Name	Color	Linetype	Lineweight
Objects	red	Continuous	0.40 mm
Border	blue	Continuous	0.60 mm
Vports	magenta	Continuous	0.20 mm

**LAYOUT
VIEWPORTS**

9. From the Layout tab of the Ribbon, pick the **Rectangular** button on the Layout Viewports panel.

10. Select the **4** option on the Command line or from the pull-down menu at the dynamic input prompt, and press **ENTER** to accept the Fit default value.

AutoCAD inserts four equally sized viewports on layer Vports.

11. Double-click inside the viewports.

Coordinate system icons appear in each of the four viewports, showing that you are now in model space.

Objects in Viewports

You will better understand the benefits of using viewports in paper space when you create an object.

1. Make **Objects** the current layer.

2. Enter the **THICKNESS** system variable and set it to **1**.

The THICKNESS system variable enables you to specify the thickness of an object in the z direction, resulting in a three-dimensional (3D) object. Parts 8 and 9 contain more information about 3D objects.

3. In the upper left viewport, draw the object shown in Fig. 25-2. Omit the dimensions.

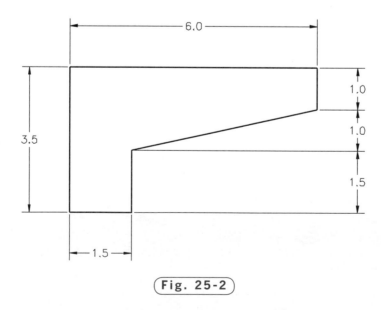

Fig. 25-2

An identical view of the object appears in all four viewports.

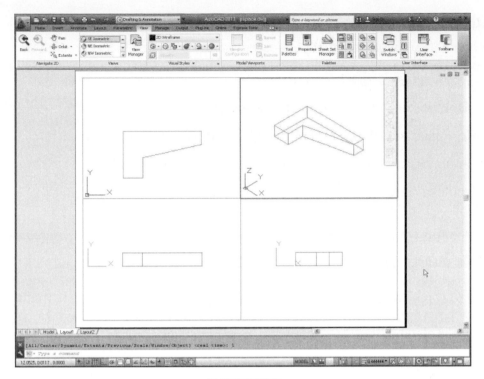

Fig. 25-3

Creating Four Individual Views

Taking advantage of AutoCAD's viewports, let's create four different views of the solid object.

1. Make the lower left viewport the current one.

2. In the Views panel of the View tab, pick the down arrow in the pull-down menu and select **Front**.

3. Enter **ZOOM** and **1**.

This scales the viewport to an apparent size of 1. Entering 2 would make it twice the size.

4. Make the lower right viewport the current one.

5. View the object from the right side by picking **Right** from the menu on the Views panel.

6. Enter **ZOOM** and **1**.

7. Make the upper right viewport current and view the object from above, from the front, and from the right by picking **SE Isometric** from the menu on the Views panel.

8. Enter **ZOOM** and **1**, and save your work.

The screen should look similar to the one in Fig. 25-3.

Editing Objects

When you are working in paper space, you cannot edit objects created in model space. Likewise, when you are in model space, you cannot edit objects created in paper space.

1. Double-click outside the four viewports.

Little appears to change except for the coordinate system icons.

2. Attempt to select the object in any of the four viewports.

As you can see, you cannot select it because it was created in model space.

Working with Paper Space Viewports

One of the big advantages of working with paper space viewports is that they allow you to print more than one view of an object on a single sheet of paper. They are also helpful for placing and arranging views on the drawing sheet.

Editing Viewports

Viewports in paper space are treated much like other AutoCAD objects. They can be moved and even erased, but not in model space. Only the views (objects) themselves can be edited in model space.

1. Enter the **UNDO** command and the **Mark** option.

2. Select one of the lines that make up one of the four viewports and move the viewport a short distance toward the center of the screen.

3. Erase one of the viewports.

4. Scale one of the viewports by **.75**.

5. Switch to model space within the layout.

6. Attempt to move, erase, or scale a viewport.

 Do not double-click a viewport because this will cause you to enter paper space.

You can't do it, because you are no longer in paper space.

7. Enter **UNDO** and the **Back** option.

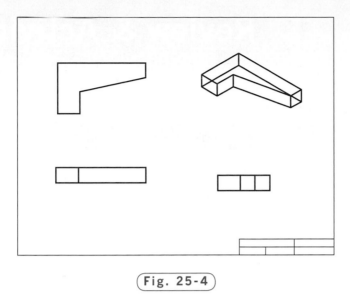

Fig. 25-4

Plotting Multiple Viewports in Paper Space

One of the benefits of paper space is multiple viewport plotting. (You cannot plot more than one model space viewport at a time, although you will learn to create multiple views in a single model space viewport in Chapter 47.)

1. Freeze layer **Vports** and make **Border** the current layer.

The lines that make up the viewports should now be invisible.

2. Switch to model space within the layout.

The viewport lines are invisible in model space also.

3. Switch to paper space and draw a border and basic title block similar to those in Fig. 25-4.

4. Plot the layout at a scale of **1 = 1**.

Positioning Viewports

It is possible that the current position of the border and views did not plot perfectly. Even if they did, make adjustments to the location of the border, title block, and views by following these steps.

1. Thaw layer **Vports**.

2. Move the individual viewports to better position them in the drawing. It is normal for them to overlap.

3. Edit the size and location of the border and title block if necessary.

4. Freeze layer **Vports**, save your work, and replot the drawing.

5. Save your drawing and exit AutoCAD.

● REVIEW QUESTIONS

Answer the following questions on a separate sheet of paper.

1. Explain the difference between model space and paper space.

2. How can using multiple viewports in model space help you construct drawings?

3. How do you make a viewport in model space the current viewport?

4. Explain how you would join two viewports in model space.

5. What option should you enter to obtain two viewports in the top half of the screen and one viewport in the bottom half of the screen?

6. If you are working in paper space and you discover that you need to change an object that was created in model space, what must you do before you can make the change? Why?

7. Explain why you may want to edit viewports in paper space.

8. What is the main benefit of plotting in paper space?

● CHALLENGE YOUR THINKING

These questions are designed to further your knowledge of AutoCAD by encouraging you to explore the concepts presented in this chapter. Answer each question on a separate sheet of paper.

1. Find out how many viewports you can have at one time in AutoCAD. Would you want to use that many? Why? Explain the advantages and disadvantages of using multiple model space viewports in your drawings.

2. Experiment with viewports created in model space and paper space. Is it possible to create more than one viewport in model space, then import the model space viewports into a paper space viewport? Explain.

● APPLYING AUTOCAD SKILLS

Work the following problems to practice the commands and skills you learned in this chapter.

1. Create each of the viewport configurations shown in Fig. 25-5.

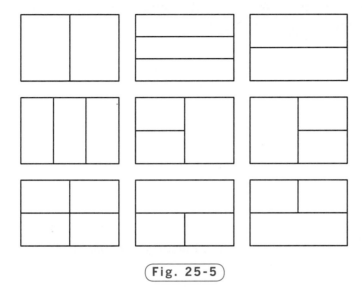

Fig. 25-5

2. Using viewports in paper space, create the top, front, right side, and SE isometric view of the security clip shown in Fig. 25-6. Create a border and title block, and plot the multiple views on a single sheet.

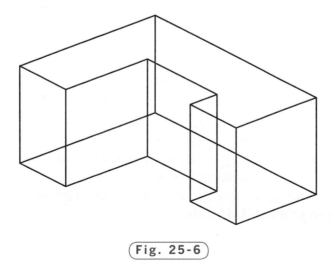

Fig. 25-6

Continued

3. Open the db_samp.dwg file in AutoCAD's Sample/Database Connectivity folder and save it in your named folder as Fremont.dwg. The current tenant of this building wants to reconfigure the office by moving a printer island to an area of unused offices.

a. Create two viewports side by side (use the Vertical option).

b. Make the right viewport active and zoom in so that the building almost fills the viewport. Then zoom in further on the printer island on the right side of the building.

c. Make the left viewport active and again zoom in so that the building almost fills the viewport. Then zoom in on the empty offices in a horizontal line at the bottom of the viewport (office numbers 6156, 6158, 6160, 6162, and 6164). This will be the new location of the printer island.

d. Erase the five offices, including the wall that lies against the outside wall.

e. Using the two viewports as necessary, move the entire printer island to the space formerly occupied by the offices. (Note: When you select the printer island, do not include the gray H or the lines above and below it. It is a structural I-beam that helps support the building.)

f. Return to a single viewport and save the drawing.

• USING PROBLEM-SOLVING SKILLS

Complete the following activities using problem-solving skills and your knowledge of AutoCAD.

1. Draw the locking receptacle (Fig. 25-7) for the alarm system to be installed in the administration building. Create Three: Right viewports in paper space, and show the top and front views. In the large right viewport, show the SW isometric view. Plot the views.

2. Draw the hexagonal locking pin that fits into the alarm system receptacle in the administration building from problem 1. The pin, shown in Fig. 25-8, is 2″ long and measures 2.5″ across the flats. Use three viewports in paper space, and show the same views that you used for the locking receptacle in problem 1. Plot the views.

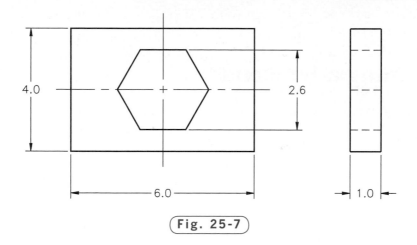

Fig. 25-7

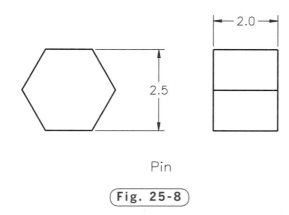

Pin

Fig. 25-8

Designing a Bookend

Creating and working with multiple viewports allows the AutoCAD user greater flexibility in preparing, viewing, and plotting drawings. This exercise requires the use of both model space and paper space.

▶ Description

Design a bookend based primarily on one or more initials of your name. Figure P5-1 shows an example of a bookend created using the initials L.B.

Create or modify the design to have a thickness (Z axis) of 2 or more inches. Actual dimensions are not critical, but they should approximate the height and width of a small book.

Keep the following suggestions in mind as you work.

• Begin by drawing everything in model space.

• Switch to paper space and set up four viewports.

• Show the plan view in the lower left viewport, the top view in the upper left viewport, the right view in the lower right viewport, and an isometric view in the upper right viewport.

• Place each view on its own layer. Assign each layer a different color.

• Zoom each viewport appropriately.

▶ Hints and Suggestions

1. Thickness gives the piece a three-dimensional look.

2. Remember as you work that lines and objects created in model space cannot be altered in paper space, and vice versa.

3. Only paper space allows adjustment and manipulation of multiple viewport plotting. Some trial and error may be required to achieve the desired look.

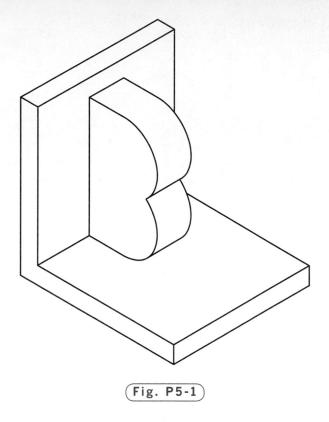

Fig. P5-1

▶ Summary Questions

Your instructor may direct you to answer these questions orally or in writing. You may wish to compare and exchange ideas with other students to help increase efficiency and productivity.

1. What factors determine the drawing setup? (How do you determine the "right" choices?) Why can't all drawings successfully use the same setup?

2. What variables can be modified on each layer?

3. Discuss some advantages of being able to draw on and individually control separate layers.

4. Compare and contrast model space with paper space.

5. What is useful about working in multiple viewports?

Chapter **26**

Basic Dimensioning

Objectives

- Set up a text style for dimensions
- Produce linear dimensions using dimensioning commands and shortcuts
- Dimension round shapes, curves, and holes
- Dimension angles
- Determine the need for and use baseline and ordinate dimensioning when appropriate

Vocabulary

baseline dimensions
datum
datum dimensions
dimension line
dimensioning
extension lines
linear dimensions
ordinate dimensions

Technical drawings lack meaning without information that communicates size. Drafters and designers use a method known as **dimensioning** to describe the size of features on a drawing. In this chapter and the two chapters that follow, you will discover the wide range of dimensioning options, settings, and styles that AutoCAD makes available to you for mechanical, architectural, and other types of drafting and design work.

Fig. 26-1 shows a drawing of a part that fits into an injection mold (mold insert). We will dimension the drawing, but first we need to prepare the drawing file.

1. Start AutoCAD and use the Quick Setup wizard to establish decimal units and a drawing area of 11″ × 8.5″.

2. Select the **Drafting & Annotation** workspace.

3. Create two new layers using the information in Table 26-1.

Table 26-1. Layers

Layer Name	Color	Linetype	Lineweight
Objects	red	Continuous	0.50 mm
Dimensions	green	Continuous	0.20 mm

4. Set layer **Objects** as the current layer.

5. Set snap at **.25** and grid at **.5**; **ZOOM All**.

6. Produce the drawing shown in Fig. 26-1 on layer **Objects**, but omit the dimensions. From the lower left corner of the object, position the center of the hole **2″** in the positive *x* direction and **1.25″** in the positive *y* direction.

7. Save your work in a file named **dimen.dwg**.

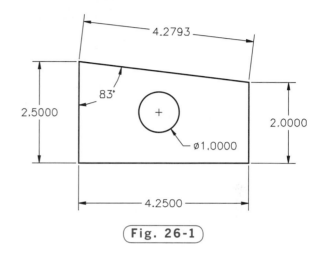

Fig. 26-1

Setting the Dimension Text Style

AutoCAD uses text styles for its dimension text. AutoCAD's romans is a suitable font for dimension text and is popular among companies that use AutoCAD. This font resembles the Gothic lettering used extensively in hand-produced drawings. For most drafting applications, a text height of .125″ (¹⁄₈″) is standard practice in industry. Some companies use taller text, such as .150″. Whichever height you use, it's important that you use a consistent text height throughout the drawing or set of drawings.

1. Create a new text style named **rom** using the **romans.shx** font. Accept the default settings for the new style.

Activate the STYLE command by entering ST or by selecting the arrow in the lower right corner of the Text panel under the Annotate tab of the Ribbon.

2. On the Annotate tab of the Ribbon, pick the arrow in the lower right corner of the Dimensions panel.

This displays the Dimension Style Manager dialog box. We will use it here, but you will learn much more about it in the following two chapters.

3. Pick the **Modify...** button.

4. Pick the **Text** tab.

5. At the right of Text style, pick the down arrow and select **rom**.

6. Pick **OK** and then pick **Close**.

Creating Linear Dimensions

DIMENSIONS

Linear dimensions are those with horizontal, vertical, or aligned dimension lines. A **dimension line** is the part of the dimension that typically contains arrowheads at each of its ends, as shown in Fig. 26-2.

1. Make layer **Dimensions** the current layer.

2. Pick the **Linear** button from the Dimensions panel on the Annotate tab of the Ribbon.

This enters the DIMLINEAR command.

3. In response to Specify first extension line origin, pick one of the endpoints of the mold insert's horizontal edge.

4. In response to Specify second extension line origin, pick the other end of the horizontal line.

5. Move the crosshairs downward to locate the dimension line 1″ away from the object and press the pick button.

The dimension's **extension lines** are the lines that extend from the object to the dimension line.

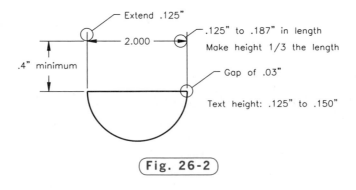

Fig. 26-2

Dimensioning Overall Size

Let's dimension the vertical edges of the object.

1. Press the spacebar to reenter the **DIMLINEAR** command.

2. This time, when AutoCAD asks for the first extension line origin, press **ENTER** or the spacebar.

The crosshairs change to a pick box.

3. Pick any point on the left edge of the object.

4. Move the crosshairs to the left and pick a point 1″ from the edge to place the dimension.

Let's dimension the other vertical edge, but this time we will use AutoCAD's Quick Dimension feature.

5. Pick the **Quick Dimension** button on the Dimensions panel or enter the **QDIM** command.

DIMENSIONS

As you can see, a pick box replaces the crosshairs. AutoCAD asks that you select geometry to dimension.

6. Pick the right edge of the object and right-click for **ENTER**.

7. Move the crosshairs to the right and pick a point for the dimension.

Notice that adding the last two dimensions required only three clicks for each dimension.

You can create more than one dimension at a time using the Quick Dimension method. However, the results are often unpredictable.

Dimensioning Inclined Edges

Let's dimension the inclined edge by aligning the dimension to the edge.

1. Pick the **Aligned** button from the flyout menu under the Linear button on the Dimensions panel.

DIMENSIONS

This enters the DIMALIGNED command.

2. Press **ENTER**, pick the inclined edge, and place the dimension.

Dimensioning Round Features

DIMENSIONS

Now let's dimension the hole. For mechanical drafting, you should use diameter dimensions for features such as holes and cylinders. Use radius dimensions for features such as fillets and rounds (rounded outside corners).

1. Pick the **Diameter** button from the flyout menu under the Aligned button on the Dimensions panel.

This enters the DIMDIAMETER command.

2. Pick a point anywhere on the hole.

The dimension appears.

3. Move the crosshairs around the circle and watch what happens.

Notice that you have dynamic control over the dimension.

4. Pick a point down and to the right as shown in Fig. 26-1 (page 357).

Dimensioning Angles

DIMENSIONS

Let's dimension the angle, as shown in Fig. 26-1.

1. Pick the **Angular** button from the flyout menu on the Dimensions panel.

This enters the DIMANGULAR command.

2. Pick both edges that make up the angle.

3. Move the crosshairs outside the mold insert and watch the different possibilities that AutoCAD presents.

4. Pick a location for the dimension arc inside the drawing as shown in the illustration.

5. Save your work. Also, save this file as a template file in your named folder. Name the file **dimen.dwt** and enter **Dimensioning Practice** for the template description.

6. Close the file.

Dimensioning Arcs

Now we will modify our injection mold insert by adding an arc. Then we will dimension the arc in two ways.

1. Pick the **New** button on the Quick Access toolbar.

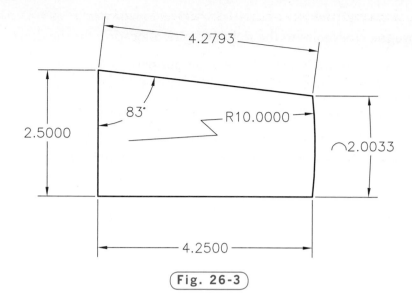

Fig. 26-3

2. Pick **Use a Template** and select **dimen.dwt** as the template file to begin a new drawing. You will need to browse to find your named folder.

3. Save the drawing and name it **dimenarc.dwg**.

4. Erase the right edge of the part, the center circle, and the two associated dimensions.

5. Set layer **Objects** as the current layer.

6. Pick the **Arc** button from the Draw panel, or enter **A** at the keyboard.

DRAW

7. Create the arc as shown in Fig. 26-3 by following this procedure:

   • Pick the lower right corner of the part as the start point of the arc.

   • Select the **End** option and pick the upper right corner of the part as the second endpoint.

   • Select the **Radius** option and move the crosshairs to the far left side of the screen, and type **10**.

8. Set layer **Dimensions** as the current layer.

9. Pick the **Arc Length** button from the flyout menu on the Dimensions panel.

This enters the **DIMARC** command.

DIMENSIONS

10. Pick the arc you just created.

The dimension appears. Notice the arc symbol at the beginning of the dimension. This indicates that the dimension is the length of an arc and not a linear dimension.

11. Move the crosshairs to the right and pick a location for the dimension.

The radius of this arc is much larger than our part. To dimension large-radius arcs and circles, we use the DIMJOGGED command.

DIMENSIONS

12. Pick the **Jogged** button from the flyout menu on the Dimensions panel.

13. Pick the arc.

14. In reply to Specify center location override, pick a point inside the part near its left edge.

15. Specify a location for the dimension line; then specify a location for the jog. Undo and recreate the jogged dimension until yours looks like the dimension in Fig. 26-3.

The jog in the dimension line indicates that the center point of the dimension is not the true center point of the arc.

16. Save your work and close the file.

Using Other Types of Dimensioning

The dimensioning process you have followed so far in this chapter works for most objects. However, other types of dimensioning are better suited for some objects and engineering drawing applications. In many cases, they produce a cleaner, less confusing drawing.

Baseline Dimensioning

Baseline dimensions are progressive, each starting at the same place, as shown in Fig. 26-4. Baseline dimensioning is most useful on machine drawings, on which precision is critical. Measuring from a single reference dimension reduces the chance of errors that can accumulate from a "stack" of dimensions.

1. Select **dimen.dwt** as the template file to begin a new drawing.

2. Save the drawing and name it **base.dwg**.

3. Erase the drawing and dimensions and create the sheet-metal drawing shown in Fig. 26-4. Place the part on layer Objects. Turn Snap on. Fill the top half of the screen, and estimate the dimensions.

DIMENSIONS

4. Make **Dimensions** the current layer.

5. Pick the **Linear** button and dimension line **A**. It is important that you pick line A's right endpoint first.

DIMENSIONS

6. Pick the **Baseline** button (you need to pick the down arrow next to the Continue button on the Dimensions panel to see the button).

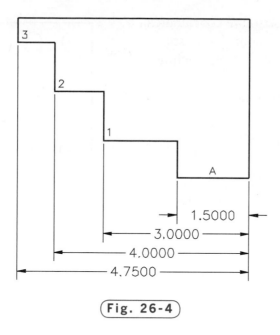

Fig. 26-4

This enters the DIMBASELINE command.

7. Pick points **1**, **2**, and **3** in order.

8. Press **ENTER** twice and save your work.

The Quick Dimension method also works with baseline dimensioning.

9. Erase the dimensions.

10. Pick the **Quick Dimension** button or enter **QDIM**.

DIMENSIONS

11. Type **ALL** to select the entire drawing and press **ENTER** twice.

Read the options AutoCAD presents.

12. Select the **Baseline** option or type **B** at the keyboard.

13. Select the **datumPoint** option or type **P** at the keyboard and pick the lower right corner of the sheet-metal part.

This establishes the point from which the baseline dimensions start.

14. Place the dimensions.

15. Save your work.

Ordinate Dimensioning

Ordinate dimensions, also known as **datum dimensions**, are similar to baseline dimensions in that both use a datum, or reference, dimension. A **datum** is a surface, edge, or point that is assumed to be exact. One basic difference between baseline dimensions and ordinate dimensions is appearance. Baseline dimensioning uses conventional dimension lines, whereas ordinate dimensions show absolute coordinates. Both types of dimensioning are especially useful when producing machine parts. Using a datum reduces the chance of error buildup caused by successive dimensions.

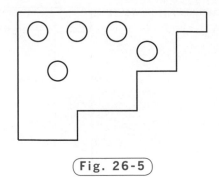

Fig. 26-5

1. In the Application Menu, pick **Save As...** to create a new file named **ordinate.dwg**.

2. Create a new layer named **Orddim**, assign the color magenta to it, make **Objects** the current layer, and freeze layer **Dimensions**.

3. Mirror the sheet-metal part to the right (see Fig. 26-5) and delete old objects. If necessary, move the part so it is in the positive x and y quadrant.

4. On the Objects layer, create a hole with a diameter of **.5**. Place the hole **.5** units down and **.5** units to the right of the top left corner.

5. Use **COPY** to create four additional holes. Place the holes as shown in Fig. 26-5.

6. Create a layer named **Center**, and on this layer add center marks to each of the holes. To do this, expand the Dimensions panel and pick the **Center Mark** button.

DIMENSIONS

Now we're ready to define the datum, or reference dimension.

7. Make **Orddim** the current layer.

DIMENSIONS

8. Pick the **Ordinate** button from the flyout menu on the Dimensions panel to enter the DIMORDINATE command.

9. In reply to the Specify feature location prompt, pick the upper left corner of the sheet-metal part.

The Xdatum option measures the absolute x ordinate along the X axis of the drawing. Also, it determines the orientation of the leader line and prompts you for its endpoint. This defines the location of the datum.

10. Select the **Xdatum** option or type **X** at the keyboard.

11. Pick a point about **1″** above the object to make the ordinate dimension appear.

12. Reenter the command and pick the upper left corner of the sheet-metal part again.

13. Select the **Ydatum** option or type **Y** at the keyboard and pick a point about **1″** to the left of the object to make the ordinate dimension appear.

The Ydatum option measures the absolute *y* ordinate along the Y axis of the drawing. This defines the location of the datum. Your drawing should now look similar to the one in Fig. 26-6, although it will not include the 3.0000 ordinate dimension.

 The ordinate values may differ in your drawing, depending on where in the drawing area you placed the drawing.

DIMENSIONS

14. Pick the **Ordinate** button.

15. Snap to the center of the hole nearest to the upper left corner, as shown in Fig. 26-6.

16. Pick a point about **1**″ from the top of the sheet-metal part. (Snap should be on.)

The X-datum ordinate dimension appears.

17. Repeat Steps 14 through 16 to create X-datum ordinate dimensions for the remaining holes.

18. Create the Y-datum ordinate dimensions on your own using Fig. 26-7 as a guide.

When you finish, your drawing should look similar to the one in Fig. 26-7.

19. Save your work and exit AutoCAD.

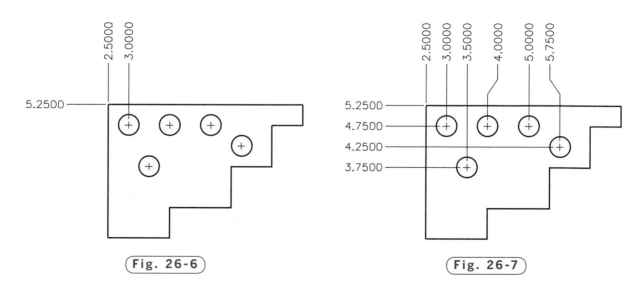

Fig. 26-6   Fig. 26-7

● REVIEW QUESTIONS

Answer the following questions on a separate sheet of paper.

1. How do you specify a text style for dimension text?

2. Describe the alternative to specifying both endpoints of a line when dimensioning the entire length of an edge.

3. Which dimension button do you use to dimension inclined lines? ...to dimension angles?

4. Which dimension buttons do you use to dimension fillets, rounds, and holes? Explain when you would use each button.

5. What does a jog in a radius dimension represent?

6. Describe baseline dimensioning. On what types of drawings is it useful?

7. Describe ordinate dimensioning. On what types of drawings is it useful?

● CHALLENGE YOUR THINKING

These questions are designed to further your knowledge of AutoCAD by encouraging you to explore the concepts presented in this chapter. Answer each question on a separate sheet of paper.

1. Experiment with altering a linear dimension using its grips. Try to change the location of the dimension line by selecting and moving a grip located at the end of an extension line that is closest to the object. Is it possible? Explain.

2. Drafters generally classify dimensions in two broad categories: size dimensions and location dimensions. What is the difference between them? Are both necessary on every drawing you create? Write a paragraph explaining your findings.

3. What is the relationship between the height of text saved in a text style and the height saved in a dimension style?

• APPLYING AUTOCAD SKILLS

Work the following problems to practice the commands and skills you learned in this chapter. Begin a new drawing for each problem. Save your work.

1–2. Create the blocks shown in Figs. 26-8 and 26-9. Place object lines on a layer named Objects and dimensions on a layer named Dimensions. Create a new text style using the romans.shx font. Approximate the location of the holes. Refer to block1.dwg and block2.dwg (These drawings are located on the OLC for this book, in the Drafting & Design Problems section.)

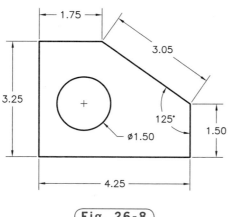

Fig. 26-8

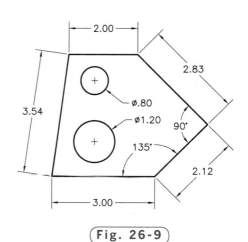

Fig. 26-9

3. Open the slide.dwg file that you created at the end of Chapter 23 and dimension it. When you finish, your drawing should look like the one in Fig. 26-10. Save the drawing as slide2.dwg.

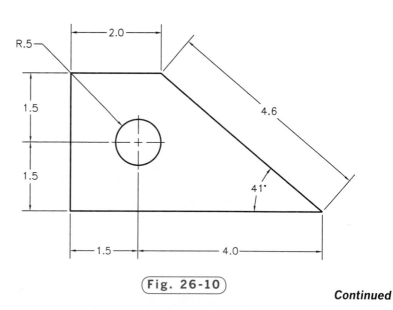

Fig. 26-10

Continued

4. Create the mounting bracket drawing shown in Fig. 26-11. Place the bracket on layer Objects. Then dimension the bracket on a layer named Dimensions. Refer to the file bracket1.dwg. (This drawing is located on the OLC for this book, in the Drafting & Design Problems section.)

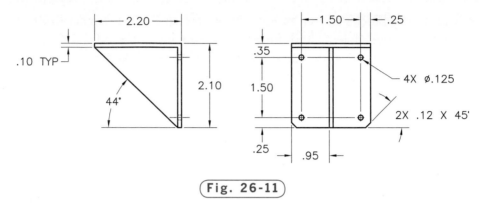

Fig. 26-11

Courtesy of Joseph K. Yabu, Ph.D., San Jose State University

5. Open the rockerarm.dwg file that you created at the end of Part 1 on page 112. Place the rocker arm on a layer named Objects. Then dimension the part on a layer named Dimensions. (See Fig. 26-12.) Save the file as rockerarmdim.dwg.

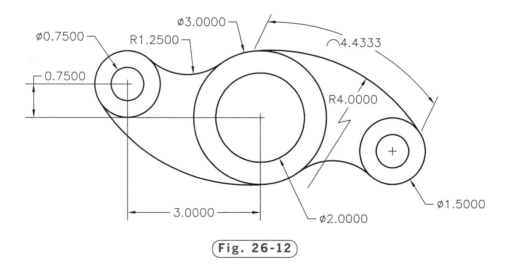

Fig. 26-12

6. Draw the front view of the shaft lock shown in Fig. 26-13. Create appropriate layers for the dimensions and linetypes. Use the romans.shx font to create a new text style named roms for the dimension text. Dimension the front view, placing the dimensions on layer Dimensions. Show all pertinent dimensions. Also, draw and dimension the right-side view. Resize the drawing area as necessary to accommodate the dimensions of the part. Refer to the file shaftloc.dwg. (This drawing is located on the OLC for this book, in the Drafting & Design Problems section.)

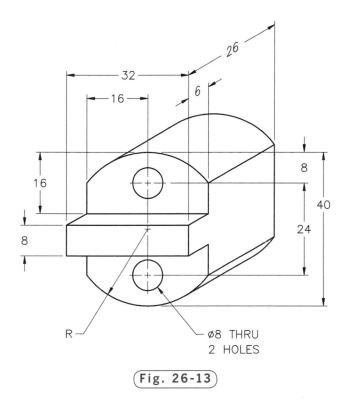

Fig. 26-13

Continued

● USING PROBLEM-SOLVING SKILLS

Complete the following activities using problem-solving skills and your knowledge of AutoCAD.

1. The architect for your firm has given you the elevation sketch shown in Fig. 26-14. Select an appropriate template file, complete the title block, and draw the front elevation; include dimensions. Since the drawing was not made by a drafter, make any changes necessary to comply with correct dimensioning practices. Save the drawing as elev1.dwg.

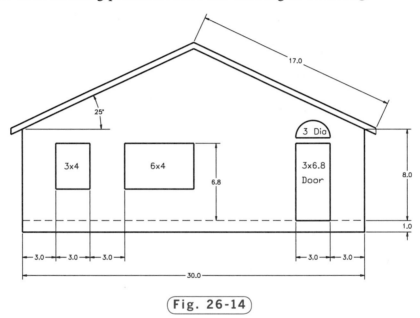

Fig. 26-14

2. The sketch of a shaft support shown in Fig. 26-15 is not to scale; however, the following description applies. The base is 2″ square and ¼″ thick. The boss (lipped area) is in the center of the base and has an outer diameter of 0.5″, a thickness of ¼″, and a through hole with a .35″ radius. The four corner holes are each .15″ in diameter and their center lines are offset from the horizontal and vertical sides by .35″. Draw and dimension two views of the shaft support.

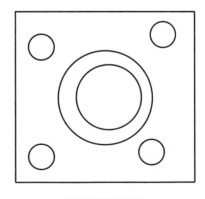

Fig. 26-15

Chapter 27

Advanced Dimensioning

Objectives

- Create a dimension style and apply it to a drawing and template file
- Adjust the dimension text, arrowheads, center marks, and scale in a dimension style
- Edit individual dimensions using a shortcut menu
- Use AutoCAD's associative dimensioning feature to update dimensions on a rescaled drawing
- Apply dimension styles to a drawing template file

Vocabulary

associative dimensions
dimension styles
leaders
spline leader

AutoCAD permits you to create **dimension styles**, which consist of a set of dimension settings. Dimension styles control the format and appearance of dimensions and help you apply drafting standards. They define the format of dimension lines, extension lines, arrowheads, and center marks. They also define the appearance and position of dimension lines and text. Using a dimension style, you can format and define the precision of the linear, radial, and angular units.

Before we can apply a dimension style to something meaningful, we need to produce a drawing. For this chapter, we will work in the AutoCAD Classic workspace.

1. Start AutoCAD and select the **tmp1.dwt** file.
2. Select the **AutoCAD Classic** workspace and close all floating toolbars and palettes.

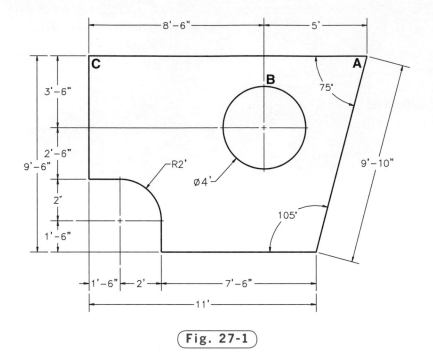

Fig. 27-1

3. On layer **Objects**, create the top view drawing of the redwood deck shown in Fig. 27-1. Do not add dimensions at this time.

The round object represents an outdoor table. It was included in the drawing to ensure that the deck is of sufficient size and shape to accommodate a table of this size.

4. Save your work in a file named **deck.dwg**.

 For best results, begin in the upper right corner and work counterclockwise to draw the edges of the deck. Add the inside rounded corner using the FILLET command.

Creating a Dimension Style

Dimension styles provide an easy method of managing the appearance of dimensions. This saves time and effort and helps you produce professional drawings. Note that drafting standards vary from industry to industry and even among companies in the same industry.

1. Display the **Dimension** toolbar. (Right-click on any docked toolbar and select **Dimension**.)

2. On layer **Dimensions**, place a horizontal dimension along the bottom edge of the deck.

As you can see, the dimension text, arrowheads, etc., are much too small. We need to make some adjustments.

3. Erase the dimension.

4. From the Dimension toolbar, pick the **Dimension Style...** button.

This button is also available in the docked Styles toolbar.

This displays the Dimension Style Manager dialog box. As you can see, Standard is the default dimension style. The sample drawing shows how the dimensions will appear using this style.

5. Pick the **New...** button.

6. For New Style Name, enter **My Style**.

7. Pick the down arrow below Start With and pick **Standard**.

This selection means that we will start with all of the settings saved in the Standard style.

8. Pick the down arrow below Use for.

You can apply a dimension style to the types of dimensions included in this list.

9. Select **All dimensions** and pick the **Continue** button.

The New Dimension Style dialog box appears, showing several tabs across the top. This is where you change the settings to store in the dimension style.

10. Pick the **OK** button.

11. Pick the **Set Current** button, pick **OK** in the pop-up window, then the **Close** button.

Notice on the Styles toolbar that My Style is now the current dimension style.

12. Save your work.

Changing a Dimension Style

Currently, My Style is identical to the Standard style. We will need to customize the new My Style dimension style for our particular drawing. This means changing the size of the arrowheads and center marks, the style and size of the text, and the overall scale of the dimensions.

DIMENSION

Arrowhead Size

1. Pick the **Dimension Style...** button and pick the **Modify...** button.
2. Pick the **Symbols and Arrows** tab unless it is already in front.

You can make adjustments to the arrowheads in the left side of the dialog box.

3. Using the up and down arrows below Arrow size, change the value to **1/8"** and watch closely how the arrowheads in the sample drawing change to reflect the new value. They may be difficult to see.

Sizing Center Marks or Center Lines

In the area titled Center marks, you can make changes to the appearance of the centers of circles. Use center marks when space is not available for complete center lines. You may also need to make changes in this area when a company style requires one or the other.

1. Pick the radio button next to Mark unless it is already selected.

Notice that the center mark appears on the sample drawing.

2. Click the down arrow until it drops to a value of **0"**.
3. Reselect the radio button next to Mark and then set the size to **1/16"**.
4. Select the **Line** radio button and watch to see how the sample drawing changes to reflect the new center lines.

This causes complete center lines to appear at the centers of circles and arcs, as shown in the sample drawing.

Tailoring Dimension Text

Let's create a new text style and height for our dimension style.

1. Pick the **Text** tab.
2. In the upper left area titled Text appearance, pick the button containing the three dots at the right of Text style.

This causes the familiar Text Style dialog box to appear.

3. Create a new style named **roms** using the **romans.shx** font and the default settings.

4. After closing the Text Style dialog box, pick the down arrow at the right of Text style and select **roms**.

Roms is now the text style for My Style, as shown in the sample drawing.

5. In the Text height edit box, change the value to 1/8″.

As a reminder, the dimension text should be .125″ to .150″ in height.

Setting the Units

1. Pick the **Primary Units** tab.

2. At the right of Unit format, change the setting to **Architectural**.

3. At the right of Precision, change the value to **0′-0″**.

The sample drawing changes to reflect the unit format and precision.

Scaling the Dimensions

1. Pick the **Fit** tab and focus your attention on the area titled Scale for dimension features.

This is where you set the scale for dimensions. The value should be the reciprocal of the plot scale. When creating tmp1.dwt, we determined that the scale would be 1/2″ = 1′, which is the same as 1/24. The reciprocal of 1/24 is 24.

2. Double-click the number in the edit box and change it to **24**.

 You could also enter **DIMSCALE** at the AutoCAD prompt and enter **24**.

There are many other options. We have changed only those needed for the dimension style we will use to dimension the redwood deck. You will learn more about dimension style options in Chapters 28 and 29.

3. Pick **OK**; pick **Close**.

4. Save your work.

Applying a Dimension Style

Now that you have made several adjustments to the dimension style, you are ready to dimension the drawing. Layer Dimensions should be the current layer.

DIMENSION

1. Begin by dimensioning the table and rounded corner. Use the **Diameter** button for the table and the **Radius** button for the filleted corner. Place the dimensions as shown in Fig. 27-1 (page 372).

Now that you have center lines, you can use them to locate the centers of the table and rounded corner using dimensions.

DIMENSION

2. Pick the **Linear** button from the Dimension toolbar.

DIMENSION

3. For the first and second extension line origins, snap to points A and B, and then place the dimension **1.5′** from the edge of the deck.

HINT

Use grid or snap to determine the 1.5′ distance.

DIMENSION

4. Pick the **Continue** button, snap to point **C**, and press **ENTER** twice to terminate the command.

HINT

When locating the center of the rounded corner with dimensions, use the end of the center lines for the extension line origins, similar to the way you picked point B.

Notice that the dimension lines for the two dimensions are aligned when Continue Dimension is used. Another way to align dimensions is with the DIMSPACE command.

5. With snap off, pick the **Linear** button and place the four short vertical dimensions to the left of the deck. Do not attempt to align the dimension lines.

DIMENSION

6. Pick the **Dimension Space** button from the Dimension toolbar.

7. In response to Select base dimension, pick the **3′-6″** dimension at the top.

8. In response to Select dimensions to space, pick the other three dimensions and press **ENTER**.

9. In response to Enter value, enter **0** and press **ENTER**.

AutoCAD aligns the four dimensions. Dimension Space can also be used to evenly space a series of dimensions that start from a common baseline. One example of this type of dimensioning is Fig. 26-4 on page 363.

10. Using similar methods, add the remaining horizontal and vertical dimensions.

11. Use the grips method of moving dimension text to move the **9'-6"** text of the vertical dimension so that it is lower than the **2'-6"** dimension beside it.

12. Add the aligned and angular dimensions.

Your dimensioned drawing of the redwood deck should now look similar to the one in Fig. 27-1 (page 372).

13. Save your work.

14. If you have access to a plotter or printer, plot the drawing.

Editing a Dimension

Using AutoCAD's grips, you can change the location of dimension lines and dimension text.

1. Pick one of the linear dimensions.

Five grip boxes appear. The floating Quick Properties panel may appear too if Quick Properties is toggled on.

2. Pick the grip box located at the center of the dimension text.

3. Move the text up and down, then back and forth, and then pick a new point.

4. Press **ESC**.

By selecting a dimension and right-clicking it, you can make other useful changes.

5. Select one of the linear dimensions and rest the crosshairs over the dimension text grip.

A shortcut menu of options appears. These options permit you to adjust the location of the dimension text.

6. Try each of the options, and pick **Reset Text Position** last.

As you can see, it's easy to tailor individual dimensions, regardless of the current dimension style.

7. Select one of the linear dimensions and right-click.

8. Rest the pointer on Dimension Style.

This allows you to save a new style or apply an existing one.

9. Select **Standard**.

The dimension uses settings stored in the Standard dimension style. The arrowheads and dimension text are there, but you can't see them because they are so small. The value of the dimension scale in Standard is 1, and we set it to 24 in My Style. This means that the text, arrowheads, and other dimensioning features are 24 times larger in My Style.

10. Change the dimension's style back to **My Style**.

11. Save your work.

Leaders

Lines with arrowheads at one end that are used to point out a particular feature in a drawing are called **leaders**. Radial and diameter dimensions usually use leaders. The plastic retainer in Fig. 27-2 provides an example of a leader.

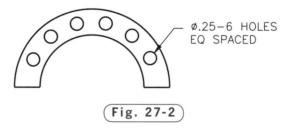

Ø.25−6 HOLES
EQ SPACED

Fig. 27-2

1. Display the **Multileader** toolbar.

2. Pick the **Multileader Style** button on the Multileader toolbar.

MULTILEADER

This displays the Multileader Style Manager dialog box. Standard is the default multileader style.

3. Pick the **New...** button, name the new style **My leader style**, and pick the **Continue** button.

4. In the Type area of the Leader Format tab, pick the down arrow next to **Straight** and select Spline.

Notice how the leader in the sample drawing changes. A **spline leader** is similar to a regular, straight leader, except the leader line can consist of a spline curve, giving you more flexibility with its shape when space is tight in a drawing.

5. Reselect the **Straight** leader type, and change the arrowhead size to **1/8″**.

6. Select the **Leader Structure** tab and specify a leader scale of **24**.

7. Pick **OK**; pick **Close**.

You have created a leader style that matches your dimension style in scale, text size, and arrowhead size. Notice on both the Multileader and the Styles toolbar that My leader style is now the current leader style.

8. Pick the **Multileader** button on the Multileader toolbar.

MULTILEADER

This enters the MLEADER command.

9. Pick point A (shown in Fig. 27-1, page 372).

10. Move the crosshairs up and to the right about 1.5′ and pick a point.

The Text Formatting toolbar appears, along with a blinking cursor in a blank text editor field.

11. In the first drop-down list, change the text style from Standard to **roms**.

12. Enter **USE REDWOOD.**

13. Pick **OK** to close the Text Formatting toolbar.

AutoCAD creates the leader. Notice the arrowhead size, the text size, and the text style of the leader match the dimensions.

14. Pick the new leader, pick the down arrow in the Multileader toolbar and select **Standard**.

The leader text is now too small to read, because text in the Standard leader style is 1/24 the size of text in My leader style.

15. Pick the **Undo** button.

16. Save your work.

Working with Associative Dimensions

By default, AutoCAD creates **associative dimensions**—dimensions that update automatically when you change the drawing by stretching, scaling, and so on.

1. Using the **SCALE** command, scale the drawing by **1.1**.

Notice that the dimensions changed automatically. This becomes useful when you need to change a drawing after you've dimensioned it.

2. Using the **STRETCH** command, stretch the rightmost part of the drawing a short distance to the right. (When selecting, use a crossing window.)

The dimensions change accordingly.

3. Undo Steps 1 and 2 and save your work.

Adding Dimension and Leader Styles to Templates

Now you know how to perform the remaining steps (12 through 14) in creating a template file (see Table 22-1 on page 297). Let's apply these steps to the tmp1.dwt template.

1. Open **tmp1.dwt**.

2. Create a new dimension style using the following information.
   - Name the dimension style **Preferred**. In the Create New Dimension Style dialog box, start with **Standard**. Apply the new style to all dimensions.

- In the Primary Units tab, set Unit format to **Architectural** and Precision to **0'-0"**.

- In the Symbols and Arrows tab, set the dimension arrow size to **1/8"**. Pick the **Line** radio button in the Center marks area and set the size of the center marks to **1/16"**.

- In the Text tab, create a new text style named **roms** using the **romans.shx** font and use the default values.

- Close the Text Style dialog box, make **roms** the style for Preferred, and set the text height at **1/8"**.

- In the Fit tab, set Scale for dimension features to **24**.

3. Pick **OK** and close the dialog boxes.

4. Create a new leader style named **Preferred leader**. Set the arrowhead size to **1/8"**, the leader scale to **24**, the text height to **1/8"**, and the text style to **roms**.

5. Save your work and close the file.

The tmp.dwt template file is now ready for use with other new drawing files.

6. Close the **Dimension** and **Multileader** toolbars.

7. Save **deck.dwg** and exit AutoCAD.

● REVIEW QUESTIONS

Answer the following questions on a separate sheet of paper.

1. Describe the purpose of dimension styles. Why might you need to create one?
2. How do you determine the dimensioning scale?
3. Explain the difference between center marks and center lines. When should you use each?
4. Explain the use of the Continue button in the Dimension toolbar.
5. When you select a dimension and right-click, AutoCAD displays a menu. Name and briefly describe the three dimension-related options that are available to you.
6. When might you use a spline leader?
7. What is the advantage of using associative dimensions?

● CHALLENGE YOUR THINKING

These questions are designed to further your knowledge of AutoCAD by encouraging you to explore the concepts presented in this chapter. Answer each question on a separate sheet of paper.

1. In this chapter, you placed object lines on one layer and dimensions on another. Discuss the advantages of placing dimensions on a separate layer. Are there any disadvantages to placing dimensions on a separate layer? Explain.
2. Describe the changes you would need to make to tmp1.dwt to create a template for a B-size sheet at the same drawing scale.

● APPLYING AUTOCAD SKILLS

Work the following problems to practice the commands and skills you learned in this chapter.

1. Begin a new drawing and establish the following drawing settings. Store it as a template file named tmp2.dwt.

   Units: Engineering
   Scale: 1″ = 10′ (or 1″ = 120″)
   Sheet size: 17″ × 11″
   Drawing Area: You determine the drawing area based on the scale and sheet size.
   Grid: 10′
   (Reminder: Be sure to enter ZOOM All.)
   Snap: 2′
   Layers: Set up as shown in Table 27-1.

   ### Table 27-1. Layers

Layer Name	Color	Linetype	Lineweight
Thick	red	Continuous	0.50 mm
Thin	green	Continuous	0.25 mm

   Create a new dimension style for the template using the following information.

   - Name the dimension style D001. In the Create New Dimension Style dialog box, start with Standard and apply the style to all dimensions.
   - Set the unit format to Engineering and the precision to 0′-0.00″.
   - Set the dimension arrow size to .125″.
   - Use center marks and set their size to .1″.
   - Create a new text style named arial using the Arial font. Use the default settings and make arial the current style for NewStyle. Set the text height at .125″.
   - Set the dimension scale to the proper value.

Continued

2–3. Use the template file you created in problem 1 to create the hotel recreation areas shown in Figs. 27-3 and 27-4.

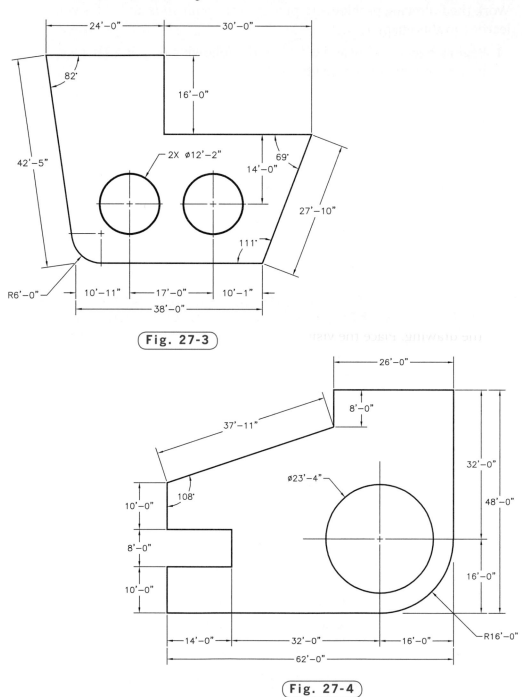

Fig. 27-3

Fig. 27-4

4. Create and identify the radii of the sheet metal elbow shown in Fig. 27-5. First draw the perpendicular lines, and then create two arcs: one with a radius of 1.5 and the other with a radius of 3. Then dimension the 90° angle and add the notes. Before you create the leaders, open the Multileader Style Manager and change the arrowheads to dots. Add the text as shown, and save the drawing as elbow1.dwg.

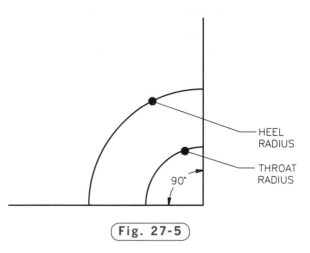

Fig. 27-5

5. Draw the front view of the shaft shown in Fig. 27-6. Use the Quick Setup method to change the drawing area to accommodate the dimensions of the drawing. Place the visible lines on a layer named Visible using the color green. Create a layer named Forces using the color red. Draw and label the four load and reaction forces, using leaders and a text style based on the romans.shx font. Create another layer named Dimensions using the color cyan. Dimension the front view, placing the dimensions on layer Dimensions.

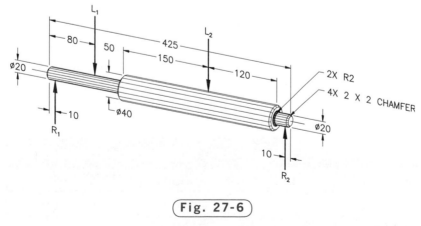

Fig. 27-6

Courtesy of Gary J. Hordemann, Gonzaga University

Continued

6. Draw and dimension the front view of the metal casting shown in Fig. 27-7. Use the following drawing and dimension settings, and place everything on the proper layer.

Units: Decimal with two digits to the right of the decimal point
Drawing Area: 11 × 8.5
Scale: 1:1
Grid: .1
Snap: .05
LTSCALE: Start with .5
Layers: See Table 27-2.
Create a new dimension style using the following information:

- Choose a name for a new dimension style that will help you remember the purpose of this dimension style. Start with Standard and apply the style to all dimensions.

- Set the unit format to Decimal and the precision to 0.00.

- Set the dimension arrow size to .125.

- Set the size of the center marks to .1 and use center lines.

- Create and name a new text style. Use romans.shx, apply default settings, and make it the current style. Set the text height at .125.

- Set the dimension scale to the proper value.

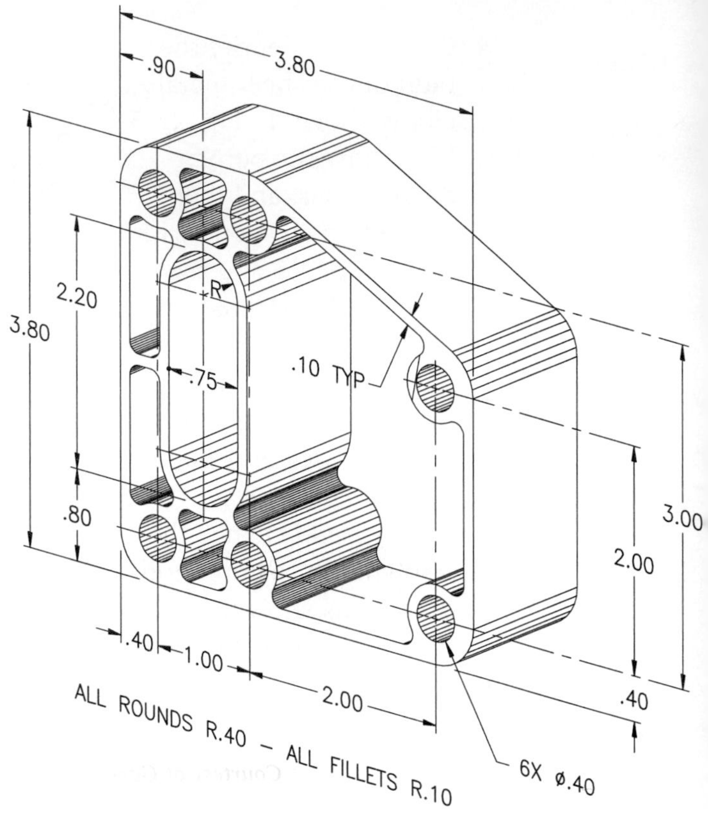

Fig. 27-7

Problem 6 courtesy of Gary J. Hordemann, Gonzaga University

Table 27-2. Layers

Layer Name	Color	Linetype	Lineweight
Visible	red	Continuous	0.50 mm
Dimensions	blue	Continuous	0.25 mm
Text	magenta	Continuous	0.30 mm
Center	green	CENTER	0.20 mm

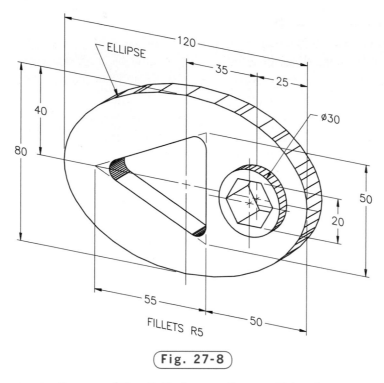

Fig. 27-8

Courtesy of Gary J. Hordemann, Gonzaga University

7. Draw and dimension the front view of the steel plate shown in Fig. 27-8.
Use the following drawing, dimension, and leader settings to set up the
metric drawing, and place everything on the proper layers. Also draw
and dimension the top and right-side views, assuming the plate and boss
thicknesses to be 10 and 5, respectively.

Units: Decimal with no digits to the right of the decimal point
Drawing Area: 280 × 216
Scale: 1:1
Grid: 10
Snap: 5
LTSCALE: Start with 10
Layers: Set up as shown in Table 27-3 on page 388.
Create new dimension and leader styles using the following
information:

- Name the styles using names of your choice. Start with Standard and
  apply the new styles to all dimensions and leaders.

- Set the unit format to Decimal and the precision to 0.

Continued

Table 27-3. Layers

Layer Name	Color	Linetype	Lineweight
Visible	red	Continuous	0.60 mm
Dimensions	blue	Continuous	0.20 mm
Text	magenta	Continuous	0.20 mm
Center	green	CENTER	0.20 mm
Hidden	magenta	HIDDEN	0.30 mm

- Set the arrow size to 3.
- Set the size of the center marks to 1.5 and use center lines.
- Create a new text style using a name of your choice. Use romans.shx, apply default settings, and make it the current style. Set the text height at 3.
- Set the scale to the proper value.

USING PROBLEM-SOLVING SKILLS

Work the following problems to practice the commands and skills you learned in this chapter.

1. Your architectural drafting trainee showed you the wall corner shown in Fig. 27-9. Notice that the trainee dimensioned the drawing in a mechanical drafting style. Redraw the corner and set the correct dimension size and units, change the arrowheads to architectural ticks (short diagonal lines), and move the dimensions above the dimension line to conform to your company's style. Save the drawing as framing1.dwg.

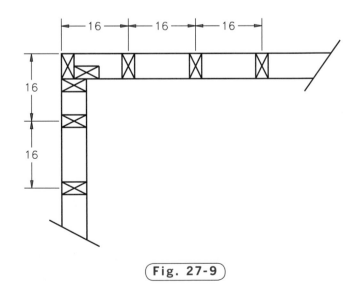

Fig. 27-9.

2. As the set designer for a theatrical company, you are creating a template for a cityscape. The director has given you a sketch of the general building landscape she would like (Fig. 27-10). Draw the template and dimension it from the datum plane. Each square equals 9 inches. Save and plot the drawing using the Extents plotting option.

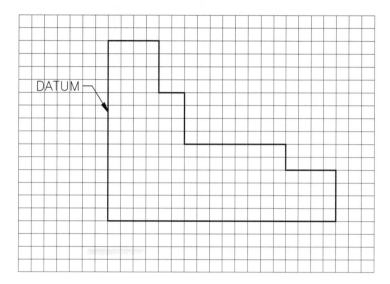

Fig. 27-10

Chapter 28

Fine-Tuning Dimensions

Objectives

- Adjust the appearance of dimension lines, arrowheads, and text
- Control the placement of dimension text
- Adjust the format and precision of primary and alternate units
- Edit a dimension's properties
- Explode a dimension

Vocabulary

alternate units
primary units

Most industries that use AutoCAD apply dimensioning differently. The architectural industry, for example, uses architectural units and fractions, whereas manufacturers of consumer and industrial products typically use decimal units. Within each industry, companies apply standards, some more rigidly than others. In manufacturing, most use a variation of ANSI or ISO standards and enforce the standard company-wide. Because thousands of companies use AutoCAD worldwide, the software must accommodate a wide range of industry and company standards. This chapter focuses on how you can fine-tune dimensions to meet a particular company's style or standard.

Dimension Style Options

AutoCAD offers a wealth of options for tailoring a dimension style to a specific need.

1. Start AutoCAD and open **deck.dwg**.
2. Select the **Drafting & Annotation** workspace.
3. On the Quick Access toolbar, select **Save As...** and enter **deck2.dwg** for the file name.
4. On the Annotate tab of the Ribbon, pick the arrow in the lower right corner of the Dimensions panel to open the Dimension Style Manager.

The current dimension style is My Style, as shown in the upper left corner of the dialog box. In the last chapter, we covered the purpose of the New... and Modify... buttons. The Set Current button permits you to make another style current. The Override... button enables you to override one or more settings in a dimension style temporarily. The Compare... button allows you to compare the properties of two dimension styles or view all the properties of one style.

Lines

1. Pick the **Modify...** button and pick the **Lines** tab unless it is already in front.

Focus your attention on the Dimension lines area. Color, Linetype, and Lineweight permit you to change the color, linetype, and lineweight of dimension lines. In most cases, it is best not to change them. Extend beyond ticks specifies a distance to extend the dimension line past the extension line when you use oblique, architectural tick, integral, and no marks for arrowheads. Baseline spacing sets the spacing between the dimension lines when they're outside the extension lines. Dim line 1 suppresses the first dimension line, and Dim line 2 suppresses the second dimension line, as shown in Fig. 28-1.

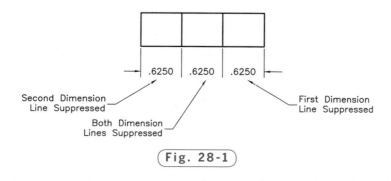

Fig. 28-1

2. Make changes to the settings in the Dimension lines area and notice how the sample drawing changes.

Focus on the area labeled Extension lines. Color, Linetype, and Lineweight allow you to change the color and lineweight of the extension lines. Extend beyond dim lines specifies a distance to extend the extension lines beyond the dimension line. Offset from origin specifies the distance to offset the extension lines from the origin points that define the dimension. Suppress prevents the display of extension lines. Fixed length extension lines allows you to specify a fixed length for all extension lines. Ext line 1 suppresses the first extension line, and Ext line 2 suppresses the second extension line. Suppressing an extension line is useful when you want to dimension to an object line. In this case, the object line would serve the same purpose as the extension line.

3. Make changes to the settings in the Extension lines area and notice how the sample drawing changes.

Symbols and Arrows

1. Pick the **Symbols and Arrows** tab.

Concentrate on the area titled Arrowheads. First sets the arrowhead for the first end of the dimension line. Second sets the arrowhead for the second end of the dimension line. In most cases, you would use the same arrowhead style for both ends of a dimension, but AutoCAD gives you flexibility in case you need to make them different. Leader sets the arrowhead for quick leaders—a leader type that is different than a multileader.

2. To the right of First, pick the down arrow and review the types of arrowheads available.

Some companies and individual architects use ticks instead of arrows at the ends of dimension lines. It would be rare to use None, but AutoCAD gives you that option. As you learned in the previous chapter, Arrow size displays and sets the size of the arrow. You also learned how to change the appearance of center marks in the Center marks area.

3. Make changes to the settings in the Arrowheads area and notice how the sample drawing changes.

Text

1. Pick the **Text** tab.

In the Text appearance area, you are familiar with Text style and Text height. Text color permits you to change the color of the dimension text. In most circumstances, it is best to leave it alone. Fraction height scale sets the scale of fractions relative to dimension text by multiplying the value entered here by the text height. Draw frame around text draws a frame around the dimension text. This option would be useful to companies that frame certain notes in a drawing.

2. Make changes to the settings in the Text appearance area and notice the changes in the sample drawing.

The Text placement area allows you to control the placement of text. Vertical controls the vertical justification of dimension text along the dimension line. Horizontal controls the horizontal justification of dimension text along the dimension line and the extension line. View Direction sets the direction of text, which in most cases will be left to right. Offset from dim line sets the text gap (distance) around the dimension text when the dimension line is broken to accommodate the dimension text. If the space between the dimension text and the end of the dimension lines becomes too large or small, use this option to adjust the space. It is rare for a company or even a government agency to require a specific distance.

3. Make changes to the settings in the Text placement area and notice the changes in the sample drawing.

The Text alignment area permits you to change the orientation (horizontal or aligned) of dimension text, whether it is inside or outside the extension lines. Horizontal places text in a horizontal position. Aligned with dimension line aligns the text with the dimension line. ISO Standard aligns text with the dimension line when text is inside the extension lines. If the text is outside the extension lines, it aligns it horizontally.

4. Experiment with the text alignment options.

Fit

1. Pick the **Fit** tab.

Focus on the Fit options area. The fit options permit you to control the placement of dimension text and arrows when there is not enough space to put them inside the extension lines.

2. Select each of the fit options and notice how they affect the placement of dimension text and arrows.

The Text placement area allows you to control the placement of text when it's not in the default position.

3. Try each of the placement options.

As you discovered in Chapter 27, the Scale for dimension features area controls the scaling of dimensions. When you pick Scale dimension to layout, AutoCAD determines a scale factor based on the scaling between the current model space viewport and paper space.

In the Fine tuning area, if you check Place text manually, AutoCAD allows you to place the text at the position you specify. If you check Draw dim line between ext lines, AutoCAD draws dimension lines between the measured points even when the arrowheads are outside the extension lines.

4. Experiment with these two options.

Primary Units

The **primary units** are those that appear by default when you add dimensions to a drawing. The term *primary* is used by AutoCAD to distinguish them from alternate dimensions, which are discussed in the next section.

1. Pick the **Primary Units** tab.

This tab establishes the format and precision of the primary dimension units, as well as prefixes and suffixes for dimension text.

You were introduced to the Linear dimensions area in Chapter 27, so you know the purpose of Unit format and Precision. Fraction format controls the format for fractions. Decimal separator sets the separator for decimal formats. Options include a period (.), comma (,), or space ( ). Round off sets rounding rules for dimension measurements for all dimension types except angular. Prefix and Suffix permit you to enter text or use control codes to display special symbols such as the diameter symbol.

2. Experiment with the settings in the Linear dimensions area.

Under Measurement scale, Scale factor establishes a scale factor for all dimension types except angular.

Zero suppression controls the suppression of zeros. Some companies do not use a leading zero, for example. One widely accepted rule of thumb is to include the leading zero in drawings dimensioned using the metric system, but to exclude it when dimensioning using the English system (feet and inches). Space (or lack of it) for the dimension may also be a consideration.

A distance of 0.2000 becomes .2000 when you check Leading. A distance of 2.0000 becomes 2 when you check Trailing. A distance of 0'-2-1/4″ becomes 2-1/4″ when you check 0 feet, and a distance of 2'-0″ becomes 2' when you check 0 inches.

Concentrate on the area labeled Angular dimensions. Units format sets the angular units format, and Precision sets the precision for the units. The Zero suppression area controls the suppression of zeros in angular dimensions.

3. Make changes to the settings in the Angular dimensions area.

Alternate Units

For some drawings, you may need to show dimensions in more than one type of unit. **Alternate units** are a second set of distances inside brackets in dimensions. The most common reason for using alternate units is to show both inches and millimeters, so millimeters are shown by default when you activate alternate units. However, the conversion factor is customizable, as you will see, so that you can change the alternate units to meet needs for specific drawings.

1. Pick the **Alternate Units** tab and experiment with the settings in this area.
2. Pick the **Cancel** button; pick the **Close** button to close the Dimension Style Manager dialog box.

Editing a Dimension's Properties

AutoCAD offers additional options for editing a dimension.

1. Select the **9′-10″** aligned dimension on the right side of the deck, right-click to display the shortcut menu, and review the editing options.

Dimension Style allows you to select from the named styles or to create a new style based on the changes you make to the dimension. Under the Precision option, you can change the linear precision of the dimension. Remove Style Overrides removes any overrides you have made to a dimension or dimension style for the selected dimension.

2. Press **ESC** to exit the shortcut menu, then rest the crosshairs on one of the arrow grips.
3. Pick **Flip Arrow** from the shortcut menu.
4. Repeat for the other arrow.
5. Right-click and select **Properties** near the bottom of the shortcut menu.

AutoCAD displays the Properties palette.

6. If information is not displayed under the General heading, click the down arrow located to the right of General.

AutoCAD provides a list of the dimension's properties.

7. Change the color and lineweight to new values of your choice.

8. Display the Text information by clicking the down arrows located to the right of the Text heading.

This displays the settings associated with the dimension text.

9. Change the height to **3/16″**.

10. Open each of the other headings, review their contents, and then close them.

As you can see, you can easily tailor the properties of an individual dimension. In most circumstances, all of the dimensions on a drawing should use a consistent style, but AutoCAD gives you the option of changing the individual elements of a particular dimension if necessary.

11. Close the Properties palette.

Exploding a Dimension

1. Attempt to erase a single element of any dimension on the drawing, such as the dimension text or an extension line.

MODIFY

AutoCAD selects the entire dimension because it treats an associative dimension as a single object.

2. Press the **ESC** key and pick the **Explode** button on the Modify panel of the Ribbon.

This enters the EXPLODE command.

3. Select any dimension and press **ENTER**.

AutoCAD explodes the dimension, enabling you to edit (move, erase, trim, etc.) the individual parts of the dimension.

4. Select the dimension.

As you can see, it is now possible to select individual parts of the dimension. Sometimes, it is necessary to explode a dimension to edit it because it is impossible to change it by any other way. Use this method, however, as the last alternative because exploded dimensions lose their associativity and you cannot control their appearance using a dimension style.

5. Undo the Explode, save your work, and exit AutoCAD.

• REVIEW QUESTIONS

Answer the following questions on a separate sheet of paper.

1. Why is it important for AutoCAD to offer so many settings that control the appearance of dimensions?

2. What are alternate units?

3. Describe a drawing situation in which you might need to rotate dimension text.

4. What is the purpose of the New option of the Dimension Style Manager dialog box?

5. How do you adjust the angle of a dimension's extension lines? Why might you choose to make this adjustment?

6. What is the fastest way of reviewing and changing the properties of an individual dimension?

7. Under what circumstances would you want to suppress one or both dimension lines? Explain.

• CHALLENGE YOUR THINKING

These questions are designed to further your knowledge of AutoCAD by encouraging you to explore the concepts presented in this chapter. Answer each question on a separate sheet of paper.

1. A drawing of a complex automotive part has been completed with AutoCAD using the English inch system. Without redimensioning the drawing, how could you change it to show metric units only?

2. Contact two different companies in your area that use CAD and find out what their company dimensioning standards are. How do the two companies compare? How might any differences be explained?

Continued

● APPLYING AUTOCAD SKILLS

Work the following problems to practice the commands and skills you learned in this chapter.

1. Create a new drawing named prb28-1.dwg. Plan for a drawing scale of 1″ = 1″ and sheet size of 11″ × 17″. After you apply the following settings, create and dimension the drawing of an aircraft door hinge component shown in Fig. 28-2. In addition to saving prb28-1.dwg as a drawing file, save it as a template file.

   Snap: .25
   Grid: 1
   Layers: Create layers to accommodate multiple colors, linetypes, and line-weights.
   Linetype scale: .5
   Font: romans.shx
   Dimension style name: Heather
   Dimension scale: 1
   Dimension text height: .16
   Arrowhead size: .16
   Center mark size: .08
   Mark with center lines? Yes

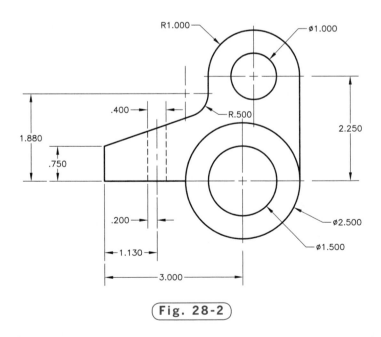

Fig. 28-2

2. Draw and dimension the front view of the oscillating follower shown in Fig. 28-3 on page 400. Use the drawing settings shown below. Draw the top and right-side views, assuming the follower to have a thickness of 8. This is a metric drawing.

Units: Decimal with no digits to the right of the decimal
Drawing area: 280 × 216
Grid: 10
Snap: 5
Linetype scale: Start with 10
Text style: Use the romans.shx font to create a new text style and set the height at 3.
Layers: Set up layers as shown in Table 28-1.

Table 28-1. Layers

Layer Name	Color	Linetype	Lineweight
Visible	red	Continuous	0.60 mm
Dimensions	blue	Continuous	0.20 mm
Text	magenta	Continuous	0.35 mm
Center	green	CENTER	0.20 mm
Hidden	magenta	HIDDEN	0.35 mm

Create two dimension styles named Textin and Textout. Use Textin for those dimensions in which the text is inside the dimension lines, and Textout for the few radial dimensions in which the text should be outside the extension lines. Use the Standard settings except for the following:

Arrowhead size: 3
Text size: 3
Distance to extend extension line beyond dimension line: 1.5
Distance the extension lines are offset from origin points: 1.5
Distance around the dimension text: 1.5
Color of dimension text: magenta

Set the dimension style to Textin and Textout as appropriate for each of the two styles.

Continued

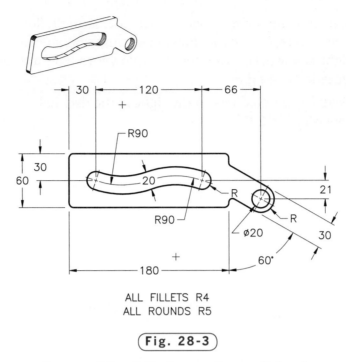

Fig. 28-3

Courtesy of Gary J. Hordemann, Gonzaga University

3. Dimension the gasket as shown in Fig. 28-4. Set the arrowheads to Integral and the text font to Times New Roman. Display alternate units below the primary values and in the same unit format. Save the drawing as ch28gasket.dwg.

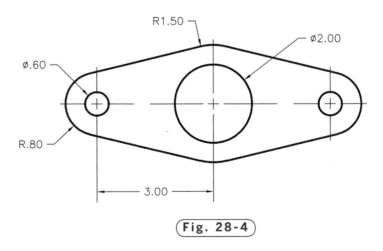

Fig. 28-4

4. Create a new drawing using the prb28-1.dwt template file and make the changes shown in Fig. 28-5. Note that the 2.250 diameter is now 2.500. Stretch the top part of the hinge component upward .250 units and let the associative dimension text change on its own.

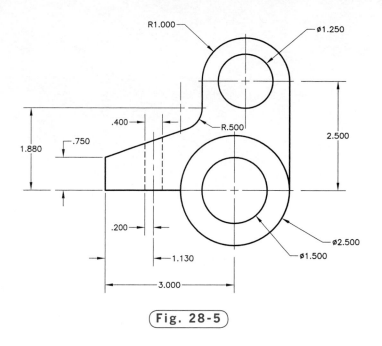

Fig. 28-5

USING PROBLEM-SOLVING SKILLS

Work the following problems to practice the commands and skills you learned in this chapter.

1. Architects are noted for their individuality. Open elev1.dwg from Chapter 26. Redimension the elevation, including text, with the following style modifications. Save the drawing as ch28elev-rev1.dwg.

Text font name: Trekker
Arrowheads: Architectural tick
Dimension lines: Extend beyond ticks 1/8
Vertical text placement: Outside

2. Dimension the shim shown in Fig. 28-6. Set the font for Times New Roman. Apply Override as necessary. Save the drawing as ch28shim.dwg.

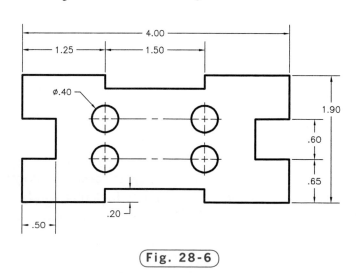

Fig. 28-6

Chapter 29

Tolerancing

Objectives

- Apply the symmetrical, deviation, limits, and basic methods of tolerancing
- Insert surface controls on a drawing
- Create geometric characteristic symbols and feature control frames that follow industry standard practices for geometric dimensioning and tolerancing
- Add material condition symbols, datum references, data identifiers, and projected tolerance zones to drawings in accordance with industry standards

Vocabulary

basic dimension
deviation tolerancing
feature control frames
geometric characteristic symbols
geometric dimensioning and tolerancing (GD&T)
limits tolerancing
material condition symbols
Maximum Material Condition (MMC)
projected tolerance zone
surface controls
symmetrical deviation
symmetrical tolerancing
tolerance

AutoCAD permits you to add tolerances to dimensions. A **tolerance** specifies the largest variation allowable for a given dimension. Tolerances are necessary on drawings that will be used to manufacture parts because some variation is normal in the manufacturing process.

This chapter applies AutoCAD's tolerancing capabilities and provides a basic introduction to **geometric dimensioning and tolerancing (GD&T)**. This is a system by which drafters and engineers define tolerances for the location of features (such as holes) of a part. The guidelines in this chapter reflect the standards that have been adopted by the American National Standards Institute (ANSI).

Methods of Tolerancing

Tolerances can be shown in more than one way, so AutoCAD provides many options for different kinds and styles of tolerances. The method used depends on the individual drawing. In some cases, you can place the tolerance information in a note or title block. However, when many different tolerances apply to different parts of the drawing, you may need to specify them individually with the dimension information.

1. Start AutoCAD and use the **dimen.dwt** template file to create a new drawing.

2. Select the **AutoCAD Classic** workspace, close all floating toolbars and palettes, and display the **Dimension** toolbar.

3. Pick the **Dimension Style** button from the toolbar and pick the **New...** button.

4. Create a new dimension style named **toler**. Start with Standard and use it for all dimensions.

5. Pick the **Text** tab and change the text height to **0.125**.

6. Pick the **Primary Units** tab and set Precision to **0.000**.

7. Pick the **OK** button; pick the **Close** button.

8. Make **toler** the current dimension style.

9. Save your work in a file named **toler.dwg**.

10. Pick the **Dimension Style** button from the toolbar and pick the **Modify...** button.

11. Pick the **Tolerances** tab to display the tolerance options.

DIMENSION

Focus your attention on the area titled Tolerance format. Notice that many of the settings are grayed out and not available. This is because Method is set to None. Method refers to the approach used to calculate the tolerance. The four methods that AutoCAD makes available, along with an example, are shown in Fig. 29-1 on page 404.

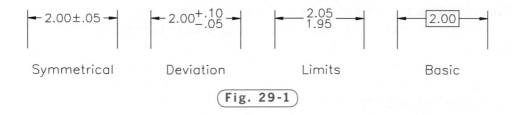

Symmetrical   Deviation   Limits   Basic

Fig. 29-1

The Symmetrical option permits you to create a plus/minus expression tolerance. AutoCAD applies a single value of variation to the dimension measurement. In other words, the upper and lower tolerances are equal. Deviation allows you to add different values for the upper and lower tolerance for cases in which the two are not equal. Limits adds a limit dimension with a maximum and a minimum value—one on top of the other. Basic creates a plain dimension inside a box. A **basic dimension** is one to which allowances and tolerances are added to obtain limits.

12. In the box to the right of Method, pick each of the options and notice how the sample drawing changes.

Precision sets the number of decimal places for tolerances. Precision varies with the tolerance precision and the numbering system you use. If you are using the Imperial system (inches, feet, etc.), the basic dimension should have the same number of decimal places as the tolerance. For example, a 1.5-inch hole with a tolerance of ±.005 should be written as 1.500 ±.005. Note that leading zeros (zeros to the left of the decimal point) are often not used with Imperial units.

Upper value sets the maximum or upper tolerance value, and Lower value sets the minimum or lower value. Scaling for height sets the current height for tolerance text. Vertical position controls text justification for symmetrical and deviation tolerances. Zero suppression works the same way as it does for primary and alternate units.

The settings in the Alternate unit tolerance area sets the precision and zero suppression rules for alternate tolerance units. Zero suppression works the same way for tolerancing as it does for primary and alternate units.

13. Experiment with each of the options in the **Tolerance** tab and notice how their settings affect the sample drawing.

14. Pick the **Cancel** button to clear all changes that you made.

Symmetrical Tolerancing

In **symmetrical tolerancing** (also called **symmetrical deviation**), the dimension specifies a plus/minus (±) tolerance in which the upper and lower limits are equal.

1. Pick the **Modify...** button and pick the **Tolerances** tab.

2. Set Method to **Symmetrical** and Precision to **0.000**.

3. Set Upper value to **0.005**.

Lower value is not available because it is the same as the upper value by definition in symmetrical tolerancing.

4. Under Zero suppression, check both **Leading** and **Trailing**, and do not change anything else.

5. Pick **OK**; pick **Close**.

Tolerances do not appear yet because the toler dimension style was created after the drawing was completed, and toler has not yet been applied to the drawing.

6. Select the drawing and dimensions.

7. Make **toler** the current dimension style and press **ESC** to remove the selection.

Your drawing should look similar to the one in Fig. 29-2.

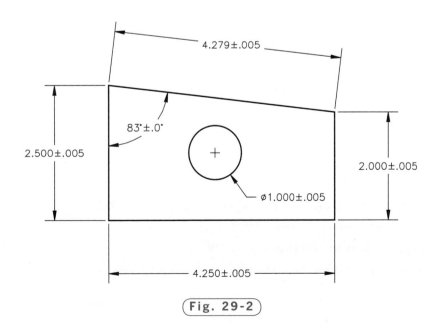

4.279±.005

83°±.0°

2.500±.005

2.000±.005

ø1.000±.005

4.250±.005

Fig. 29-2

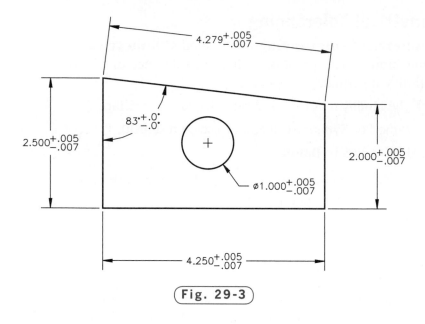

Fig. 29-3

Deviation Tolerancing

Sometimes the upper and lower tolerances for an object are unequal. AutoCAD's **deviation tolerancing** method allows you to set the upper and lower values separately.

1. Open the **Dimension Style Manager**, pick the **Modify...** button, and display the **Tolerances** tab of the Modify Dimension Style dialog box.

2. Change Method to **Deviation**, set Lower value to **0.007**, and leave everything else the same. Pick **OK**.

3. Close the Dimension Style Manager and notice how the drawing changes.

Your drawing should now look similar to the one in Fig. 29-3.

Limits Tolerancing

In **limits tolerancing**, only the upper and lower limits of variation for a dimension are shown. The basic dimension is not shown.

1. Open the **Dimension Style Manager** and display the **Tolerances** tab in the Modify Dimension Style dialog box.

2. Change Method to **Limits** and pick **OK**.

3. Close the Dimension Style Manager and notice how the drawing changes.

Your drawing should now look similar to the one in Fig. 29-4.

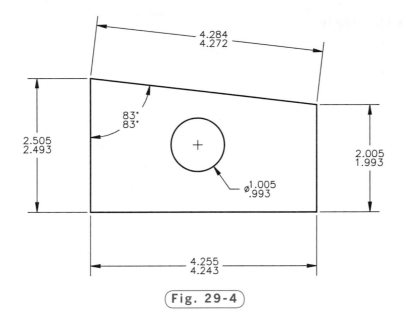

Fig. 29-4

Basic

When you select the Basic method, AutoCAD draws a box around the dimension. As you may recall, a basic dimension is one to which allowances and tolerances are added to obtain limits. Therefore, you should use the Basic method to present theoretically exact dimensions.

1. In the Dimension Style Manager, display the **Tolerances** tab of the Modify Dimension Style dialog box.

2. Change Method to **Basic** and pick **OK**.

3. Close the Dimension Style Manager and notice how the drawing changes.

GD&T Practices

With AutoCAD, you can create geometric characteristic symbols and feature control frames that follow industry standard practices for geometric dimensioning and tolerancing (GD&T). **Geometric characteristic symbols** are used to specify form and position tolerances on drawings. **Feature control frames** are the frames used to hold geometric characteristic symbols and their corresponding tolerances.

Surface Controls

When a feature control frame is not associated with a specific dimension, it usually refers to a surface specification regardless of feature size. For this reason, such frames are often known as **surface controls**.

In Chapter 27, you created leaders using the MULTILEADER command. To create leaders for surface controls, you must use the LEADER command.

1. Create a new layer named **GDT** and assign the color magenta to it. Make it the current layer and freeze layer **Dimensions**.
2. Enter **LEADER** at the keyboard.
3. In reply to **specify leader start point**, pick the midpoint of the inclined line.

 Use the Midpoint object snap.

4. Pick a second point about 1 unit up and to the right of the first point to specify the second point of the leader line.
5. Pick a third point to the right of the second point.
6. Press **ENTER** twice to reach the **Annotation Options** and then select the **Tolerance** option.

The Geometric Tolerance dialog box appears, as shown in Fig. 29-5.

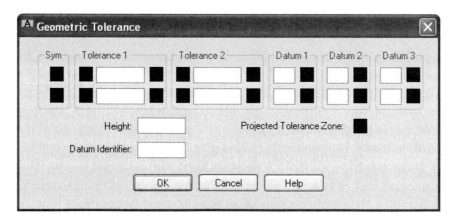

Fig. 29-5

This dialog box permits you to create complete feature control frames by adding tolerance values and their modifying symbols.

Focus your attention on the area labeled Sym.

7. Pick the top black box under **Sym**.

This displays the Symbol dialog box, as shown in Fig. 29-6. The dialog box contains 14 geometric symbols that are commonly used in GD&T.

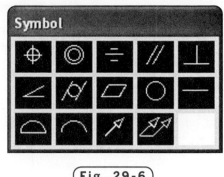

Fig. 29-6

8. Pick the flatness symbol (the third symbol on the second row).

AutoCAD adds the symbol to the black box you picked under Sym.

9. Under **Tolerance 1**, in the top white box, type **.020** for the tolerance value.

10. Pick **OK**.

The feature control frame appears as a part of the leader, as shown in Fig. 29-7. The feature control frame specifies that every point on the surface must lie between two parallel planes .020 apart.

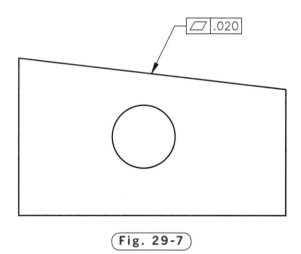

Fig. 29-7

Geometric Characteristic Symbols

To associate geometric symbols with specific dimensions, you must first create the dimension using the appropriate dimensioning command. Then you use the TOLERANCE command to add the tolerancing information.

1. Pick the **Diameter** button from the Dimension toolbar and pick a point anywhere on the hole.

2. Pick a second point to position the dimension as shown in Fig. 29-8 on page 410.

Let's change the method of tolerance from Basic to Limits.

3. Select the new dimension, right click, and pick **Properties** from the shortcut menu.

DIMENSION

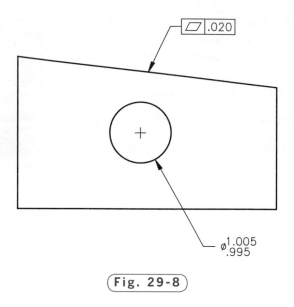

Fig. 29-8

4. In the Properties palette, click the down arrow at the right of Tolerances, change Tolerance display to **Limits**, and close the Properties palette.

5. Press **ESC** to remove the selection.

DIMENSION

Your drawing should now look similar to the one in Fig. 29-8.

6. Pick the **Tolerance...** button from the Dimension toolbar.

This enters the TOLERANCE command and displays the Geometric Tolerance dialog box.

7. Under Sym, pick the top black box and pick the true position symbol located in the upper left corner of the Symbol box.

8. Under Tolerance 1, pick the black box located in the upper left corner.

The diameter symbol appears.

9. Pick it again, and again, causing it to disappear and reappear.

As you can see, this serves as a switch, toggling the diameter symbol on and off.

10. In the top white box under Tolerance 1, type **.010** for the tolerance value.

Material Condition Symbols

The black box located at the right of the white text box specifies material condition. **Material condition symbols** are used in GD&T to modify the geometric tolerance in relation to the produced size or location of the feature.

In this case, we will use the symbol for **Maximum Material Condition (MMC)**. MMC specifies that a feature, such as a hole or shaft, is at its maximum size or contains its maximum amount of material.

11. At the right of the same white box, pick the black box.

This displays the Material Condition dialog box, as shown in Fig. 29-9.

12. Pick the first symbol, which is the symbol for Maximum Material Condition.

The symbol appears in the Geometric Tolerance dialog box.

Material Condition

Fig. 29-9

Datum References and Datum Modifiers

AutoCAD allows you to specify datum reference information on three different levels (primary, secondary, and tertiary).

1. In the top white boxes under Datum 1, Datum 2, and Datum 3, enter **A**, **B**, and **C**, respectively.

You can also associate a material condition with any or all of the datum references. To do this, you would pick the black box to the right of the datum boxes and select the appropriate material condition symbol. For the current drawing, do not use material condition symbols.

Datum Identifier near the bottom left of the dialog box allows you to enter a feature symbol to identify the datum.

2. In the box located to the right of Datum Identifier, enter **-D-**.

The Geometric Tolerance dialog box on your screen should now look like the one in Fig. 29-10.

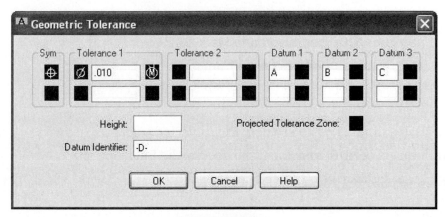

Fig. 29-10

The black and white boxes in the second row permit you to create a second feature control frame. You would need to use a second feature control frame when the feature has two geometric characteristics that are of dimensional importance.

3. Pick the **OK** button and position the feature control frame as shown in Fig. 29-11.

4. Save your work.

See Appendix D for more information about geometric symbols.

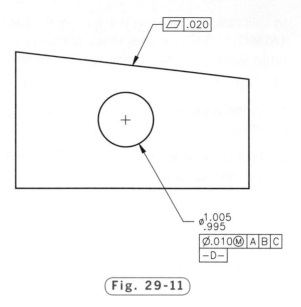

Fig. 29-11

Projected Tolerance Zone

You can specify projected tolerances in addition to positional tolerances to make a tolerance more specific. The **projected tolerance zone** controls the height of the extended portion of a perpendicular part. To display a projected tolerance, you would pick the black box next to Projected Tolerance Zone in the Geometric Tolerance dialog box. Then, under the Tolerance 1 area of the dialog box, enter a value for the height. Doing so creates a projected tolerance zone in the feature control frame.

Experiment with placing a projected tolerance zone in a feature control frame by following the steps below.

1. Enter the **UNDO** command and **Mark** option.

2. Create another feature control frame.

3. Pick the black box next to Projected Tolerance Zone and enter a height in the Height box.

4. Place a dimension to see how the projected tolerance appears.

5. Enter the **UNDO** command and specify the **Back** option.

6. Close the **Dimension** toolbar and exit AutoCAD.

● REVIEW QUESTIONS

Answer the following questions on a separate sheet of paper.

1. What is the purpose of adding tolerances to a drawing?

2. When is it necessary to include tolerances on a drawing?

3. Give an example for each of the following tolerancing methods.

   a. Symmetrical

   b. Deviation

   c. Limits

4. Which tolerancing method should you choose in AutoCAD if the upper tolerance value of a dimension is different from the lower value?

5. What is a surface control? How can you create a surface control in AutoCAD?

6. What is a geometric characteristic symbol?

7. Explain how you would include material condition symbols in feature control frames.

8. How would you include a second feature control frame below the first one?

● CHALLENGE YOUR THINKING

These questions are designed to further your knowledge of AutoCAD by encouraging you to explore the concepts presented in this chapter. Answer each question on a separate sheet of paper.

1. With AutoCAD, you can quickly produce feature control frames, complete with geometric characteristic symbols. Investigate ways of editing them.

2. When a second feature control frame is added to a dimension, how does this affect the meaning of the first feature control frame?

3. Investigate material condition symbols and their meanings. In what way do material conditions affect dimensions?

Continued

● APPLYING AUTOCAD SKILLS

Work the following problems to practice the commands and skills you learned in this chapter.

1. Create the two-view drawing of the plug according to the dimensions shown in Fig. 29-12. Use a drawing area of 100,80, and name the drawing plug.dwg. Dimension it as shown in Fig. 29-13 to show the runout of the plug. (*Runout* is the form and location of a feature relative to a datum.) Notice that the limits are different on the two holes.

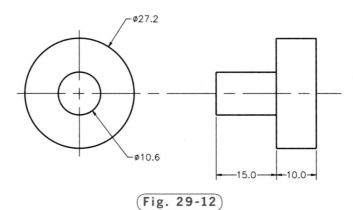

Fig. 29-12

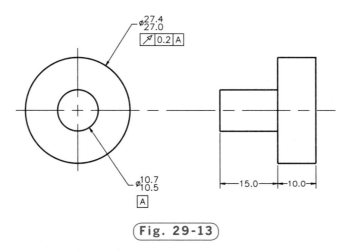

Fig. 29-13

2. Create a new drawing named ch29gasket.dwg and draw the gasket shown in Fig. 29-14. Dimension the gasket using limits tolerancing, and specify an upper value of .02. Create a feature control frame to show a true position diameter tolerance of .025 at maximum material condition.

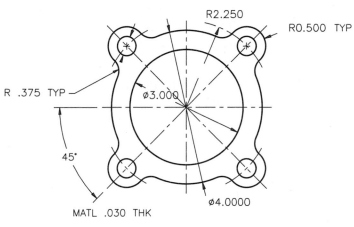

4X ⌀0.500 EQUALLY SPACED

R2.250

R0.500 TYP

R .375 TYP

⌀3.000

45°

MATL .030 THK

⌀4.0000

UNLESS OTHERWISE SPECIFIED ALL
DIMENSIONS ARE IN INCHES

Fig. 29-14

3. Create a new drawing named hinge.dwg and draw the aircraft door hinge component shown in Fig. 29-15. Dimension the hinge using symmetrical tolerancing and specify tolerances as shown.

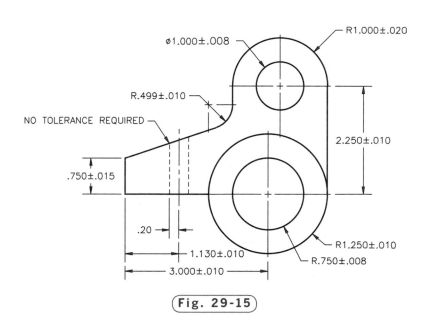

R1.000±.020

⌀1.000±.008

R.499±.010

NO TOLERANCE REQUIRED

2.250±.010

.750±.015

.20

1.130±.010

R1.250±.010

R.750±.008

3.000±.010

Fig. 29-15

Continued

● USING PROBLEM-SOLVING SKILLS

Complete the following activities using problem-solving skills and your knowledge of AutoCAD.

1. The drawing you dimensioned in Chapter 26, slide2.dwg, will be used in a new CAD textbook to illustrate dimensioning styles. Save the drawing with a new name of slide3.dwg. Create two new layers named Symmetrical and Deviation. Freeze the layer on which the current dimensions were drawn. Make the Symmetrical layer current and redimension the drawing with tolerances expressed in the symmetrical style. Then freeze Symmetrical, make Deviation current, and redimension using the deviation style. The tolerance on the angle is ±15 units, on the radius +0.02° and −0.00°, and all others ±.05 unit. Plot the drawing twice to show the two different dimensioning methods.

2. Your editor also wants an example of GD&T with limits dimensioning. Save slide3.dwg as slide4.dwg. Redimension the drawing to maintain the 3.0-unit vertical side perpendicular to the 5.5-unit base within 0.02 units. Save the drawing.

Prototype Testing Expert

Traditional prototype testing is a long and expensive proposition. First, the physical prototype has to be built at full size using the materials to be used in the final product. Then the prototype has to be tested—a process that often involves expensive testing devices and sometimes the actual destruction of the prototype.

Now, using computers and high-end CAD modeling and testing software, companies can carry out much of their testing on virtual prototypes before a physical prototype is created.

Courtesy of Objet Ltd.

Advantages of Virtual Prototypes

Testing a virtual prototype has many advantages. It reduces costs and shortens design time by allowing designers to eliminate design flaws without having to build another prototype each time a major change is made to the product.

But testing virtual prototypes also increases safety. For example, in the illustration, air flow around a hovering fighter plane is being analyzed using a computer simulation. The pattern of air flow caused by the jet's movable exhaust nozzles is critical. If hot air from the exhaust is sucked into the plane's engines, the plane may crash. Using a physical prototype for this test would put pilots at serious risk.

Testing Experts

In addition to understanding the software, prototype testers should have a solid background in math and perhaps even engineering in the field in which they are testing. They must have an eye for detail to catch the slightest indication of possible design flaws.

▶ Career Activities

1. Choose a manufacturer of a vehicle (car, plane, train, space vehicle, etc.) and find out how the company tests prototypes for new designs.

2. Research job openings in this field. What skills would you need to become a prototype tester?

Calculations

Objectives

- Find the exact coordinates of points
- Calculate the distance between specific points in a drawing
- Calculate area and circumference of objects in a drawing
- Use AutoCAD's online geometry calculator to place objects at precise points in a drawing
- Use **QuickCalc** to perform arithmetic, algebraic, and trigonometric calculations
- Display information from AutoCAD's drawing database on objects and entire drawings
- Divide an object into equal parts
- Place markers or points at specified intervals on an object

Vocabulary

area
circumference
delta
drawing database
integer
perimeter
vector

With the power of the computer, AutoCAD can perform measurement and calculation tasks that would be too time-consuming to do manually. An example is calculating miles of a chain-link fence on a drawing.

The drawing in Fig. 30-1 shows an apartment complex with parking lots, streets, and trees. With AutoCAD, you can calculate the square footage of the parking lot and the distance between parking stalls on such a drawing.

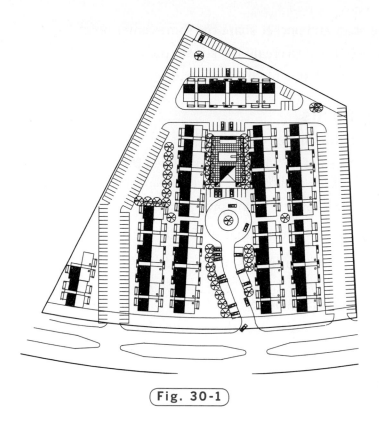

(Fig. 30-1)

Performing Calculations

AutoCAD includes commands that help you find the coordinates of specific points on a drawing, calculate the distance between two points, calculate the area of an object, and perform other geometric calculations.

1. Start AutoCAD and start a new drawing using the **dimen.dwt** template file.

2. Select the **Drafting & Annotation** workspace.

3. Erase the drawing and dimensions.

4. Create the drawing of the end view of a shaft with a square pocket machined into it (Fig. 30-2). Use the following guidelines:

   • Create the end view on the layer named Objects.

   • Use the sizes shown, but omit dimensions.

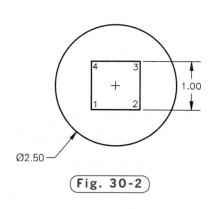

(Fig. 30-2)

- The shaft and pocket share the same center point.
- Use RECTANG to create the square pocket.

5. Save your work in a file named **calc.dwg**.

Locating Points

Let's find the coordinates of point 1.

UTILITIES

1. Pick the **ID Point** button on the expanded Utilities panel or type on the Home tab of the Ribbon **ID** at the keyboard to enter the ID command.

2. In response to Specify point, pick point **1**.

AutoCAD displays the coordinates of point 1.

3. Determine the coordinates of point 2.

Calculating Distances

The DIST command calculates the distance between two points.

UTILITIES

1. On the Utilities panel, select the **Distance** button.

AutoCAD enters the **DIST** command.

2. Pick points **1** and **3**.

In addition to the distance, AutoCAD calculates angle and delta information. If you have studied geometry, you may be familiar with the use of the term **delta** to mean "change." Delta X is the distance, or change, in the X direction, and Delta Y is the distance in the Y direction from one point to another.

3. Enter **X** to exit the DIST command.

Calculating Area

The AREA command determines the area and perimeter of several different object types. **Area** is the number of square units (inches, acres, miles, etc.) needed to cover an enclosed two-dimensional shape or surface. **Perimeter** is the distance around a two-dimensional shape or surface.

UTILITIES

1. Select the **Area** button from the flyout menu of the Utilities panel.

AutoCAD enters the AREA command.

2. Select the **Object** option or enter **O** and pick the square pocket.

AutoCAD displays the area and perimeter of the pocket. Suppose we want to know the area of the end of the shaft minus the pocket.

3. Press **ESC,** select the **Area** button again, and select the **Add area** option or enter **A**.

This puts the command in Add mode.

4. Select the **Object** option and pick the shaft.

AutoCAD displays 4.9087 for the area and 7.8540 for the circumference. (**Circumference** is the distance around a circle.) Notice that AutoCAD is still in Add mode and is asking you to select objects.

5. Press **ENTER**.

6. Select the **Subtract** option.

Now the command is in Subtract mode.

7. Select the **Object** option, select the pocket, and press **ENTER**.

AutoCAD subtracts the area of the pocket from the area of the end of the shaft and displays the result (3.9087).

8. Press **ESC** to terminate the command.

Using the Geometry Calculator

With the CAL command, AutoCAD offers a geometry calculator that evaluates vector, real, and integer expressions. A **vector** is a line segment defined by its endpoints or by a starting point and a direction in 3D space. An **integer** is any positive or negative whole number or 0. You can use the expressions in any AutoCAD command that requests points, vectors, or numbers.

1. Enter the **CAL** command.

2. Type **(3*2)+(10/5)** at the keyboard and press **ENTER**.

As in basic algebra, the parts in parentheses are calculated first. The asterisk (*) indicates to multiply and the forward slash indicates to divide the numbers.

AutoCAD calculates the answer as 8 and displays the answer in the Command Line window. You can also use CAL with object snaps to calculate the exact placement of a point. For example, suppose you wanted to add a feature to the shaft that required starting a polyline halfway between the center of the shaft and the lower left corner of the machined pocket.

3. Enter the **PLINE** command.

4. Enter **'CAL**. (Note the leading apostrophe.)

By preceding the command with an apostrophe, you have entered the command transparently. In other words, when you complete the CAL command, the PLINE command will resume.

5. Type **(cen+end)/2** and press **ENTER**.

6. Pick any point on the shaft.

7. Pick either line near point 1.

AutoCAD calculates the midpoint between the shaft's center and point 1 and places the first point of the polyline at that location.

8. Press **ENTER** to terminate the PLINE command.

9. Undo the polyline and save your work.

Using QuickCalc

The QUICKCALC command allows you to perform a full range of mathematical and trigonometric calculations, as well as determine graphic information such as the location of a point or the length of a line.

1. Select the **Open...** button from the Quick Access toolbar and open the **deck.dwg** file.

2. Pick the **Quick Calculator** button from the Utilities panel, or type **QUICKCALC** at the keyboard.

UTILITIES

This displays the Quick Calculator palette, as shown in Fig. 30-3. (If the calculator buttons are not displayed on your screen, pick the down arrow button in the lower right corner of the palette.) The buttons across the top of the palette allow you to find the coordinates of a point, the distance between two points, the angle of a line, and the intersection point of two lines.

3. Pick the **Get Coordinates** button and select the upper right corner of the deck.

The *x,y,z* coordinates of the point are displayed.

4. Pick the **Clear** button to clear the calculator's display.

Fig. 30-3

5. Pick the **Distance Between Two Points** button.

6. Pick the upper left and lower right corners of the deck.

The distance between the two corners of the deck is displayed: 14′-6-3/8″.

(AutoCAD may display the decimal inch equivalent: 174.413302.)

7. Pick the **Clear** button.

Now focus your attention on the QuickCalc Number Pad.

8. Experiment with the calculator by entering mathematical equations, just as you would with an ordinary calculator.

 You may have to drag the bottom line of the calculator down to enlarge the window so that you can see the entire calculator.

Notice that the equations and solutions are stored in a history list at the top of the QuickCalc palette. More functions are located below the Number Pad in the Scientific area.

9. Pick the arrow to the right of Scientific and review the trigonometric and algebraic functions available there.

Now focus your attention on the Units Conversion area. QuickCalc allows you to obtain values for different units of measurement.

10. Pick the arrow to the right of Units Conversion to display the Units Conversion area.

11. Convert an area of 15 square feet to square meters.

 To access the Square feet and Square meters conversion options, you must first select Area in the Units type drop-down box.

12. Close the **QuickCalc** palette and close the **deck.dwg** file without saving.

Displaying Database Information

AutoCAD maintains an internal database on every drawing. The **drawing database** contains information about the types of objects in the drawing, defined layers, and the current space (model or paper). It also contains specific numerical information about individual objects and their placement in the drawing. This information can be useful to drafters and programmers using AutoCAD for a variety of tasks.

PROPERTIES

Listing Information About Objects

1. Pick the **List** button from the expanded Properties panel.

2. Pick any point on the shaft and press **ENTER**.

The AutoCAD Text window displays the object type, layer, space, center point, radius, circumference, and area.

3. Close the window.

4. Select the shaft.

5. Right-click and pick **Properties** from the shortcut menu.

AutoCAD displays most of the same information, but does not give the space. The benefit of this list is that you can make changes to the object, such as changing its circumference and area.

6. Pick the text field to the right of Area.

Notice the calculator button that appears on the right side of the Area line. This provides you with another way to open QuickCalc.

7. Pick the calculator button to open the QuickCalc palette, then close it.

8. In the Properties palette, change the area of the shaft to **3**.

AutoCAD redraws the shaft with an area of 3 units.

9. Close the Properties palette and undo Step 8.

Editing Objects Mathematically

You can apply the same power that AutoCAD uses to calculate coordinates, areas, and circumferences to edit objects with mathematical precision. You can use this power to divide objects into equal parts and to measure exact intervals along a line or arc.

Dividing an Object into Equal Parts

The DIVIDE command divides an object, such as the end of the shaft, into a specified number of equal parts.

1. Expand the Draw panel in the Home tab of the Ribbon, then pick the **Divide** button.

DRAW

This enters the DIVIDE command.

2. Select the shaft.

3. Enter **20** for the number of segments.

It appears as though nothing has happened. Something did happen: the DIVIDE command divided the end of the shaft into 20 equal parts using 20 points; you just can't see them. Here's how to display them.

4. Turn off snap and enter the **LINE** command.

5. Right-click on the **Object Snap** button on the status bar and pick the **Node** object snap from the shortcut menu. (Node is used to snap to the nearest point.)

6. Move the crosshairs along the shaft and snap to one of the nodes.

7. Snap to the center of the shaft.

8. Snap to another node point on the shaft.

9. Terminate the **LINE** command.

10. Save your work.

Displaying the Points

AutoCAD's DDPTYPE command is used to control the appearance of points. Let's use it to make the points on the shaft visible.

1. Expand the Utilities panel and pick the **Point Style...** button, or enter **DDPTYPE.**

AutoCAD displays the dialog box shown in Fig. 30-4.

2. Pick the first point style in the second row—the one with a circle and a point at its center—and pick **OK**.

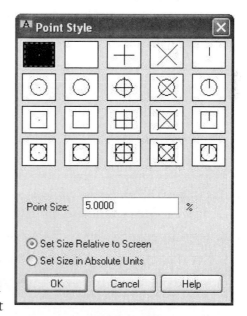

UTILITIES

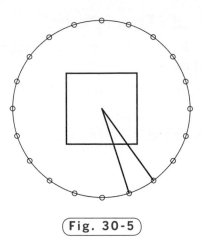

Fig. 30-5

AutoCAD places a small circle at each of the 20 equally spaced points, as shown in Fig. 30-5.

3. Display the same dialog box (press the spacebar).

4. Experiment with other point styles and adjust the point size.

5. Display the Point Style dialog box again and set the style to the last one in the top row.

6. Pick **OK** to close the dialog box.

Placing Markers at Specified Intervals

The MEASURE command is similar to DIVIDE except that MEASURE does not divide the object into a given number of equal parts. Instead, the MEASURE command allows you to place markers along the object at specified intervals.

DRAW

1. On the expanded Draw panel, pick the **Measure** button.

This enters the MEASURE command.

2. Select one of the two lines.

3. Enter **.3** units.

AutoCAD adds points spaced .3 units apart.

4. Further experiment with MEASURE.

5. Set the point style to a single dot.

6. Save your work and exit AutoCAD.

● REVIEW QUESTIONS

Answer the following questions on a separate sheet of paper.

1. Which AutoCAD commands are used to find coordinate points?
2. What information is produced with the AREA command?
3. What information is produced with the LIST command?
4. How do you calculate the perimeter of a polygon?
5. How do you find the circumference of a circle?
6. Explain how you control the appearance of points.
7. Explain the difference between the DIVIDE and MEASURE commands. Under what conditions would you use each of these commands?

● CHALLENGE YOUR THINKING

These questions are designed to further your knowledge of AutoCAD by encouraging you to explore the concepts presented in this chapter. Answer each question on a separate sheet of paper.

1. Describe one drafting situation in which you might need to use the DIVIDE command and one in which you would prefer to use the MEASURE command.
2. Investigate the difference between AutoCAD's points created when you use the POINT command and those created when you use the DIVIDE and MEASURE commands. Write a paragraph comparing and contrasting the two methods.

Continued

● APPLYING AUTOCAD SKILLS

Work the following problems to practice the commands and skills you learned in this chapter.

Draw the views of the fasteners in Figs. 30-6 through 30-8 at the sizes indicated. Omit text and dimensions. Use the appropriate commands to find the information requested. Write their values on a separate sheet of paper. Refer to the files calcprb1.dwg, calcprb2.dwg, and calcprb3.dwg located on the OLC in the Drafting & Design Problems section. These files correspond to Figs. 30-6, 30-7, and 30-8, respectively.

1. Draw the top view of the Phillips-head screw shown in Fig. 30-6 according to the dimensions shown. Then find the following information:

   • location of point A

   • distance between points A and B

   • area of the polygon

   • perimeter of the polygon

2. Draw the top view of the nut shown in Fig. 30-7 according to the dimensions shown. Then find the following information:

   • distance between A and B

   • distance between B and C

   • area of hole

   • circumference of hole

   • area of nut

   • perimeter of nut

3. What information does AutoCAD list for arc A on the screw shown in Fig. 30-8?

4. List the information AutoCAD provides for line B on the screw.

5. On the screw shown in Fig. 30-8, divide line B into five equal parts. Make the points visible.

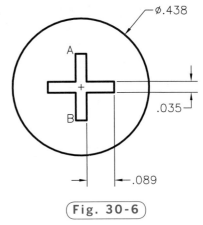

Fig. 30-6

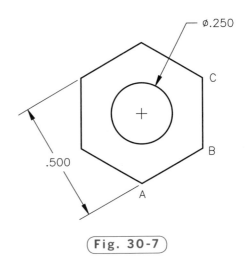

Fig. 30-7

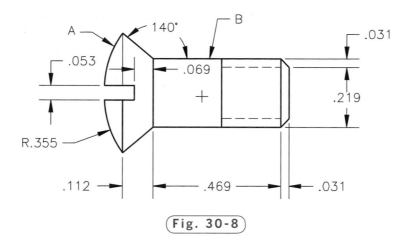

Fig. 30-8

6. On arc A in Fig. 30-8, place markers along the arc at intervals of .02 units. If the markers are invisible, make them visible.

7. Create the floor plan shown in Fig. 30-9 on page 430 according to the dimensions given. Then find the information requested below. This floor plan is also available in the Drafting & Design Problems section of the OLC for this book as flplan.dwg.

Find the following information in square feet:
- area of showroom carpet
- area of entry clay tile
- area of bathroom roll tile

As you may know from your math courses, 1 square yard contains 9 square feet. Calculate the area of the carpet, entry clay tile, and bathroom roll tile as square yards using the Units Conversion area of the QuickCalc palette.
- square yards of carpet
- square yards of clay tile
- square yards of roll tile

Calculate the distance between opposite corners of the following areas:
- showroom
- entry area
- bathroom

Continued

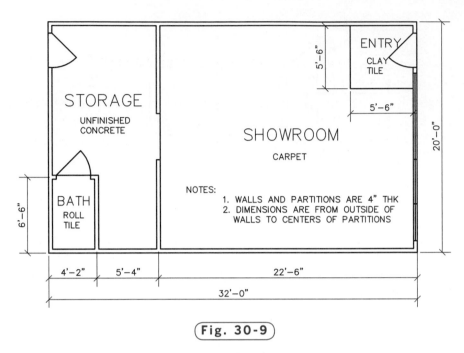

Fig. 30-9

Problem 7 courtesy of Mark Schwendau, Kishwaukee College

8. Create the top view of the nut shown in Fig. 30-10 according to the dimensions given. Then find the information requested below. This drawing is also available as 1-8UNC2B.dwg in the Drafting & Design Problems section of the OLC for this book.

Find the following information:

- distance between A and B
- area of the minor diameter of the thread
- area of the major diameter of the thread
- circumference of the minor diameter
- circumference of the major diameter
- area of the top surface of the nut
- area of the top surface of the nut minus the minor diameter
- area of the top surface of the nut minus the major diameter
- Calculate the average of the last two answers using the QUICKCALC command.

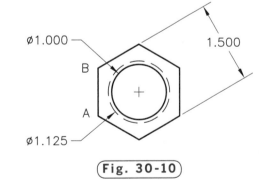

Fig. 30-10

● USING PROBLEM-SOLVING SKILLS

Complete the following activities using problem-solving skills and your knowledge of AutoCAD.

1. Create a new drawing using the Quick Setup wizard. Specify a drawing area of 300′ × 200′. Draw two horizontal lines of different lengths. Place one near the top of the drawing area and the other near the bottom. Then connect their ends to form a four-sided polygonal area. Determine the area and perimeter of the polygon.

   The area you created represents a real-estate plot. As a paralegal for an attorney, you have been asked to divide this plot into five equal lots. Determine how to do this and carry out your plan. (*Hint:* consider using the DIVIDE command and the Node object snap.) To avoid future conflict, the attorney also wants you to verify the size of each lot. Determine the area and perimeter of each lot, and prepare a statement for the attorney explaining why the areas are the same, but the perimeters are not. Save the drawing as realestate.dwg.

2. You work for a company that designs and manufactures custom clocks for wealthy international clients. You have been assigned to design a clock for a client whose office is decorated in ultramodern style. Research the ultramodern style to determine what characteristics the clock should have to fit in with the client's current decor. Then design the clock in an ultramodern style using the following customer specifications:

   • The client does not care whether the clock is a floor-standing or desk model, but she does not want a wall-mounted clock.

   • The clock must have a mark at each minute in the hour as well as at each hour.

   • The clock must have a large face so that it can be seen easily from any point in the client's 20′ × 30′ office.

Part 6 Project | Applying Chapters 26–30

Designer Eyewear

Part 6 reviewed the extensive dimensioning and calculating abilities of AutoCAD. In this exercise, you will be asked to design and draw an object and then to dimension it as specified. To review, traditional dimensions generally inform of size and location. Tolerance dimensions also yield information about characteristics or features of the item (*e.g.*, How round is the hole? How flat is the surface?).

▶ Description

Your task is to design a pair of eyeglasses. The actual style (appearance) and size are up to you. Whether latest fashion fad or very old-fashioned, the eyeglasses should be designed to be appropriate for the human face. (Feel free to inquire whether your instructor will allow a design for some other creature, real or imagined.) In any event, the chosen design must be revealed in two or three views and fully dimensioned. Of particular interest will be dimensioning information concerning the minimum and maximum tolerance permitted for the frame and lenses (such as ±.01). The nose pads or nosepiece may be separate or part of the frame itself. You may wish to measure your own eyeglasses or those of a family member or friend, or you may have someone actually measure your face to gain some insight with respect to realistic sizes. Figs. P6-1 and P6-2 may also give you some design ideas.

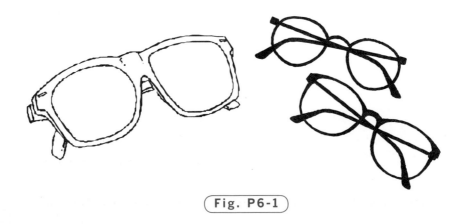

Fig. P6-1

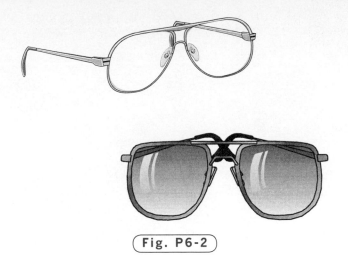

Fig. P6-2

▶ Hints and Suggestions

1. Consider showing a sketch of your idea to the instructor for feedback before investing significant time and effort into the three-view CAD version. Guidance gained early is likely to save more time than constructive criticism given upon completion.

2. Don't spend too much time on exact sizes for the eyeglasses, but once drawn, dimension details thoroughly and accurately.

3. Include dimensions for the angles of the earpieces (end pieces) as well as radii for the curvature of the lenses.

4. Consider drawing one-half of the eyeglasses; then mirror the drawing to complete the pair.

▶ Summary Questions

Your instructor may direct you to answer these questions orally or in writing. You may wish to compare and exchange ideas with other students to help increase efficiency and productivity.

1. Why might accurate dimensions be needed for one drawing and not for another?

2. Name some purposes for which tolerance dimensioning (min/max) might be considered of vital importance.

3. Once created, a dimension style can be a great timesaver. In what ways might it save time?

4. How do you determine which dimensions are needed on a drawing?

Chapter 31

Groups

Objectives

- Create a group
- Add and delete objects from a group
- Edit the group name and description
- Make a group selectable or unselectable
- Reorder the objects in a group

Vocabulary

group
selectable

A **group** in AutoCAD is a set of objects. A circle with a line through it, for instance, could become a group. Groups save time by allowing you to select several objects with just one pick. This speeds up editing operations such as moving, scaling, and erasing these objects.

Creating a Group

The GROUP command permits you to create a selection set of objects.

1. Start AutoCAD and open the drawing named **calc.dwg**.

2. Select the **AutoCAD Classic** workspace and close all floating toolbars and palettes.

3. Using **Save As...**, create a new drawing named **groups.dwg** and erase all objects.

4. Create the drawing of the pulley wheel shown in Fig. 31-1 using the following guidelines:

   - Create a layer named **Hidden** and assign the **HIDDEN** linetype and a light gray color to it. Place the hidden line on this layer.

- Place the remaining objects on layer Objects.
  - Omit the dimensions, including the center line.
  - Set **LTSCALE** to **.5**.
5. Enter the **GROUP** command.

Read the options AutoCAD presents.

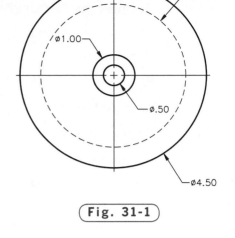

6. Select the **Name** option, type **wheel** in upper- or lowercase letters for the group's name, and press **ENTER**.

7. Select the **Description** option, type **Cast aluminum** for the group's description, and press **ENTER**.

Fig. 31-1

8. In response to the Select objects prompt, select all four objects and press **ENTER**.

In the Command Line window, notice the message **Group "WHEEL" has been created**.

9. Pick any one of the objects.

Notice that AutoCAD selects all four circles and places a rectangular bounding box around them.

10. Move the group a short distance and then press **ESC**.

11. Lock layer Hidden.

12. Pick one of the objects again and then move the group a short distance.

The gray hidden-line object did not move because it is on a locked layer.

13. Undo the move and unlock layer Hidden.

14. Enter the **UNGROUP** command and pick any object in the group WHEEL.

As you can see, the UNGROUP command allows you to explode the group. The four circles are now separate, ungrouped objects.

15. Undo the UNGROUP command.

Changing Group Properties

The GROUPEDIT command dialog box offers several features for changing a group's definition and behavior. For example, you can add and delete objects from the group or change the group's name.

Adding and Deleting Objects

1. Enter the **GROUPEDIT** command.

2. Pick any of the objects in the group WHEEL.

3. Select the **Remove objects** option.

4. Pick the gray hidden line.

AutoCAD removes this object from the group.

5. Press **ENTER**.

6. Press the spacebar to reenter the GROUPEDIT command and pick the largest circle in the group WHEEL.

Notice that the hidden-line circle is not highlighted; it is no longer part of the group.

7. Select the **Add objects** option, pick the gray hidden-line circle, and press **ENTER**.

This adds the circle back to the group.

Changing the Group Name

1. Reenter the **GROUPEDIT** command and pick any of the objects in group WHEEL.

2. Select the **REName** option and change the group's name to **PULLEY**.

Making a Group Selectable

The Object Grouping dialog box offers all the group commands we've covered thus far, as well as additional options.

1. Enter the **CLASSICGROUP** command.

This displays the Object Grouping dialog box, as shown in Fig. 31-2.

2. Select **PULLEY** in the Group Name list box.

3. In the Group Name: edit box under Group Identification, change the group name from pulley back to **wheel**.

4. Pick the **Rename** button.

5. In the Description edit box, change the group description from Cast aluminum to **Cast magnesium**.

6. Pick the **Description** button (just above the **OK** button).

This causes the group name to update, as indicated at the bottom of the dialog box.

7. In the Change Group area, pick the **Selectable** button.

In the upper right area under Selectable, notice that Yes changed to No. **Selectable** means that when you select one of the group's members (i.e., one of the circles), AutoCAD selects all members of the group.

8. Pick the **OK** button and select one of the objects in the group.

As you can see, the group is no longer selectable. Grips appear only for the specific object you select.

9. Reenter the **CLASSICGROUP** command, pick **WHEEL** in the Group Name list box, pick the **Selectable** button again, and pick **OK**.

The group is selectable once again.

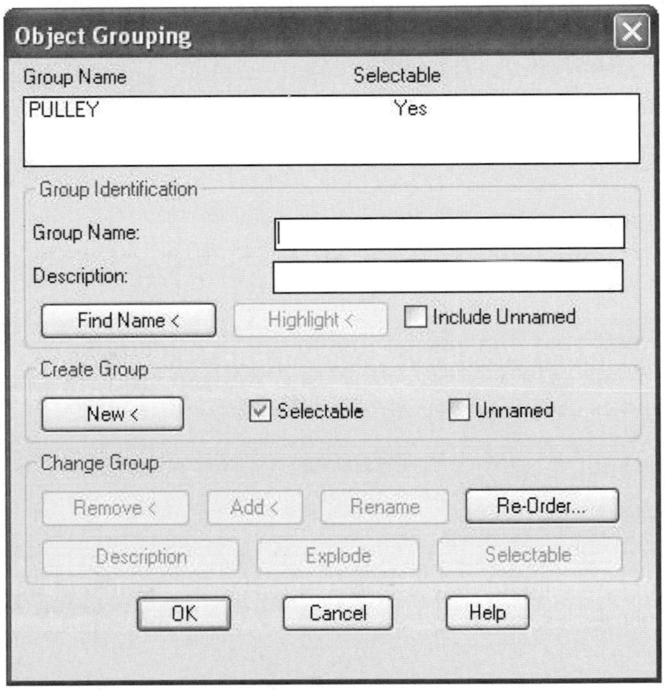

Fig. 31-2

Reordering the Objects

AutoCAD numbers objects in the order in which you select them when you create the group. In the Order Group dialog box, AutoCAD allows you to change the numerical sequence of objects within a group. Reordering can be useful when you are creating tool paths for computer numerical control (CNC) machining.

1. Enter the **CLASSICGROUP** command and select the group **WHEEL**.

2. Pick the **Re-Order...** button to make the **Order Group** dialog box appear, as shown in Fig. 31-3.

If there were more than one group in the drawing, you would be able to re-order the groups in the Order Group dialog box.

3. Pick **OK**; pick **OK** again.

4. Save your work and exit AutoCAD.

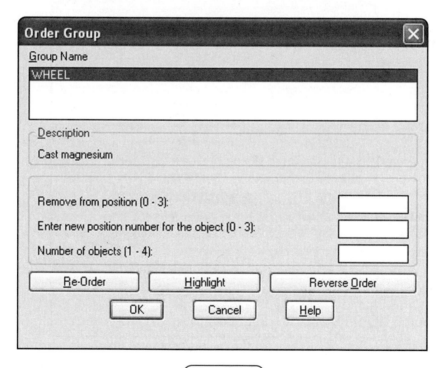

Fig. 31-3

• REVIEW QUESTIONS

Answer the following questions on a separate sheet of paper.

1. What is a group?
2. How can groups help AutoCAD users save time?
3. When creating a new group, what is the purpose of entering the group description?
4. Explain how you would add and delete objects from a group.
5. Is it possible to remove the selectable property from a group but still retain the group definition in the drawing? Explain.
6. Under what circumstances might you need to re-order the objects in a group?

• CHALLENGE YOUR THINKING

These questions are designed to further your knowledge of AutoCAD by encouraging you to explore the concepts presented in this chapter. Answer each question on a separate sheet of paper.

1. Experiment with object selection using a group name. Note that you can enter group, followed by the group name, at any Select objects prompt for moving, copying, scaling, etc. When might this feature be useful? Explain.
2. Under what circumstances might you want to make a group unselectable? Explain.

• APPLYING AUTOCAD SKILLS

Work the following problem to practice the commands and skills you learned in this chapter.

1. Perform the following tasks.

    a. Load calc.dwg from the previous chapter and use Save As... to create a file named stop.dwg. If the points are visible, make them invisible.

Continued

b. Create a group named WIDGET. Enter New product design for the description. The group should consist of all four objects.

c. Change the name of the group to WATCH and the group description to Stop watch design. Remove the two lines from the group.

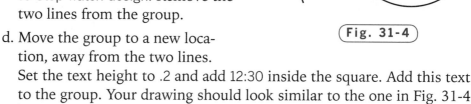

Fig. 31-4

d. Move the group to a new location, away from the two lines. Set the text height to .2 and add 12:30 inside the square. Add this text to the group. Your drawing should look similar to the one in Fig. 31-4.

• USING PROBLEM-SOLVING SKILLS

Complete the following activities using problem-solving skills and your knowledge of AutoCAD.

1. The exterior trim design of your company's administration building has not yet been approved. Open db_samp.dwg in AutoCAD's Sample folder and save it as AdmBldg.dwg in the folder with your name. Make a group of each of the three exterior entrances (two sides in turquoise and the front in yellow). Move these groups to various locations to find a better arrangement. You may also remove the five-sided structure at the rear of the building to be replaced by a possible entry. Save the design you think is the best, and be prepared to explain your choice.

2. Your company has been hired to create a sales brochure for a new bicycle. The bicycle will be available in several colors, and your customer wants the brochure to show the bicycle in each color. Research typical bicycle shapes and designs. Then create a bicycle design of your own. Use commands such as PLINE, SOLID, and FILL to make the body (frame) of the bicycle solid green. Save the drawing as bikesales.dwg. Then, using the GROUP and COPY commands, create additional copies of the bicycle and change the body colors to show a version in six colors that you think would look good on a bicycle. Use the entire Color palette to make your color selections.

Chapter 32

Blocks

Objectives

- Create and insert blocks
- Rename, explode, and purge blocks
- Organize and insert blocks using the **DesignCenter** and **Tool Palettes Window**
- Add commands to a tool palette
- Insert a drawing file into the current drawing
- Create a drawing file from a block
- Copy and paste objects from drawing to drawing

Vocabulary

block
copy and paste
purge

With AutoCAD, you seldom need to draw the same object twice. Using blocks, you can define, store, retrieve, and insert symbols, components, and standard parts without the need to recreate them. A **block** is a collection of objects that you can associate together to form a single object. Blocks are especially useful when you are creating libraries of symbols and components.

Blocks are similar to groups, but the two are different. A block definition is stored as a single, selectable object in a drawing file that you can insert, scale, and rotate. A group is a selection set of objects.

Working with Blocks

The BLOCK command allows you to combine several objects into one and then store and retrieve it for later use.

1. Start AutoCAD and start a new drawing from scratch using **Imperial** units.
2. Select the **Drafting & Annotation** workspace.
3. Draw the top view of a flathead screw as shown in Fig. 32-1. Omit the dimensions.
4. Save your work in a file named **blks.dwg**.

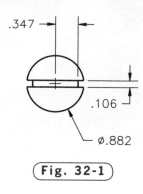

Fig. 32-1

Creating a Block

BLOCK

1. Pick the **Create** button from the Block panel on the Home tab of the Ribbon, or enter the **BLOCK** command at the keyboard.

This displays the Block Definition dialog box, as shown in Fig. 32-2.

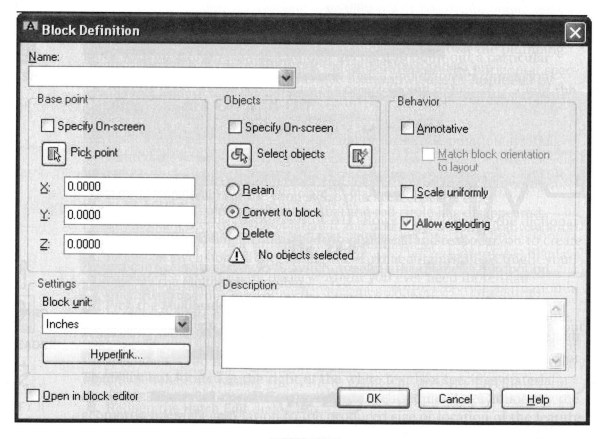

Fig. 32-2

2. In the Base Point area, pick the **Pick point** button.

3. Snap to the center point of the head of the screw.

This defines the base point for subsequent insertions of the screw head.

4. In the Objects area, pick the Convert to block radio button and then pick the **Select objects** button.

5. Select the entire screw head and press **ENTER**.

In the Objects area, AutoCAD displays 6 objects selected. The Retain option retains the selected objects as distinct objects in the drawing after you create the block. Convert to block converts the selected objects to a block instance in the drawing after you create the block. Delete deletes the selected objects from the drawing after you create the block. (The block definition remains in the drawing database.)

6. In the Settings area, review the options available for the Block unit.

7. Set Block unit at **Unitless;** do not add a description.

8. In the Name text field, type **flat** and pick the **OK** button.

You have just created a new block definition.

Inserting a Block

BLOCK

1. Pick the **Insert** button on the Block panel of the Ribbon, or enter the **INSERT** command.

This displays the Insert dialog box, as shown in Fig. 32-3. Flat is listed after Name because it is the only block definition in the drawing. Picking the down arrow would list additional blocks, if any existed, in alphabetical order.

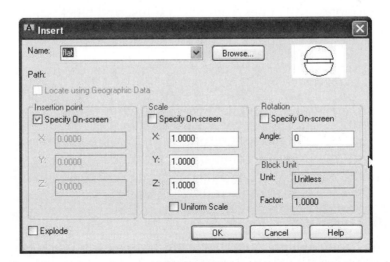

Fig. 32-3

The Insertion point area permits you to enter specific coordinates for the block's insertion. Specify On-screen, which should be checked, allows you to specify the insertion point on-screen. The Scale area allows you to scale the block now or on the screen when you insert it. The Rotation area enables you to specify a rotation angle in degrees now or when you insert the block. The Block Unit area is not selectable; it displays the unit and scale factor used when the block was created. The Explode check box explodes the block and inserts the individual parts of the block.

2. Under Scale, pick the **Specify On-screen** check box.

3. Under Rotation, pick the **Specify On-screen** check box.

4. Pick the **OK** button.

AutoCAD locks the insertion point of the block onto the crosshairs.

5. Pick an insertion point anywhere on the screen.

6. Move the crosshairs up and down, then back and forth.

As you can see, you can drag the X and Y scale factors.

7. Enter **1.5** for the X scale factor and **1.5** for the Y scale factor.

You can also drag the rotation into place, or you can enter a specific degree of rotation.

8. Enter **45** for the rotation angle.

9. Insert additional instances of the block using different settings in the Insert dialog box.

Editing an Inserted Block

1. Try to erase one of the lines from the first inserted block; then press **ESC**.

You cannot erase part of a block because a block is a single object.

MODIFY

2. Pick the **Explode** button from the Modify panel, pick the inserted block, and press **ENTER**.

3. Now try to erase a line from the inserted block.

As you can see, exploding a block returns it to its component parts. Exploding a block with the EXPLODE command is the same as checking the Explode check box in the Insert dialog box when inserting the block.

Renaming a Block

The RENAME command lets you rename previously created blocks.

1. Enter **RENAME** at the keyboard.

This enters the RENAME command and displays the Rename dialog box. Under Named Objects, the dialog box lists all of the object types in AutoCAD that you can rename.

2. Pick **Blocks**.

AutoCAD lists the only block, flat, under Items.

3. Pick **flat** and then enter **flathead** in the edit box at the right of Rename To.

4. Pick the **Rename To** button.

Flathead is now listed under Items.

5. Pick the **OK** button.

6. Save your work.

In the future, if you need to rename blocks, dimension styles, layers, linetypes, text styles, UCSs, viewports, or views, use the RENAME command.

Purging Blocks

The PURGE command enables you to selectively delete, or **purge**, any unused named objects, such as blocks. Purging named objects reduces drawing file size.

1. Select **Drawing Utilities** from the Application Menu, then the **Purge** command, or enter **PURGE** at the keyboard.

AutoCAD displays the Purge dialog box and shows the different object types that you can purge. If the drawing database contained unused blocks, the Purge button would not be grayed out. In the future, you may choose to purge unused blocks, groups, dimension styles, layers, and other object types.

2. Pick the **Close** button.

DesignCenter and Tool Palettes

The DesignCenter is a palette that allows you to browse, find, and insert content, such as blocks and hatches. The Tool Palettes Window offers tabbed areas that provide an easy method of organizing, sharing, and placing blocks and hatches.

Inserting Blocks from DesignCenter

1. Pick the **DesignCenter** button on the Palettes panel in the View tab of the Ribbon.

This displays the DesignCenter palette.

PALETTES

2. Click each of the tabs to familiarize yourself with DesignCenter.

3. Pick the **Open Drawings** tab and double-click **Blocks** beneath blks.dwg.

Notice that the flathead block was added automatically to DesignCenter.

4. Click and drag the block named **flathead** from DesignCenter into the drawing area.

As you can see, DesignCenter offers a quick and easy alternative for inserting blocks.

5. Right-click and drag the **flathead** block into the drawing area and pick **Insert Block...** from the shortcut menu.

This displays the Insert dialog box.

6. Make sure that Specify On-screen is checked under Insertion point, Scale, and Rotation and pick **OK**.

7. Pick an insertion point.

8. Enter **.75** for the X and Y scale factors.

9. Enter **135** for the rotation angle.

10. After inserting the block, auto-hide DesignCenter to make it smaller, but leave it open.

PALETTES

Inserting Blocks from a Tool Palette

1. Pick the **Tool Palettes Window** button on the Palettes panel.

2. Right-click the title bar of the Tool Palettes Window and select **New Palette** from the shortcut menu.

3. In the text box, type **Fasteners** and press **ENTER**.

This creates a new (but empty) tool palette named Fasteners.

4. From DesignCenter, click and drag the block named **flathead** to the new palette.

The flathead block appears in the tool palette.

5. Close DesignCenter.

6. From the new tool palette, click and drag the **flathead** block into the drawing area and release the pick button.

As you can see, this is a fast and easy way of inserting blocks.

7. Insert another **flathead** block into the drawing area from the tool palette; then insert another.

8. Attempt to right-click and drag the **flathead** block from the palette into the drawing area.

As you can see, this is not an option. The Tool Palettes Window does not give you the option of inserting blocks using the Insert dialog box, similar to right-clicking and inserting them from DesignCenter.

Adding Commands to a Tool Palette

AutoCAD makes it easy to add commands to a tool palette.

1. Display the Draw toolbar by selecting **Toolbars** on the User Interface panel of the View tab, then selecting **Draw** from the cascading menu below AutoCAD.

2. Right-click in the **Fasteners** tool palette and pick **Customize Palettes...** from the menu.

This displays the Customize dialog box.

3. With the Customize dialog box open, click and drag the **Polyline** button from the Draw toolbar to the Fasteners tool palette.

The Polyline button becomes part of the palette. Notice that a small black triangle appears next to it. The button is now a flyout button—a set of Draw commands are nested under this single button.

4. Click the black triangle next to the Polyline button on the Fasteners tool palette.

As you can see, the buttons for eight of the commands from the Draw toolbar are displayed on the flyout button.

5. Pick the **Close** button in the Customize dialog box.

6. Display the **Dimension** and **Modify** toolbars.

7. Right-click in the **Fasteners** tool palette and pick **Customize Palettes...** from the menu once again.

8. Click and drag the **Linear Dimension** button from the Dimension toolbar to the Fasteners tool palette.

9. Click and drag the **Erase** button from the Modify toolbar to the Fasteners tool palette.

10. Close the Customize dialog box.

11. Try each of the commands that you added to the Fasteners tool palette.

You can also arrange the order of tools on a tool palette and rearrange the order of tabs in the Tool Palettes Window.

12. On the Fasteners tool palette, click and drag the **Erase** button up to the top of the palette.

13. Right-click on the **Fasteners** tab and pick **Move Up** or **Move Down** to rearrange the order of tabs in the Tool Palettes Window.

14. Right-click in an open area of the Fasteners tool palette, select **Delete Palette** from the menu, and pick **OK** to confirm that you want to delete the tool palette.

AutoCAD deletes the tool palette.

15. Close the **Tool Palettes Window**, close the **Draw**, **Dimension**, and **Modify** toolbars, and save your work.

Blocks and Drawing Files

It is possible to insert a drawing (DWG) file as if it were a block. This can be useful when you want to use part or all of another drawing in the current drawing. Also, it's possible to create a drawing file from a block. This can be especially helpful when you need to transport a block to another computer.

BLOCK

Inserting a Drawing File

1. Begin a new drawing from scratch using **Imperial** units.
2. Pick the **Insert** button from the Block panel and pick the **Browse...** button from the dialog box.
3. Find and open the folder with your name.
4. Open the file named **toler.dwg**.

Toler appears in the Name box.

5. Uncheck Specify On-screen under Insertion point, Scale, and Rotation.
6. Pick the **Explode** check box and pick **OK**.

Toler.dwg appears.

7. Pick one of the objects from the toler.dwg drawing.

This is possible because we selected the Explode check box to make the components of the block selectable individually.

8. Close the current drawing without saving.

The drawing named blks.dwg (containing the flathead screws) is now the current drawing.

BLOCK DEFINITION

Creating a Drawing File from a Block

WBLOCK, short for Write BLOCK, writes (saves) objects or a block to a new drawing file.

1. Pick the **Write Block** button from the fly-out menu on the Block Definition panel of the Insert tab, or enter **WBLOCK** at the keyboard.

AutoCAD displays the Write Block dialog box.

2. Under Source, pick the **Block** radio button.

3. Click in the box at the right of Block.

4. Pick **flathead**.

The Base point and Objects areas are grayed out because both base point and objects are a part of the block definition.

5. Review the information listed under Destination, but do not change it.

Notice that the proposed file name is flathead.

6. At the right, pick the button containing the three dots, select the folder with your name, and pick **Save**.

7. Pick the **OK** button to close the Write Block dialog box.

AutoCAD creates a new file named flathead.dwg.

8. Pick the **Open** button from the docked Quick Access toolbar.

9. Find **flathead.dwg** and preview it, but do not open it.

10. Pick the **Cancel** button.

Copying and Pasting Objects

AutoCAD's Windows-standard **copy and paste** feature provides an alternative to using the WBLOCK and INSERT commands. Copying and pasting can be a faster approach when you want to transfer a block or set of objects to another drawing on the same computer. This approach does not work when you need to transfer a block or set of objects to another computer.

1. On the Clipboard panel of the Home tab, pick the **Copy Clip** button to enter the **COPYCLIP** command.

CLIPBOARD

2. Pick a couple of objects and press **ENTER** to copy the objects to the Windows Clipboard.

3. Save your work.

4. Create a new drawing from scratch using **Imperial** units.

5. On the Clipboard panel, pick the **Paste** button to enter the **PASTECLIP** command.

CLIPBOARD

6. Move the crosshairs and notice that the objects are attached to it.

7. Pick an insertion point.

8. Exit AutoCAD without saving the current drawing.

● REVIEW QUESTIONS

Answer the following questions on a separate sheet of paper.

1. Briefly describe the purpose of blocks.

2. Explain how the INSERT command is used.

3. How can you list all defined blocks contained within a drawing file?

4. A block can be inserted with or without selecting the Explode check box. Describe the difference between the two.

5. Describe the function of the WBLOCK command.

6. When would WBLOCK be useful?

7. How can you rename blocks?

8. Why would you want to purge unused blocks from a drawing file?

9. What is an advantage to using the Tool Palettes Window to organize and insert blocks?

10. When copying an object from one drawing to another, why might you prefer the copy and paste method over the WBLOCK and INSERT approach?

● CHALLENGE YOUR THINKING

These questions are designed to further your knowledge of AutoCAD by encouraging you to explore the concepts presented in this chapter. Answer each question on a separate sheet of paper.

1. An electrical contractor using AutoCAD needs many electrical symbols in his drawings. He has decided to create blocks of the symbols to save time. Describe at least two ways the contractor can make the blocks easily available for all his AutoCAD drawings. Which method would you use? Why?

2. If you were to copy a block from one drawing and paste it into another, would AutoCAD recognize the pasted object as a block? Explain.

● APPLYING AUTOCAD SKILLS

Work the following problems to practice the commands and skills you learned in this chapter.

1. Open range.dwg, which you created in the "Applying AutoCAD Skills" section of Chapter 17. Make a block of the range and reinsert it into the drawing at a scale factor of .25, with a rotation of 90°. Save the drawing as range2.dwg.

2. Begin a new drawing named livroom.dwg. Draw the furniture representations shown in Fig. 32-4 and store each as a block. Then draw the living room outline. Don't worry about exact sizes or locations, and omit the text. Insert each piece of furniture into the living room at the appropriate size and rotation angle. Feel free to create additional furniture and to use each piece of furniture more than once. This file is also available on the OLC for this book, in the Drafting & Design Problems section, as livroom.dwg.

3. After creating the blocks in problem 2, write two of them (of your choice) to a file using WBLOCK. Store them in the folder with your name.

4. Copy the block of the easy chair and paste it into another drawing.

5. Explode the PLANT block and erase every fourth arc contained in it. Then store the plant again as a block.

6. Rename two of the furniture blocks.

7. Purge unused objects.

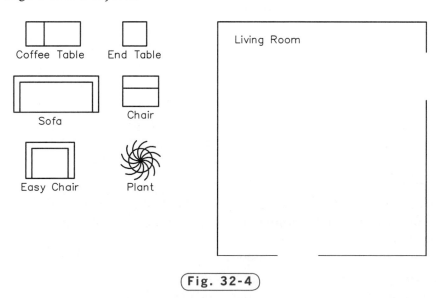

Coffee Table   End Table

Sofa   Chair

Easy Chair   Plant

Living Room

Fig. 32-4

Continued

8. Create a new drawing named revplate.dwg. Create the border, title block, and revisions box using the dimensions shown in Fig. 32-5.(Do not include the dimensions.) Make a block of the revisions box using the insertion point indicated. Insert the block in the upper right corner of the drawing. This drawing is also available as revplate.dwg on the OLC for this book, in the Drafting & Design Problems section.

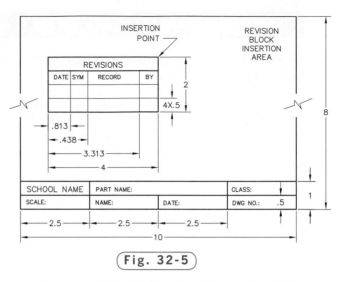

Fig. 32-5

Courtesy of Mark Schwendau, Kishwaukee College

● USING PROBLEM-SOLVING SKILLS

Complete the following problems using problem-solving skills and your knowledge of AutoCAD.

1. Your company wants to make the revisions box you created in problem 8 available to all the designers in the company. Open revplate.dwg from your named folder, create a block of the revisions box, and save it as a DWG file. Be sure to give the file a descriptive name so that the designers will know what is in the file.

2. Draw the electric circuit shown in Fig. 32-6 as follows: Draw the resistor using the mesh shown; then save it as a block. Draw the circuit, inserting the blocks where appropriate. Grid and snap are handy for drawing the resistor and circuit. Finish the circuit by inserting small donuts at the connection points. Add the text. The letter omega (Ω), which is used to represent the resistance in ohms, can be found under the text style GREEKC (character W).

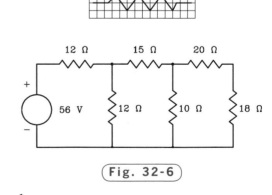

Fig. 32-6

Chapter 33

Dynamic Blocks

Objectives

- Create a dynamic block
- Assign actions to parameters
- Insert a dynamic block in a drawing
- Create a set of constraining values for a dynamic block
- Use a constrained dynamic block in a drawing

Vocabulary

constrain
dynamic blocks
parameters
value set

In the last chapter, you were introduced to blocks. By creating and using blocks, AutoCAD users avoid drawing the same object more than once. This can be extremely useful, because many objects in architecture, manufacturing, and engineering are used repeatedly and need to be drawn again and again in many drawings.

AutoCAD allows you to take advantage of the convenience of blocks even further with dynamic blocks. A **dynamic block** is a block with certain characteristics or **parameters**, such as size or location, that can be changed after you insert the block into a drawing. With dynamic blocks, you first define the parameters that you want to be changeable using AutoCAD's block editor. Once the block has been created, you only need the one dynamic block to draw a range of sizes or lengths of a part. For example, you could create a dynamic block for drawing a commonly used fastener, such as a hex nut, and vary the size of the nut when you insert the block into the drawing.

Creating a Dynamic Block

Let's create a dynamic block for a simple shape: a bed. We'll start with the shape of a twin-size bed and add parameters and actions that will allow its length and width to be stretched into a bed of any size.

DRAW

1. Start AutoCAD and start a new drawing from scratch using **Imperial** units.

2. Select the **Drafting & Annotation** workspace.

3. Using the **LIMITS** command, set the drawing area to **120 × 100** and **ZOOM All**.

4. Use the **LINE** command to create the top view of a twin-size bed using the dimensions shown in Fig. 33-1. Do not include the dimensions in your drawing. These are the standard dimensions (in inches) for a twin-size bed.

5. Name the drawing **bed.dwg**.

BLOCK

6. Pick the **Create** button from the Block panel or enter the **BLOCK** command to display the Block Definition dialog box.

7. Name the block **Bed**, pick the lower left corner of the bed for the base point, and select the entire bed to be included in the block.

8. Near the bottom of the dialog box, check the Open in block editor box.

9. Pick **OK** to close the dialog box and open the new block in the block editor.

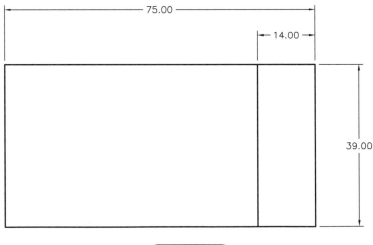

(Fig. 33-1)

Assigning Parameters to the Block

There are two ways to work with blocks in the block editor. You can use either the Block Authoring Palettes window, which appears on the left side of the block editor, or the buttons on the Block Editor tab that is automatically displayed on the Ribbon when the Block Editor is opened. The following steps use both methods.

1. In the Block Authoring Palettes window, select the **Parameters** tab if it is not already selected.

2. From the available options, pick the **Linear** parameter button.

3. Pick the bottom left corner of the bed as the start point and the bottom right corner of the bed as the endpoint. Be sure to snap accurately to these two points.

4. Pick a point inside the bed for the label location, as shown in Fig. 33-2.

Notice that the parameter is automatically labeled Distance1. Also notice the yellow icon with an exclamation mark near the lower left corner of the bed. This symbol means that the parameter is not currently associated with any action. You will set up the actions later. First, let's create a second parameter.

5. From the Block Authoring Palettes window, pick the **Linear** button again. This time, pick the lower left corner for the start point and the upper left corner for the endpoint. Place the label inside the bed, as shown in Fig. 33-3.

The second parameter is automatically named Distance2. However, you can rename parameters to make them easier to identify.

6. Select the first **Distance1** parameter you created, right-click to display the shortcut menu, and pick **Rename Parameter**.

7. Highlight the word **Distance1**, enter **Length** for the new name, and press **ENTER**.

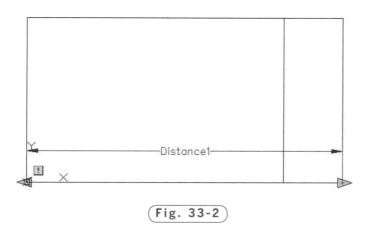

Fig. 33-2

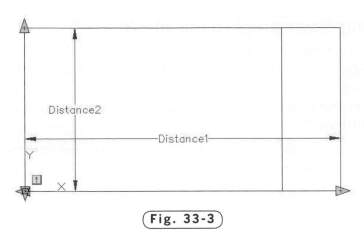

Fig. 33-3

8. Select the **Distance2** parameter and rename it **Width**.

Because we will stretch only one end of each parameter, let's change the grip display.

9. Pick the **Length** parameter, right-click, and move the pointing device to **Grip Display**.

10. Pick a Grip Display of **1** from the cascading menu.

Notice that the closed blue arrow on the left endpoint disappears.

11. Change the Grip Display of the Width parameter to **1**.

12. Save your work.

Assigning Actions to the Parameters

The two parameters that you have established will allow you to adjust both the length and the width of the dynamic block named Bed. Next, we will set up the actions to be associated with each parameter.

1. Expand the Action Parameters panel on the Block Editor tab of the Ribbon and select the **Show All Actions** button.

> You can also define an action using the Actions tab of the Block Authoring Palettes window.

ACTION PARAMETERS

2. Pick the down-facing arrow below the Move button on the Action Parameters panel, and select the **Stretch** action button from the flyout menu.

3. Select the **Length** parameter.

Notice that a red symbol appears at the lower right corner of the bed.

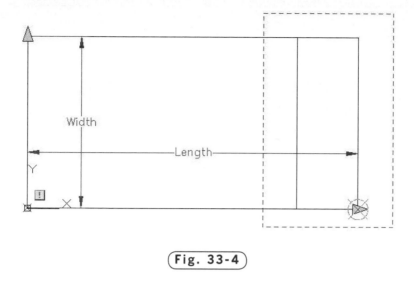

Fig. 33-4

4. Pick the point in the lower right corner of the bed as the parameter point to associate with the action.

5. In reply to Specify first corner of stretch frame, use a standard crossing window (right to left selection). Place the window as shown in Fig. 33-4. Be sure to include the interior line that represents the bedcovers in the crossing window.

6. At the Select Objects prompt, use another crossing window to select roughly the same area. Press **ENTER** when you are finished.

A Stretch action icon appears to the right of the parameter point. Let's rename the label for our new action.

7. Pick the **Stretch** action icon, right-click, and pick **Rename Action**.

8. Rename the action **Stretch Length**.

Next, we will define the action needed to change the width of the bed.

9. Pick the **Stretch** button again and select the **Width** parameter.

10. Pick the endpoint at the upper left corner as the parameter point to associate with the action.

11. For the stretch frame, place a crossing window across the upper half of the bed, as shown in Fig. 33-5.

ACTION PARAMETERS

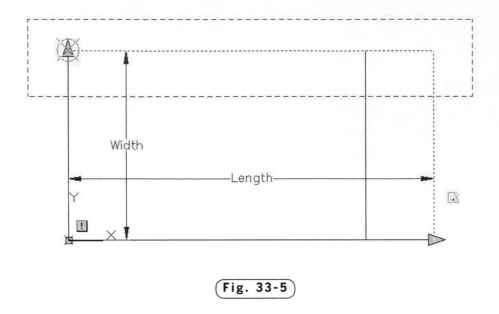

Fig. 33-5

12. At the Select objects prompt, use another crossing window to select roughly the same area. Press **ENTER** when you are finished.

13. Rename the action to **Stretch Width**.

The Length and Width parameters are now completely defined, so the alert icons for the parameters have disappeared. Your block should now look similar to the one in Fig. 33-6.

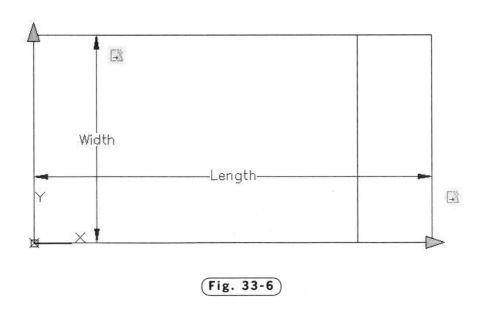

Fig. 33-6

14. Pick the **Close Block Editor** button on the Close panel of the Ribbon to return to the AutoCAD window.

15. When prompted, save changes to the Bed block.

CLOSE

Using Dynamic Blocks

The Bed block should now appear in AutoCAD's drawing area. Now we will change the size of the inserted block by changing its parameters.

1. Select the **Bed** block.

Notice the arrows that appear instead of grip boxes at the upper left and lower right corners of the bed. These arrows indicate changeable parameters.

2. Pick the arrow at the lower right corner of the bed and notice that it becomes active (red) just as a grip would.

The Length parameter appears and an edit box appears to show the current length of the bed, which is 75 inches.

3. Move the crosshairs and notice that the length of the bed changes dynamically.

4. At the keyboard, enter a new length of **80** and press **ENTER** to change the length of the bed to 80 inches.

The bed updates automatically. This is the standard length for a queen-size, king-size, or extra-long twin-size bed. Let's change the Width dimension to create a queen-size bed, which has a standard width of 60 inches.

5. Pick the arrow at the upper left corner of the bed.

6. Enter a new width of **60**.

Once again, the **Bed** block automatically updates when you press ENTER.

7. Press **ESC** to deselect the block.

Constraining Parameters

It may have occurred to you that it is possible to change the dynamic Bed block to any length and width using the dynamic block defined in the previous procedure. In real life, beds are usually made to standard dimensions, as shown in Fig. 33-7.

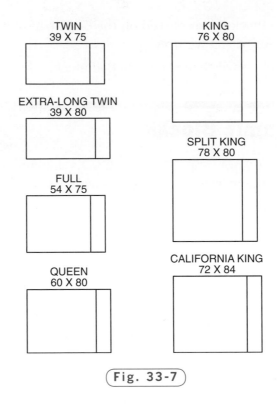

TWIN
39 X 75

KING
76 X 80

EXTRA-LONG TWIN
39 X 80

SPLIT KING
78 X 80

FULL
54 X 75

QUEEN
60 X 80

CALIFORNIA KING
72 X 84

Fig. 33-7

Creating a Value Set

AutoCAD makes it easy to **constrain**, or limit, the choices that can be used for a given parameter. You can do this by creating a **value set** for the parameter. For example, the three standard lengths for a bed are 75, 80, and 84 inches. The following steps constrain the Length parameter of the Bed block using a value set that contains these three values.

1. Double-click the block to display the Edit Block Definition dialog box.

2. Select **Bed** and pick **OK** to enter the block editor.

3. Select the **Length** parameter.

4. Right-click and select **Properties** from the shortcut menu.

This displays the properties of the linear parameter in the Properties palette.

5. Near the bottom of the Properties palette, find the Value Set section and click to activate the box next to Dist type. (You may need to scroll down to find this section.)

6. Pick the down arrow and then pick **List** from the options that appear.

Notice that the next row, Dist value list, now shows 75.0000. This is the value you used when you originally created the block, and it becomes the default value.

7. Pick to activate the Dist value list row and pick the three dots to the right of the edit box.

This displays the Add Distance Value dialog box.

8. In the Distances to add edit box, enter **80** and pick the **Add** button.

The new value of 80.0000 appears in the window below the edit box.

9. With the cursor still active in the Distances to add edit box, enter **84** and pick the **Add** button to add 84.0000 to the list.

10. Pick **OK** to close the dialog box.

11. Press **ESC** to deselect the Length parameter.

The standard widths for beds are 39, 54, 60, 72, 76, and 78. Let's constrain the Width parameter to these values.

12. With the Properties palette still open, pick the **Width** parameter.

13. In the Dist type row in the Properties palette, change the value to **List**.

14. Click to activate the Dist value list row and pick the button with three dots to display the Add Distance Value dialog box.

Once again, the default value appears in the window.

15. Add the other standard width values: **54**, **60**, **72**, **76**, and **78**. Pick **OK** when you are finished.

16. Close the **Properties** palette and save the file. If AutoCAD displays a message asking whether you want to save the drawing and update the existing block definition in the drawing, pick **Yes**.

17. Close the block editor.

Using a Constrained Block

Constrained blocks can be altered on-screen in the same way as unconstrained dynamic blocks. The only difference is that, when you pick one of the parameter arrows, small tick marks appear to mark the valid values for the parameter.

1. Select the **Bed** block to display the parameter arrows.

2. Pick the lower right parameter arrow so that it turns red.

Just below the arrow, notice the faint tick marks on either side of the current bed length. See Fig. 33-8.

You may recall that the bed is currently configured with dimensions for a queen-size bed: 60 × 80 inches. The 80 inches is one of the three standard values. The two ticks mark the other two values allowed for the Length

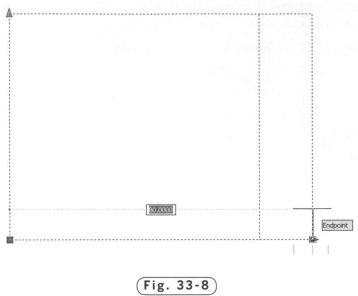

Fig. 33-8

parameter: 75 and 84. Let's change the Bed block to represent a full-size bed (54 × 75 inches).

3. Move the cursor to the left and notice that a tentative line appears at the 75-inch tick mark.

4. Pick any point while that line is present to reset the length of the bed to 75 inches.

5. Pick the upper left parameter arrow to activate the **Width** parameter and notice the tick marks representing allowed values.

The current value of 60 shows up in the edit box for reference.

6. Move the cursor down until the edit box shows a value of 54 and pick any point to change the width of the bed to 54 inches.

The block is now configured to represent a full-size bed.

If you enter an undefined value into the edit box for a constrained parameter, AutoCAD automatically chooses the defined value that is closest to the value you entered. For example, if you entered a value of 63 for the Width parameter, the width of the bed would change to 60, which is the closest value defined in the value set.

7. Save your work and exit AutoCAD.

• REVIEW QUESTIONS

Answer the following questions on a separate sheet of paper.

1. Define *dynamic block* in your own words.
2. Briefly describe the process of creating a dynamic block.
3. How can you rename a parameter or an action?
4. What is a value set? How would you create a value set to constrain a parameter for a dynamic block?
5. Why is careful selection of a dynamic block's base point important?

• CHALLENGE YOUR THINKING

These questions are designed to further your knowledge of AutoCAD by encouraging you to explore the concepts presented in this chapter. Answer each question on a separate sheet of paper.

1. One dynamic block action is visibility, which allows you to make a block invisible. This is somewhat similar to freezing a layer to make it invisible. When would each method of hiding drawing elements be preferable?
2. Review the Parameter Sets tab of the Block Authoring Palettes. Which parameter set would we have used to create the adjustments of the bed's length?

• APPLYING AUTOCAD SKILLS

Work the following problems to practice the commands and skills you learned in this chapter.

1. Open the bed.dwg file and save it as hotelplan.dwg. Create a hotel floor plan of your choosing, making a variety of rooms and room sizes. Place beds in each of the rooms. In about half of the rooms, insert two twin-size beds. In most of the others, insert a single king-size bed. In the remaining rooms, install two queen-size beds. Save your work.

Continued

2. Open the file named flathead.dwg that
you created in Chapter 32. Create
a block named dynflathead with the
screw head's center as the base point.
In the block editor, create a polar
parameter with the center of screw
head as the base point. Specify a
Grip Display of 1. Assign a scale action
to the parameter. Your dynamic block
should look similar to the one in
Fig. 33-9. Save the dynamic block and
close the block editor. Insert several of
the blocks into the drawing area. Scale
the blocks to a variety of sizes. Use
the keyboard to enter numeric values

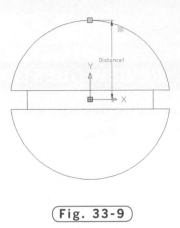

Fig. 33-9

for the diameter on several, then scale the others by dragging the grip
with the cursor. Your finished drawing should look similar to Fig. 33-10.
Save your work as dynflathead.dwg.

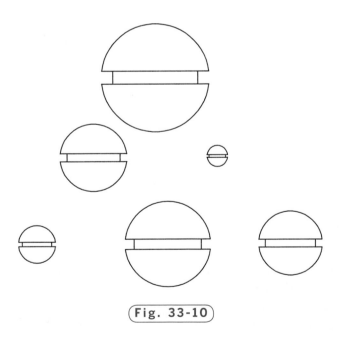

Fig. 33-10

• USING PROBLEM-SOLVING SKILLS

Complete the following activities using problem-solving skills and your knowledge of AutoCAD.

1. Open the chair.dwg file, which you created in Chapter 15. Delete all but one of the chairs. Make a block named chair, using the midpoint on the chair back as the base point. Save the drawing as dynchair.dwg. Create a rotation parameter and action. Rename the parameter Chair Angle. Your dynamic block should look like the one in Fig. 33-11. Insert six of the chair blocks into the drawing. Use dynamic rotation to rotate the individual chairs to the angles specified in Fig. 33-12.

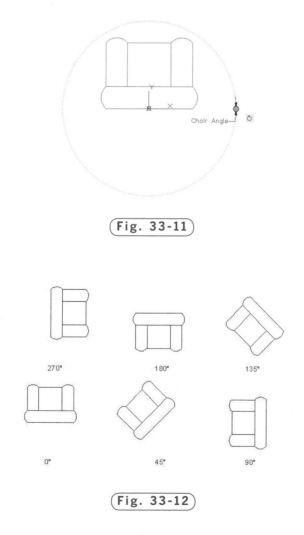

Fig. 33-11

Fig. 33-12

Continued

2. Create the hex bolt shown in Fig. 33-13. Approximate the size and shape of the hex head. Then make a dynamic block of the bolt, creating a parameter and action to stretch the length of the bolt's shank. Create a value set of the following shank lengths: 1.00, 1.25, 1.50, 1.75, 2.00, 2.50, and 3.00 inches. Place seven blocks in the drawing, and use the constraints of the value set to create one bolt of each length. Your finished drawing should look similar to Fig. 33-14. Save the file as dynbolt.dwg.

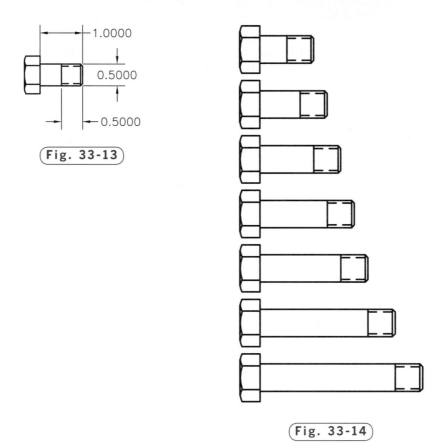

Fig. 33-13

Fig. 33-14

Chapter 34

Symbol Libraries

Objectives

- Create a library of symbols and details
- Insert symbols and details using a symbol library
- Insert layers, dimension styles, and other content from drawings using **DesignCenter**

Vocabulary

symbol library

Figure 34-1 shows a collection of electrical substation schematic symbols in an AutoCAD drawing file. Each of the symbols was stored as a single block and given a block name. (In this particular case, numbers were used for block names rather than words.) The crosses, which show the blocks'

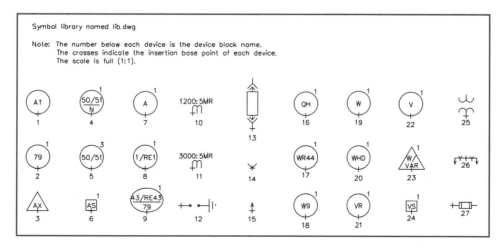

Fig. 34-1

insertion base points, and the numbers were drawn on a separate layer and frozen when the blocks were created. They are not part of the blocks; they are used for reference only. A drawing file such as this that contains a series of blocks for use in other drawings is known as a **symbol library**.

After the symbols were developed and stored in a drawing file, DesignCenter was used to insert the symbols into a new drawing for creation of the electrical schematic shown in Fig. 34-2.

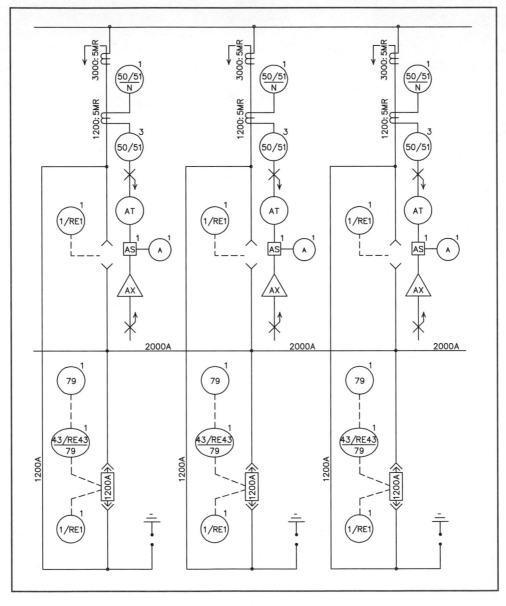

Fig. 34-2

In this example, the blocks were then inserted into their proper locations, and lines were used to connect them. Because these blocks containing the symbols were available, about 80% of the work was complete before the drawing was started. This is the primary advantage of grouping blocks in symbol libraries.

Creating a Library

Let's step through a simple version of the procedures just described. Keep in mind that a symbol library may be nothing more than a drawing file that contains a collection of blocks for use in other drawings.

1. Start AutoCAD and start a new drawing using the **tmp1.dwt** template file.

2. Select the **Drafting & Annotation** workspace.

3. Using the **LIMITS** command, set the drawing area to **24'** × **14'** and **ZOOM All**.

4. Create the schematic representations of tools shown in Fig. 34-3. Set snap at **3″**. Construct each tool on layer 0. Omit the text.

5. Save your work in a file named **lib1.dwg**.

6. Create a block from each of the tools. Use the following information:

   • Use the names shown in Fig. 34-3.

   • Pick the lower left corner of each tool for the insertion base point.

   • Under Objects, pick the Convert to block radio button.

   • For Block unit, use inches.

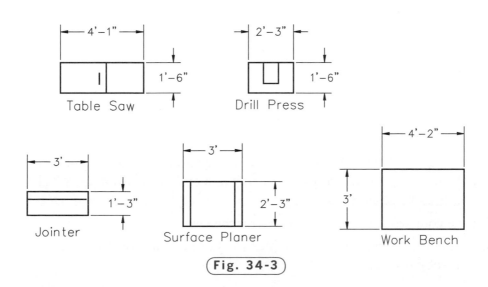

Fig. 34-3

- The block names adequately describe the blocks, so do not add a description for each block.
- Uncheck the Open in block editor box.

7. Save your work and close **lib1.dwg**.

A block can be made up of objects from different layers, with different colors and linetypes. The layer, color, and linetype information of each object is preserved in the block. When the block is inserted, each object is drawn on its original layer, with its original color and linetype, no matter what the current drawing layer and object linetype are.

A block created on layer 0 and inserted onto another layer inherits the color and linetype of the layer on which it is inserted and resides on this layer. Therefore, it is important to create blocks on layer 0 in most cases. Other options exist, but they can cause confusion.

Using a Symbol Library

We're going to use the new lib1.dwg file to create the workshop drawing shown in Fig. 34-4. With DesignCenter, we can review and insert blocks efficiently.

1. Start a new drawing using the **Quick Setup** wizard and the following information:
- Use **Architectural** units.
- Make the drawing area **22′ × 17′**.
- Set snap at **6″** and grid at **1′**.
- Be sure to **ZOOM All**.

2. Create a new layer named **Objects**, assign the color red to it, and make it the current layer.

3. Use the **POLYLINE** command to create the outline of the workshop as shown in Fig. 34-4. Make the starting and ending width **4″** and make it nearly as large as possible.

4. Save your work in a file named **workshop.dwg**.

5. In the Insert tab of the Ribbon, pick the **DesignCenter** button on the Content panel.

CONTENT

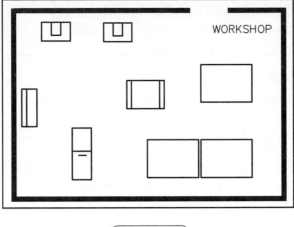

Fig. 34-4

DesignCenter appears. Notice the buttons along the top of the DesignCenter window.

6. Pick the **Load** button located in the upper left corner.

The Load dialog box appears.

7. Find the folder with your name and open it unless it is already open.

8. Find and open the drawing file named **lib1.dwg**.

This displays several types of lib1.dwg content in DesignCenter. As you can see, one of them is Blocks.

9. Double-click the **Blocks** icon or select **Blocks** below lib1.dwg in the Folder List.

The icons of the five blocks you created appear.

Inserting Blocks

1. Right-click and drag the jointer into the drawing area and release the right button.

2. Pick **Insert Block...** from the shortcut menu.

The Insert dialog box appears.

3. Under Insertion point and Rotation, check Specify On-screen, but do not check this option under Scale.

4. Uncheck the Explode check box and pick the **OK** button.

5. Insert the block in the position shown in Fig. 34-4.

6. Insert the remaining blocks and save your work.

 If you do not need to rotate the block when you insert it, you can save steps by left-clicking and dragging the block into position in the drawing.

You can insert symbols and details that are stored as blocks from any drawing file. AutoCAD provides several good examples.

1. Pick the **Load** button in DesignCenter to display the Load dialog box.

2. Find and open AutoCAD's **Sample** folder, then the **en-us** folder, and then open the **DesignCenter** folder.

3. Find the file named **House Designer.dwg** and open it.

4. Double-click **Blocks** in DesignCenter.

5. Insert a 36″ right-swing door into the doorway of the workshop. (You may need to edit the size of the opening.)

6. Experiment with the blocks in some of the other drawing files located in the DesignCenter folder, such as Home - Space Planner.dwg.

7. Erase any blocks that are not appropriate for this drawing, and save your work.

Inserting Other Content

As you may have noticed, DesignCenter permits you to drag other content into the current drawing. Examples include layouts, text styles, layers, and dimension styles.

1. Pick the **Load** button in DesignCenter and open the **lib1.dwg** file.

2. Double-click **Dimstyles**.

The Preferred and Standard dimension styles appear because they are stored in lib1.dwg.

3. Left-click and drag **Preferred** into the drawing area.

AutoCAD adds the Preferred dimension style to the current drawing.

4. Review the dimension styles in the Dimensions panel of the Annotate tab on the Ribbon.

As you can see, Preferred was indeed added to the current drawing.

5. Pick the **Up** button (the button that has a folder with an up arrow in it) once in DesignCenter and double-click **Layers**.

This displays the layers contained in lib1.dwg.

6. Click and drag **Center** to the drawing area.

7. Check the list of layers in the current drawing.

Center is now among them. So, you see, it is very easy to add content such as blocks, dimension styles, layers, and text styles from another drawing to the current drawing.

Other Features

The buttons at the top of the DesignCenter window offer several Windows-standard features. The Tree View Toggle button displays and hides the tree view.

1. Pick the **Tree View Toggle** button (fourth button from the right).

When this button is selected, it splits the window, with the left side displaying an Explorer-style tree view of your computer's files, folders, and devices.

2. Toggle the **Tree View Toggle** button until the display of the tree view is present.

The Favorites button displays the contents of the Favorites folder. You can select a drawing or folder or another type of content and then choose Add to Favorites. AutoCAD creates a shortcut to that item, which is then added to the AutoCAD Favorites folder. The original file or folder does not actually move. Note that all of the shortcuts you create using DesignCenter are stored in the AutoCAD Favorites folder.

The Search button displays the Search dialog box. You can enter search criteria and locate drawings, blocks, and nongraphic objects within drawings.

The Preview button (third selection from the right) displays a preview of the selected item in a pane below the content area if a preview image has been saved with the selected item. If not, the Preview area is empty.

The Description button displays a description of the selected item in a pane below the content area. When creating a new block, you are given the opportunity to enter a description; this is where you can view this information.

3. Save your work, close **DesignCenter**, and exit AutoCAD.

● REVIEW QUESTIONS

Answer the following questions on a separate sheet of paper.

1. What is the primary purpose of creating a library of symbols and details?
2. When you create a symbol library, on which layer should you create and store the blocks? Why?
3. What types of content does DesignCenter display?
4. From DesignCenter, how do you insert a dimension style into the current drawing?
5. Is it possible to use DesignCenter to insert content from other AutoCAD drawings into the current drawing? Explain.

● CHALLENGE YOUR THINKING

These questions are designed to further your knowledge of AutoCAD by encouraging you to explore the concepts presented in this chapter. Answer each question on a separate sheet of paper.

1. Identify an application for creating and using a library of symbols and details. Decide what symbols the library should include. Discuss your idea with others and make changes according to their suggestions.
2. Why are blocks, rather than groups, used in symbol libraries? Can you think of any applications for which you could use groups as well as blocks in a symbol library? Explain.
3. Explore reasons it might be useful to add shortcuts to the Favorites folder that appears in DesignCenter. How might adding shortcuts to libraries of symbols and details be especially helpful?

• APPLYING AUTOCAD SKILLS

Work the following problems to practice the commands and skills you learned in this chapter.

1. Based on steps described in this chapter, create an entirely new symbol library specific to your area of interest. For example, if you practice architectural drawing, create a library of doors and windows. First create a drawing template. If you completed the first "Challenge Your Thinking" item, you may choose to create the symbol library you planned.

2. After you have completed the library of symbols and details, begin a new drawing and insert the blocks. Create a drawing using the symbol library.

3. The logic circuit for an adder is shown in Fig. 34-5. Draw the circuit by first constructing the inverter and the AND gate as blocks. Use DONUT for the circuit connections.

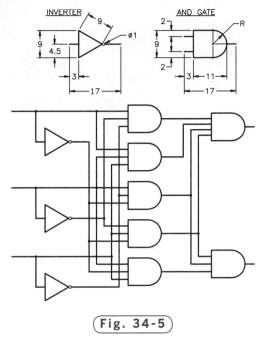

Fig. 34-5

Courtesy of Gary J. Hordemann, Gonzaga University

• USING PROBLEM-SOLVING SKILLS

Complete the following activities using problem-solving skills and your knowledge of AutoCAD.

1. As a building contractor, you want to make a customized library from AutoCAD's existing libraries to suit your own needs. Open DesignCenter, select the Sample folder, and then pick the DesignCenter folder. Find out what blocks are most suitable for a building contractor. Browse through AutoCAD's drawing libraries and select the blocks that are most appropriate. Adjust the scale so that they fit on the screen. Save the drawing as contractor.dwg.

2. As a drafter for an electronics firm, you need a library of electronic symbols. Consult reference books to determine the correct schematic symbols for a PNP transistor, NPN transistor, resistor, capacitor, diode, coil, and transformer. Draw the symbols and make blocks of them to create an electronics symbol library. Save it as electsymbol.dwg.

Chapter 35

Attributes

Objectives

- Create fixed and variable attributes
- Store attributes in blocks
- Edit individual attributes

Vocabulary

attribute tag
attribute values
attributes
fixed attributes
variable attributes

In AutoCAD, **attributes** consist of text information stored within blocks. The information describes certain characteristics of a block, such as size, material, model number, cost, etc., depending on the nature of the block. The advantage of adding attribute information to a drawing is that it can be extracted to form a report such as a bill of materials. The attribute information can be made visible, but in most cases, you do not want the information to appear on the drawing. Therefore, it usually remains invisible, even when you plot the drawing.

The electrical schematic shown in Fig. 35-1 contains attribute information, even though you cannot see it. (The numbers you see in the components are not the attributes.)

Figure 35-2 shows a zoomed view of one of the components after the attribute information has been made visible. In this example, the attribute information is displayed near the top of the component.

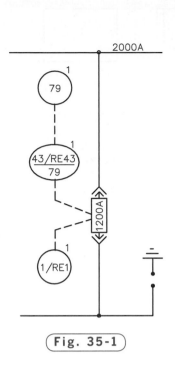

Fig. 35-1

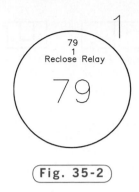

Fig. 35-2

DESCRIPTION	DEVICE	QUANTITY/UNIT
Recloser Cut-out Switch	43/RE43/79	1
Recloser Relay	79	1
Lightning Arrester	--	3
Breaker Control Switch	1/RE1	1
1200 Amp Circuit Breaker	52	1

Fig. 35-3

All of the attributes contained in this schematic were compiled into a file and placed into a program for report generation. The report (bill of materials) in Fig. 35-3 was generated directly from the electrical schematic drawing.

Fixed Attributes

Attributes can be fixed or variable. **Fixed attributes** are those whose values you define when you first create the block. You will learn about variable attributes later in this chapter.

Creating Attributes

Attributes are created, or defined, using the ATTDEF command.

1. Start AutoCAD and open the drawing named **lib1.dwg**.

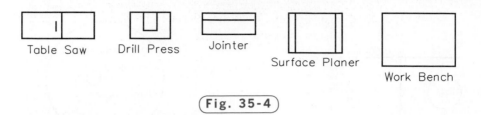

Table Saw   Drill Press   Jointer

Surface Planer

Work Bench

Fig. 35-4

It should look similar to the one in Fig. 35-4, but without the text.

If you do not have lib1.dwg on file, create it using the steps outlined in Chapter 34.

2. Select the **AutoCAD Classic** workspace and close all floating toolbars and palettes.

Let's assign information (**attribute values**) to each tool so that we can later generate a bill of materials. We'll design the attributes so that the report will contain a brief description of the component, its model, and the cost.

3. Zoom in on the table saw. It should fill most of the screen.

4. From the Draw pull-down menu, select **Block** and **Define Attributes...**, or enter **ATTDEF** (short for "attribute definition") at the keyboard.

This enters the ATTDEF command and displays the Attribute Definition dialog box, as shown in Fig. 35-5.

5. In the Mode area, check **Invisible** and **Constant**.

Invisible specifies that attribute values are not displayed when you insert the block. Constant gives attributes a fixed value for all insertions of that block. You will learn more about these two items when you insert blocks that contain attributes.

6. In the Attribute area, type the word **description** (in upper- or lowercase letters) in the box at the right of Tag.

The **attribute tag** identifies each occurrence of an attribute in the drawing. Once again, this will become clearer after you create and use attributes.

7. In the box at the right of Default, type **Table Saw** (exactly as you see it here).

Table Saw becomes the default attribute value.

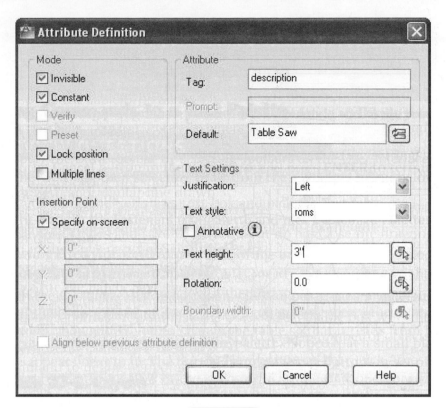

Fig. 35-5

8. Under Text Settings, enter the following information:

- Justification: **Left**
- Text Style: **roms**
- Height: **3″**
- Rotation: **0.0**

9. Pick the **OK** button.

10. In reply to Specify start point, pick a point inside the table saw, near the top left.

The word DESCRIPTION appears on the table saw.

11. Press the spacebar to repeat the ATTDEF command.

12. This time, enter **model** for the Tag and **1A2B** for the Default value.

13. Pick the **Align below previous attribute definition** check box in the lower left corner of the dialog box and pick **OK**.

The word MODEL appears below DESCRIPTION.

14. Repeat Steps 11 through 13, but enter **cost** for the tag and **$625.00** for the value. Be sure to pick the **Align below previous attribute definition** check box.

15. Save your work.

Table 35-1. Attribute Information

Description	Model	Cost
Drill Press	7C-234	$590.00
Jointer	902-42A	$750.00
Surface Planer	789453	$2070.00
Work Bench	31-1982	$825.00

You are finished entering the table saw attributes. Let's assign attributes to the remaining tools.

16. Zooming and panning as necessary, assign attributes to the remaining tools. Use the information shown in Table 35-1.

17. ZOOM All and save your work.

Be sure to pick the Align below previous attribute definition check box in the Attribute Definition dialog box when adding the second and third attributes to each tool. This will save time and will automatically align the second and third attributes under the first one.

Storing Attributes

MODIFY

Let's store the attributes in the blocks.

1. Pick the **Explode** button from the docked Modify toolbar and explode each of the tools.

Exploding the blocks permits you to redefine them using the same name.

DRAW

2. Pick the **Make Block** button from the docked Draw toolbar.

3. Pick the down arrow at the right of Name and pick **Table Saw**.

4. Pick the **Pick point** button and pick the lower left corner of the table saw for the insertion base point.

5. Pick the **Select objects** button, select the table saw and its attributes, and press **ENTER**.

6. Remove the check mark in the Open in Block editor box.

7. Pick the **OK** button.

8. When AutoCAD asks if you want to redefine Table Saw, pick **Redefine**.

This redefines the Table Saw block.

9. Repeat Steps 2 through 8 to redefine each of the remaining tool blocks.

10. Save your work.

All of the tools in lib1.dwg have been redefined and now contain attributes. When you insert these tools into another drawing, the attributes will insert also. You will learn more about attribute extraction and bills of materials in Chapter 36.

Displaying Attributes

Let's display the attribute values using the ATTDISP (short for "attribute display") command.

1. Enter **ATTDISP** and select the **On** option.

You should see the attribute values, similar to those shown in Fig. 35-6.

2. Reenter **ATTDISP** and select the **Normal** option.

The attribute values should again be invisible.

Fig. 35-6

Variable Attributes

Thus far, you have experienced the use of fixed attribute values. With **variable attributes**, you have the freedom of changing the attribute values as you insert the block. Let's step through the process.

1. Using the **tmp1.dwt** template file, begin a new drawing.

2. Zoom in on the lower left quarter of the display, make layer **0** the current layer, and set the snap resolution to **2"**.

3. Draw the architectural window symbol shown in Fig. 35-7. The dimensions not given are 2" in length. Do not place dimensions on the drawing. (The symbol represents a double-hung window for use in architectural floor plans.)

4. Save your work in a file named **window.dwg**.

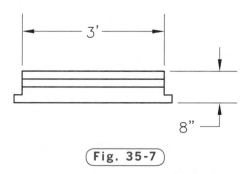

Fig. 35-7

5. Enter the **ATTDEF** command and check **Invisible**, but Constant, Verify, and Preset should be unchecked.

6. For the tag, type the word **type**.

7. In the Prompt edit box, type **What type of window?**

8. In the Default edit box, type **Double-Hung**.

9. Under Text Settings, enter the following information.
   - Justification: **Center**
   - Text Style: **roms**
   - Height: **3″**
   - Rotation: **0.0**

10. Pick the **OK** button and pick a point over the top center of the window, leaving space for two more attributes.

The word TYPE appears.

11. Redisplay the dialog box, and leave the attribute modes as they are.

12. Type **size** for the attribute tag, **What size?** for the attribute prompt, and **3′ × 4′** for the attribute value.

13. Pick the **Align below previous attribute definition** check box, and pick **OK**.

The word SIZE appears on the screen below the word TYPE.

14. Repeat Steps 11 through 13 using the following information.
   - Tag: **manufacturer**
   - Prompt: **What manufacturer?**
   - Value: **Andersen**

15. Store the window symbol and attributes as a block. Name it **DH** (short for "double-hung") and pick the lower left corner for the insertion base point.

16. Pick **OK** to close the Block Definition dialog box.

The Edit Attributes dialog box appears.

17. We do not want to edit the attributes at this time, so pick the **Cancel** button and save your work.

18. Enter **ATTDISP** and **On**.

Did the correct attribute values appear?

Inserting Variable Attributes

1. Enter the **INSERT** command and select the block named **DH** in the Insert dialog box.
2. Uncheck **Explode** if it is checked.
3. Make sure **Specify On-screen** is checked under Insertion point (Scale and Rotation should be unchecked) and pick **OK**.
4. Pick an insertion point at any location.
5. In reply to What manufacturer?, press **ENTER** to accept Andersen.
6. In reply to What size?, enter **3′ × 5′**.
7. In reply to What type of window?, press **ENTER** to accept Double-Hung.

The block and attributes appear.

Editing Attributes

ATTEDIT, short for "attribute edit," allows you to edit attributes in the same way as you edited the size attribute of the double-hung window. Your drawing should currently look similar to the one in Fig. 35-8. The exact location of the attributes is not important.

1. Enter the **ATTEDIT** command and pick one of the two blocks.

This displays the Edit Attributes dialog box.

2. Change Andersen to **Pella** and pick **OK**.

Andersen changes to Pella.

3. Save your work and close the file named **window.dwg**.

The lib1.dwg file again becomes the current drawing.

Double Hung
3'x5'
Andersen

Fig. 35-8

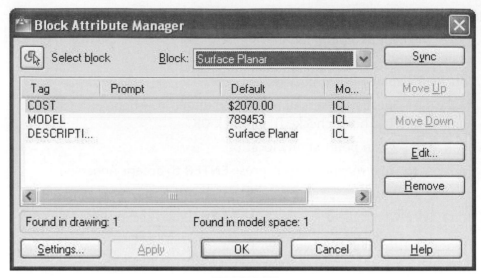

Fig. 35-9

Block Attribute Manager

AutoCAD's Block Attribute Manager permits you to edit the attribute definitions in blocks, remove attributes from blocks, and change the order in which you are prompted for attribute values when inserting a block.

1. Enter the **BATTMAN** command.

This displays the Block Attribute Manager dialog box, as shown in Fig. 35-9.

2. Pick the down arrow located at the right of Block near the top of the dialog box and pick **Surface Planer**.

3. Select the line that includes COST and $2070.00 and pick the **Edit...** button.

This displays the Edit Attribute dialog box.

4. Pick each of the tabs and review the information and options in each one and then pick the **Cancel** button.

5. Pick **Cancel** and exit AutoCAD without saving.

● REVIEW QUESTIONS

Answer the following questions on a separate sheet of paper.

1. Explain the purpose of creating and storing attributes.
2. Briefly define each of the following commands.
   a. ATTDEF
   b. ATTDISP
   c. ATTEDIT
   d. BATTMAN
3. What are attribute tags?
4. What are attribute defaults?
5. Explain the attribute modes Invisible and Constant.

● CHALLENGE YOUR THINKING

These questions are designed to further your knowledge of AutoCAD by encouraging you to explore the concepts presented in this chapter. Answer each question on a separate sheet of paper.

1. Discuss the advantages and disadvantages of using variable and fixed attribute values. Why might you sometimes prefer one over the other?
2. Brainstorm a list of applications for blocks with attributes. Do not limit your thinking to the applications described in this chapter. Compare your list with lists created by others in your group or class. Then make a master list that includes all the ideas.

Continued

● APPLYING AUTOCAD SKILLS

Work the following problems to practice the commands and skills you learned in this chapter.

1. Load the livroom.dwg drawing containing the furniture representations you created in Chapter 32. If this file is not available, create a similar drawing. Outline a simple plan for assigning attributes to each of the components in the drawing. Create the attributes and redefine each of the blocks so the attributes are stored within the blocks.

2. The small hardware shop in which you work is computerizing its fasteners inventory. The shop carries both U.S. and metric sizes. All of the available fasteners are saved as blocks in the DesignCenter folder in AutoCAD's Sample folder. Place the blocks from Fasteners – Metric.dwg and Fasteners – US.dwg into a drawing named hardware.dwg. Place them in an orderly and logical fashion. Then assign attributes to each fastener to describe it. Use attribute tags of Type and Size. For the individual attribute values, use the information given in the name of each block.

3. Look in a computer hardware catalog or on the Internet to find the sizes and basic shapes of several printers from different manufacturers. Draw the basic shape of at least five different printers. Create attributes to describe the printers in terms of type (inkjet, laser, etc.), brand, and cost. Save the drawing as printers.dwg.

4. Refer to schem.dwg and schem2.dwg located on the OLC for this book, in the Drafting & Design Problems section. Display and edit the attributes contained in the drawing files.

● USING PROBLEM-SOLVING SKILLS

Complete the following problems using problem-solving skills and your knowledge of AutoCAD.

1. Attributes are used for many reasons besides creating bills of materials. For example, you can use attributes to track items in a collection of anything from model cars to unique bottles to action figures. For this problem, suppose you are a dealer in and collector of Avon® bottles that have a transportation theme. You have been asked to participate in a display of various collections being sponsored by the local library. Display space is limited, so you need to make a layout to determine your specific needs. Create a library of schematic symbols for the car-shaped bottles shown

Table 35-2. Collectable Attributes

Bottle	Contents	Price
1937 Cord	Wild Country	$12.00
1951 Studebaker	Spicy	$8.00
1988 Corvette	Spicy	$8.00
Silver Deusenberg	Oland	$13.00
1933 Pierce Arrow	Deep Woods	$10.00

in Table 35-2. The cars vary in size from 2″ × 5″ for the Corvette and Studebaker to 2-1/2″ × 8″ for the Pierce Arrow and 3″ × 9″ for the Cord and Deusenberg. Assign the attributes as shown in Table 35-2. Save the drawing as cars.dwg.

2. You are setting up sprinklers for new landscaping for a residential plot. Create a new drawing with architectural units and a drawing area of 300′ × 200′. Draw the plot boundaries with corners at the following absolute coordinates:

45′,170′

235′,140′

210′,45′

35′,25′

The house will be a rectangle of 60′ × 32′. The northwest corner of the house is at absolute coordinates 90′, 112′. Your client wants a sprinkler system with enough sprinkler heads to cover the entire property. Each sprinkler head can be set to turn a full 360° or any number of degrees greater than 30°. When set to turn 360°, each sprinkler can cover a radial area of approximately 30′. Decide on the number of sprinklers needed, their placement on the property, and the angle (number of degrees) that each should be set to cover. Then create the sprinkler heads, load them from your library, or load them from DesignCenter. Place them in the locations you have determined. On a separate layer, draw the range of each sprinkler to show the client that the sprinkler heads will indeed cover the entire property.

Finally, define attributes with the following tags: Location, Angle, and Direction. Assign these attributes to each sprinkler head in your drawing and give them the appropriate values according to the plan you have developed. Save the drawing as sprinklers.dwg.

Architectural Drafter

What type of house do you live in? Perhaps it's an older home, where floorboard creaks and groans are part of its charm. Or maybe it is tall, narrow, and practically bumps shoulders with your neighbor's house, like many homes located in the city. If you live in the country, your home may sprawl across your property, surrounded by plenty of grass and shade trees.

Although many people are responsible for the actual construction of homes and other buildings, architectural drafters play an important role in the planning stages of a building's design.

Climbing the Architectural Ladder

Architectural drafters are employed by engineering firms, corporate development departments, and architectural and construction companies. Apprenticeships are sometimes offered for students interested in entering this field. Architectural drafters may begin as tracers or junior drafters. After gaining experience, they may advance to senior drafting or checking positions.

Paying Attention to Detail

Architectural drafters must be able to visualize objects in two and three dimensions. They must also exhibit good eye-hand coordination, which is a skill necessary to create fine, detailed drawings. The ability to pay attention to detail is a critical characteristic for an architectural drafter to possess; just one minor design flaw can result in a major problem, such as a lopsided doorway or an imperfect foundation. For information on student involvement in architecture, visit the American Institute of Architecture Students (AIAS) Web site.

Courtesy of Autodesk, Inc.

▶ Career Activities

1. What types of detail do you think architectural drafters must be aware of when creating drawings? Compare your list with a classmate's.

2. Has employment in this field been steady over the past few years? What are the forecasts for future employment?

Chapter 36

Bills of Materials

Objectives

- Extract attribute data from blocks for report generation
- Create a bill of materials

Vocabulary

comma-delimited file
data extraction
xref

After finishing the attribute assignment process (Chapter 35), you are ready to create a report such as a bill of materials. The first step in this process involves **data extraction** (gathering the attribute data and placing it into an AutoCAD table or an electronic file that can be read by a computer program).

1. Start AutoCAD and begin a new drawing using the **tmp1.dwt** template file.
2. Select the **Drafting & Annotation** workspace and **ZOOM All**.
3. Create the four walls of a workshop (fill the upper two-thirds of the drawing area) and insert each of the blocks on layer Objects as shown in Fig. 36-1 on page 490.
4. Save your work in a file named **extract.dwg**.

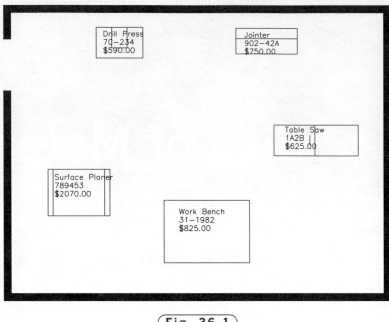

Fig. 36-1

5. Enter **ATTDISP** and **On**.

Each of the tools contains attributes, as shown in Fig. 36-1.

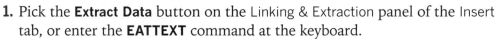

Data Extraction

The EATTEXT command permits you to extract the attributes from the extract.dwg file using the Data Extraction wizard.

LINKING & EXTRACTION

1. Pick the **Extract Data** button on the Linking & Extraction panel of the Insert tab, or enter the **EATTEXT** command at the keyboard.

This starts the Data Extraction wizard, as shown in Fig. 36-2. There are eight pages in the Data Extraction wizard.

2. Make sure the radio button for Create a new data extraction is selected and pick **Next**.

3. Save your data extraction as **extract.dxe** in the file with your name on it.

4. On page 2 of the wizard, pick the Select objects in the current drawing radio button, because we want to extract attributes from blocks in the current drawing. The current drawing, extract.dwg, should be listed in the Drawing files and folders area.

5. Pick the **Settings...** button before continuing. Review the options available in the Additional Settings dialog box.

The Extraction settings area allows you to select different blocks to be included in the extraction. The extract.dwg file does not contain any external references (xrefs), so it does not matter whether the second

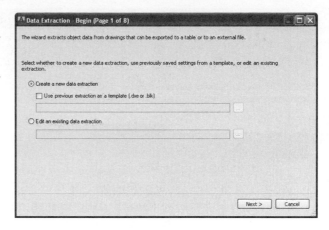

Fig. 36-2

and third boxes are checked. An **xref** is a drawing that is attached (linked) to the current drawing. Any changes made to the attached drawing are reflected in the current drawing.

The Extract from area allows you to extract blocks from the entire drawing or only blocks from the model space. Again, it does not matter which radio button is selected, because extract.dwg contains no blocks in layout space.

6. Pick **Cancel** to exit the Additional Settings dialog box.

7. Pick the Select objects button to the right of the Select objects in the current drawing radio button, select all five blocks in the drawing, and press **ENTER**.

8. Pick **Next** to go to the next page of the wizard.

Selecting and Reviewing the Attributes

Page 3 of the Data Extraction wizard displays the object names that you selected, as shown in Fig. 36-3 on page 492. The Objects area on top lists all the items you selected. The Display options at the bottom of the dialog box allows you to choose the object types to display.

1. Uncheck **Display all object types** and pick the radio button next to **Display blocks only**.

The object list now contains only the five blocks.

2. Make sure the boxes for all five blocks are checked and pick **Next**.

Page 4 of the wizard asks which properties of the selected blocks you want to extract.

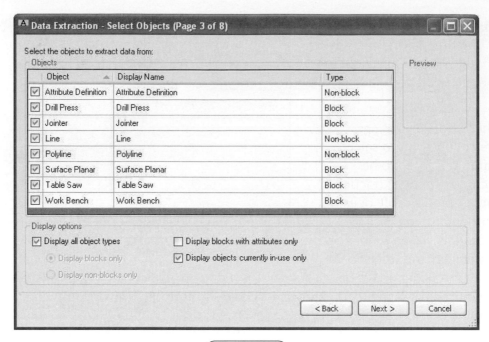

Fig. 36-3

First, focus on the Category filter area on the right of the dialog box. All the categories are checked, so all the properties available are listed on the left.

3. Uncheck all the check boxes under Category filter except the box in front of **Attribute**.

Now only the three attribute properties are listed under Property.

4. Select (highlight) the three attributes and pick **Next**.

Page 5 of the wizard allows you to refine the data in the data extraction set. We want to rearrange the columns of the table into a more logical order for a bill of materials.

5. Click in the **COST** column heading and drag it to the far right side of the table.

6. Right-click on the **DESCRIPTION** column heading and select **Sort Ascending**.

7. Uncheck the box in front of **Show count column**.

The preview of your table should now look similar to the table in Fig. 36-4. As you can see, you have many options for arranging the rows and columns of your table.

8. Pick the **Next** button.

Page 6 of the wizard allows you to choose output options for the data extraction. You can insert the table in a drawing, output the data to an external file, or both. You have the option of creating a Microsoft Excel® (XLS) file, a comma-delimited (CSV) file, a Microsoft Access® (MDB) file, or a text extract (TXT) file. A **comma-delimited file** is a popular format for importing data into databases and other programs.

9. Check the box next to **Insert data extraction table into drawing** and pick **Next**.

Page 7 of the Data Extraction wizard allows you to select a style for the attribute table. You can modify an existing table style or create a new one.

10. Pick **Next** to accept the Standard table style.

Page 8 of the Attribute Extraction wizard is displayed.

Placing the Attribute Table in a Drawing

1. Pick **Finish**.

The table attaches to the crosshairs so that you can place it in the drawing.

2. Pick a point below the workshop in response to Specify insertion point.

As you can see, our bill of materials is much too small for the drawing.

3. Type **SCALE**, select the table, and use a scale factor of **20**.
4. Move the table so it is centered below the walls of the workshop.
5. Save your work and exit AutoCAD.

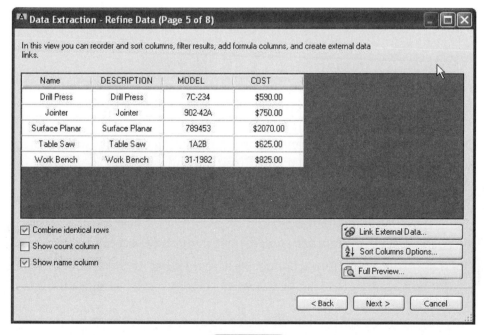

Fig. 36-4

● REVIEW QUESTIONS

Answer the following questions on a separate sheet of paper.

1. Describe the purpose of AutoCAD's EATTEXT command.

2. Describe the process of creating a report or bill of materials from a drawing that contains blocks with attributes.

3. Name the file types into which you can store the attributes using the Attribute Extraction wizard.

● CHALLENGE YOUR THINKING

These questions are designed to further your knowledge of AutoCAD by encouraging you to explore the concepts presented in this chapter. Answer each question on a separate sheet of paper.

1. Find out more about comma-delimited files. After you have extracted a CSV file, explore and list options for opening and printing its contents.

2. Research template (BLK) files. Write a short paragraph explaining what they do, what they contain, and why they are useful.

● APPLYING AUTOCAD SKILLS

Work the following problems to practice the commands and skills you learned in this chapter.

1. Open the livroom.dwg file from Chapter 32, which contains several pieces of furniture. Add attributes to the furniture and create blocks. Using the Data Extraction wizard, create a bill of materials and place it in the drawing.

2. Using the attribute data file created in problem 1, insert its contents into Microsoft Excel® or another spreadsheet if you have access to one.

3. Open sprinklers.dwg from Chapter 35. Create a list of the sprinkler heads for the landscaping contractor to use as reference when the sprinkler system is installed.

4. Refer to schem.dwg and schem2.dwg located on the OLC for this book, in the Drafting & Design Problems section. Using the procedure outlined in this chapter, create a bill of materials from each of the two drawings.

● USING PROBLEM-SOLVING SKILLS

Complete the following activities using problem-solving skills and your knowledge of AutoCAD.

1. Open cars.dwg from Chapter 35. Using the schematic blocks, create a display to fit in a 2-1/2' × 3' display space. Before you begin, you may want to consult antique car references if you are unfamiliar with any of the car models. Knowing what the cars look like will help you position them attractively in the display. Then extract the attribute data and create a list to distribute to potential customers.

2. Open printers.dwg from "Applying AutoCAD Skills" problem 3 in Chapter 35. Extract the attribute data from the printer blocks you created and create a table of the printer attributes. Display the table and all of the blocks in the drawing.

Applying Chapters 31–36

Computer Assembly Library

One rule of thumb for CAD drafters and designers is not to draw anything twice. Although that may sound strange at first, what it really means is to use CAD commands and functions to reproduce the same details again and again without having to redraw them. You are already familiar with the MIRROR, COPY, and ARRAY commands. To extend the capabilities of these and other commands, AutoCAD also permits you to block two or more objects so that you can manipulate them as one. These blocks of objects may also be assigned characteristics or attributes, which can be retrieved and listed later if desired.

▶ Description

A business that assembles computers has decided that it wishes to automate its advertising and billing practices. To do so, the CAD drafter has been assigned the task of drawing basic symbols to represent the various parts of the computer. Figure P7-1 shows possible configurations for CPUs, hard disk drives, and DVD drives as an example, and Fig. P7-2 shows a possible symbol library based on these configurations.

CPUs	
2.2 Gigahertz	$125
2.5 Gigahertz	$150
3.0 Gigahertz	$200
3.3 Gigahertz	$300
3.8 Gigahertz	$350
Hard Drives	
320 Gigabyte	$60
900 Gigabyte	$90
1.0 Terabyte	$120
7.5 Terabyte	$400
DVD-ROM Drives	
24x	$35
48x	$60
Read/Write	$200

Fig. P7-1

Creating the Symbol Library

Using the examples shown in Figs. P7-1 and P7-2, create a symbol library that contains various components of a computer. Do not limit the library to just the three parts shown in Figs. P7-1 and P7-2. Do research if necessary to find out what other components should be included in the symbol library and the configurations for each component. After you have created the symbols, assign appropriate attributes to them, as shown in Fig. P7-1, block the individual symbols, and save the symbol library.

Creating the Drawing

Create a new drawing and use the symbol library to configure three computers—a low-cost model, a high-end "supermodel," and a middle-of-the-road model. Extract the attributes and create a list of materials needed to assemble the three computers.

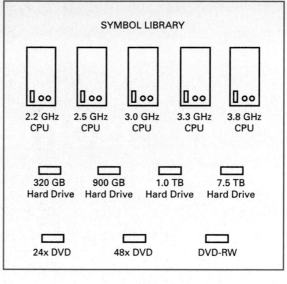

Fig. P7-2

▶ Hints and Suggestions

1. Create all blocks on layer 0.
2. Create only one symbol for each type of component (CPU, hard disk drive, etc.). Assign attributes to it. Then assign parameters to create a dynamic block that can be used to represent all of the models in that category.

▶ Summary Questions

Your instructor may direct you to answer these questions orally or in writing. You may wish to compare and exchange ideas with other students to help increase efficiency and productivity.

1. Distinguish between groups and blocks.
2. Describe the function of a symbol library.
3. What might be an advantage of maintaining a library of dynamic blocks?
4. What factors might determine the attributes assigned to a block?
5. Why is creating a bill of materials from a set of blocks a timesaving technique? Explain.

Isometric Drawing

Objectives

- Set up a drawing file for an isometric drawing
- Create an isometric drawing
- Dimension an isometric drawing correctly

Vocabulary

isometric drawing
isometric planes
pictorial representation

An **isometric drawing** is a type of two-dimensional drawing that gives a three-dimensional (3D) appearance. Isometric drawings are often used to produce a **pictorial representation** of a machined part, injection-molded part, architectural structure, or some other physical object. Pictorial representations are useful because they provide a realistic view of the object being drawn, compared to orthographic views. The drawing in Fig. 37-1 is an example of an isometric drawing created with AutoCAD.

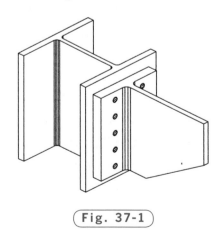

Fig. 37-1

Setting Up an Isometric Drawing

Because isometric drawings are two-dimensional, all of the lines in an isometric drawing lie in a single plane parallel to the computer screen. The drawing achieves the 3D effect by using three axes at 120° angles, as shown in Fig. 37-2. In AutoCAD, isometric drawing is accomplished by changing the SNAP style to Isometric mode.

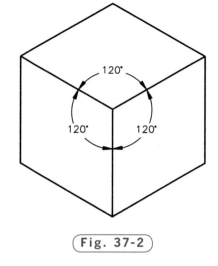

Fig. 37-2

Isometric Planes

As you can see from Fig. 37-2, the 120° angles produce left, right, and top imaginary drawing planes. These three planes are known as the **isometric planes**.

1. Start AutoCAD and begin a new drawing from scratch using **Imperial** units.

2. Select the **AutoCAD Classic** workspace and close all floating toolbars and palettes.

3. Set the drawing limits to **15,12**.

4. Right-click on the **Snap** or **Grid** button on the status bar and select **Settings...** from the shortcut menu.

5. Set the grid to **1** unit and the snap resolution to **.5** unit.

6. In the Snap type area of the dialog box, set the Grid snap to **Isometric snap.**

7. Pick **OK** to close the Drafting Settings dialog box.

You should now be in Isometric drawing mode, with the crosshairs shifted to one of the three isometric planes.

You can also enter isometric snap mode with the SNAP command. Enter S for Style and I for Isometric.

Toggling Planes

When you produce isometric drawings, you will use more than one of the three isometric planes. To make your work easier, you can shift the crosshairs from one isometric plane to the next.

1. Press the **CTRL** and **E** keys and watch the crosshairs change.
2. Press **CTRL** and **E** again, and again.

As you can see, this toggles the crosshairs from one isometric plane to the next.

 You can also toggle the crosshairs using the ISOPLANE command. Enter ISOPLANE and then select the Left, Top, or Right option to change the isometric plane. The F5 key also toggles among the three planes.

Creating an Isometric Drawing

Let's construct a simple isometric drawing.

1. Create a layer named **Objects**, assign a color of your choice to it, and make it the current layer.
2. Enter the **LINE** command and draw the aluminum block shown in Fig. 37-3. Make it **3 × 3 × 5** units in size.
3. Save your work in a file named **iso.dwg**.
4. Alter the block so that it looks similar to the one in Fig. 37-4 using the **LINE**, **BREAK**, and **ERASE** commands.
5. Further alter the block so that it looks similar to the one in Fig. 37-5. Use the **ELLIPSE** command to create the holes.

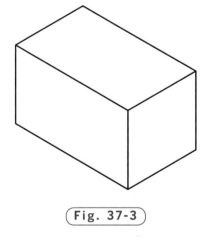

(Fig. 37-3)

 Since the ellipses are to be drawn on the three isometric planes, toggle the crosshairs to the correct plane. Then choose the ELLIPSE command's Isocircle option. You will need to change the snap spacing to .25.

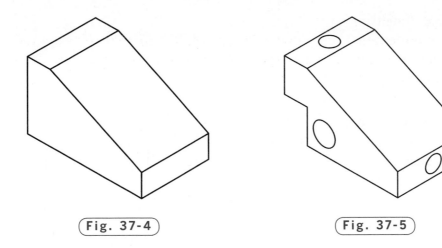

Fig. 37-4

Fig. 37-5

Dimensioning an Isometric Drawing

To dimension an isometric drawing properly, you must rotate the dimensions so that they show more clearly what is being dimensioned. To do this, you can use the Oblique option of the DIMEDIT command. Compare the two drawings shown in Fig. 37-6. The drawing at the right is the result of applying oblique to the drawing at the left.

Let's dimension the aluminum block.

1. Create a new layer named **Dimensions**, assign a different color to it, and make it current.

2. In the lower left corner of the block, draw one horizontal and one vertical extension line as shown in Fig. 37-7.

3. Display the **Dimension** toolbar and use the **Aligned** button to create the dimensions as shown in Fig. 37-7.

Your drawing should now look similar to the one in Fig. 37-7.

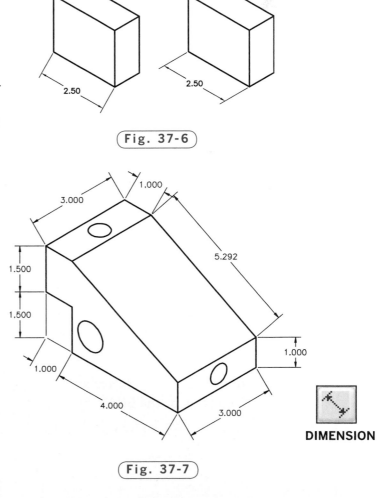

Fig. 37-6

Fig. 37-7

DIMENSION

Notice that dimensions are not well-oriented. This is because you have not yet applied oblique to them.

DIMENSION

4. From the **Dimension** toolbar, pick the **Dimension Edit** button to enter the **DIMEDIT** command.

5. Select the **Oblique** option.

6. Select the 3.000 dimension at the top of the block and press **ENTER**.

7. Enter **–30** for the obliquing angle.

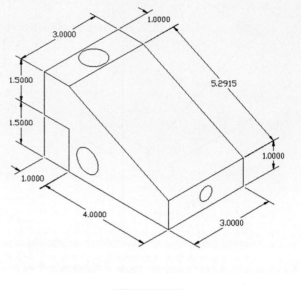

Fig. 37-8

The dimension rotates –30° to take its proper position.

8. Use **DIMEDIT** to correct the appearance of the remaining dimensions so that they match those in Fig. 37-8.

Enter an obliquing angle of 30 or –30, depending on the dimension you are editing. Use grips if necessary to adjust the positions of the dimension lines.

9. Select the 1.000 dimension at the top of the block. Right-click and select **Properties** from the shortcut menu, and change the right arrow (it may be Arrow 1 or Arrow 2) to **None** for a cleaner appearance.

10. Use the **Dimension Space** button to align the two sets of dimensions on the left and lower left sides of the drawing.

When you are finished, your drawing should look like the one in Fig. 37-8.

11. Return to AutoCAD's standard drawing format using the Drafting Settings dialog box or the **SNAP** command.

12. Save your work and exit AutoCAD.

Chapter 37 Review & Activities

● REVIEW QUESTIONS

Answer the following questions on a separate sheet of paper.

1. Describe two methods of changing from AutoCAD's standard drawing format to isometric drawing.

2. Describe how to change the isometric crosshairs from one plane to another.

3. Explain how to create accurate isometric circles.

4. How do you correct the appearance of dimensions used on an isometric drawing?

5. How do you change from isometric drawing to AutoCAD's standard drawing format?

● CHALLENGE YOUR THINKING

These questions are designed to further your knowledge of AutoCAD by encouraging you to explore the concepts presented in this chapter. Answer each question on a separate sheet of paper.

1. Discuss the purposes of isometric drawings. When and why are they used?

2. If you were to dimension the holes in the aluminum block created in this chapter, how would you go about it?

Continued

● APPLYING AUTOCAD SKILLS

Work the following problems to practice the commands and skills you learned in this chapter.

1–3. Create the machined parts shown in Figs. 37-9 through 37-11 using AutoCAD's isometric capability. Estimate their sizes.

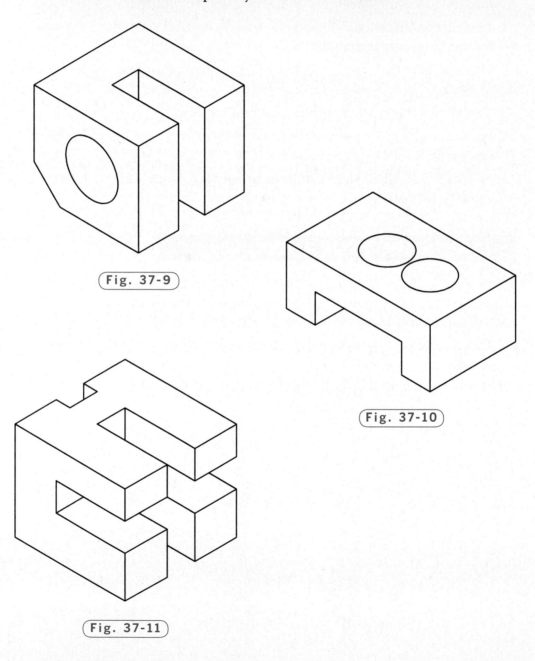

Fig. 37-9

Fig. 37-10

Fig. 37-11

4. Draw the isometric solid according to the dimensions shown in Fig. 37-12A. Center the hole in the top surface of the part. Specify a diameter of 10.61. Your finished drawing should look like the one in Fig. 37-12B.

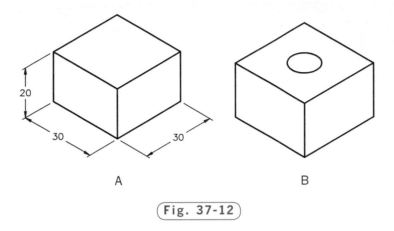

A

B

Fig. 37-12

5. Create the cut-away view of the wall shown in Fig. 37-13.

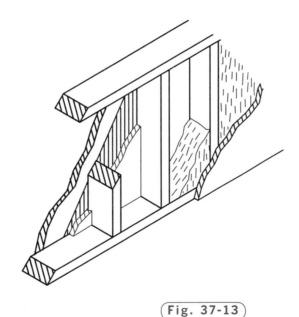

Fig. 37-13

Courtesy of Autodesk, Inc.

Continued

6. A drawing of a guide block is shown in Fig. 37-14. Draw a full-scale isometric view using the SNAP and ELLIPSE commands. An appropriately spaced grid and snap will facilitate the drawing.

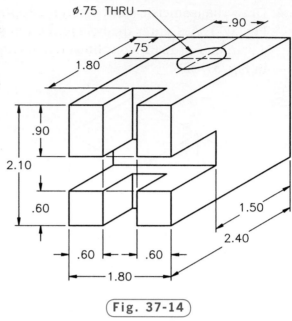

Fig. 37-14

Courtesy of Gary J. Hordemann, Gonzaga University

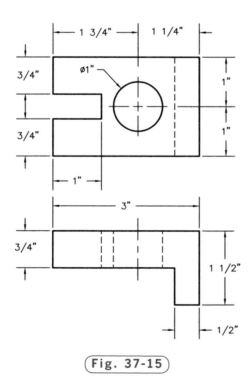

Fig. 37-15

7. Accurately draw an isometric representation of the orthographic views shown in Fig. 37-15. Draw and dimension the isometric according to the dimensions provided.

8. An exploded, sectioned orthographic assembly is shown in Fig. 37-16. Draw the assembly as a full-scale exploded isometric assembly. Use isometric ellipses to represent the threads, as shown in the isometric pictorial.

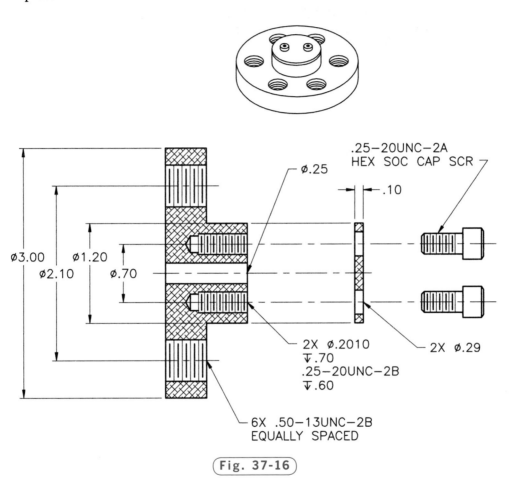

Fig. 37-16

Courtesy of Gary J. Hordemann, Gonzaga University

Continued

● USING PROBLEM-SOLVING SKILLS

Complete the following problems using problem-solving skills and your knowledge of AutoCAD.

1. The design engineer for the quality assurance division of your company asked you for an isometric drawing of the dovetail block shown in Fig. 37-17, which is used in a milling machine setup. Draw the block, dimension it, and save it as ch37dove.dwg.

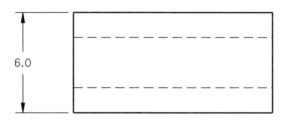

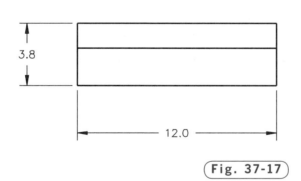

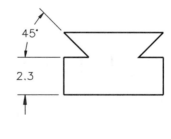

Fig. 37-17

2. The engineer also needs an isometric drawing of the slide shown in Fig. 37-18 for the same milling machine setup. This one, however, does not need to be dimensioned. Save the drawing as ch37slide.dwg.

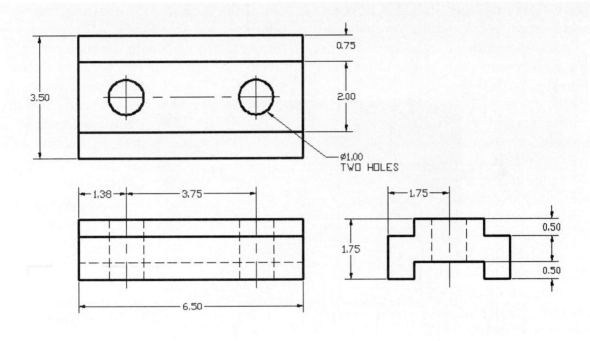

Fig. 37-18

3D Civil Designer

Civil design and drafting is a highly varied occupation. Civil designers work on highways, harbor facilities, railroads, airports, and even city plans and site developments. The actual creation of documents such as plats, operations maps, and contour maps, however, is often a highly time-consuming and sometimes frustrating endeavor. A change in one view or type of document necessitates changes in multiple other documents.

© PhotoLinkGetty Images/RF

With today's 3D civil software, however, designers can use engineering models directly to produce documents, annotations, and tables that update automatically when the engineering model changes. The software can even enforce drafting standards and company styles. Never again will a designer or drafter have to make change after change manually to civil documents. Model-driven documents are always up-to-date.

Advantages of 3D Civil Software

The most obvious result of using 3D modeling software to keep design and drafting documents up-to-date automatically is a huge time savings. Many companies are finding that their land development projects can be done in 25% less time than projects based on simpler CAD products that do not interact with the engineering model. The use of 3D civil software can literally shave years off projects compared to traditional, manual drafting methods.

Education

A 3D civil designer or drafter needs the same basic education needed by all civil drafters. In addition, a 3D designer needs to be comfortable working with and extracting data from 3D models prepared by engineers. A solid background in science and math is also helpful.

▶ Career Activities

1. If possible, arrange to see the plat for your town or city. How was it created? Explain how it could benefit from 3D civil CAD software.

2. Research civil design and drafting jobs in your area. What is the job outlook?

The Third Dimension

Objectives

- Create a basic 3D model
- Apply several 3D viewing options
- View and control a model interactively in 3D space
- Become familiar with the **3D Modeling** workspace
- Learn to use the **SteeringWheel** and **ViewCube** tools

Vocabulary

arcball
camera
clipping plane
compass
extrusion thickness
ground plane
hidden line removal
navigation bar
parallel projection
perspective projection
roll
Z axis

Three-dimensional (3D) drawing and modeling has become popular due, in large part, to improvements in hardware and software. AutoCAD's 3D capabilities allow you to produce realistic views of car instrument panels, snow skis, office buildings, and countless other products and structures. A key benefit of creating 3D drawings is to communicate better a proposed design to people and groups inside and outside a company. Also, the availability of 3D data speeds many downstream processes such as engineering analysis, digital and physical prototyping, machining, and mold-making. In

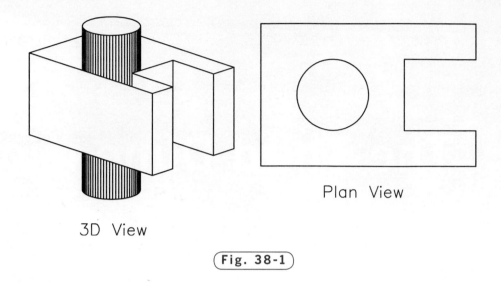

3D View

Plan View

Fig. 38-1

the field of architecture, 3D modeling can help companies persuade customers to move ahead with a proposed construction project such as a shopping mall.

This chapter introduces AutoCAD's 3D capabilities. You will create a simple 3D model, learn to view it from any point in space, use many different viewing options, and learn the different visual styles available for 3D objects in AutoCAD.

An example of a model generated in AutoCAD is shown on the left side of Fig. 38-1. The model on the right is the same object, viewed from the top. The model consists of a section of steel rod and a sheet-metal shroud that fasten into an aircraft fuel control system. The model shows how the shroud protects a wiring harness that runs through the square notch in the shroud next to the rod. The shroud protects the wires from the heat of the rod. This prevents the plastic insulation from melting and causing an electrical short.

Creating a Basic 3D Model

To work in three dimensions in AutoCAD, you will use a third axis on the Cartesian coordinate system. This axis, called the **Z axis**, shows the depth of an object, adding the third dimension. (In this context, the X axis shows the width of the object, and the Y axis shows the length or height.) The Z axis runs through the intersection of the X and Y axes at right angles to both of them.

Let's draw a steel rod and sheet-metal shroud similar to those shown in Fig. 38-1.

1. Start AutoCAD and create a new drawing using the following information.

   - Use the **Quick Setup** wizard.

   - Set the units to **Decimal**.

   - Set the area to **12″ × 9″**.

   - Select the **Drafting & Annotation** workspace.

   - • **ZOOM All**.

2. Create a **1″** grid and a **.5″** snap.

3. Create a new layer named **Objects**, assign the color magenta to it, and make it the current layer.

4. Enter the **ELEV** command and set the new default elevation at **1″** and the new default thickness at **3″**.

An elevation of 1″ means the base of the object will be located 1″ above a plane of 0 on the Z axis. The thickness of 3″ means the object will have a thickness of 3″ in the Z direction. This is called the **extrusion thickness**.

5. Draw the top (plan) view of the sheet-metal shroud using the **LINE** command. Construct it as shown on the right in Fig. 38-1. Omit the rod at this time. The width and length of the shroud are 4″ × 6″, respectively. The square notch measures 2″ × 2″.

6. Save your work in a file named **3d.dwg**.

Viewing and Hiding

Let's view the object in 3D.

1. Select the View tab of the Ribbon and focus on the Views panel.

2. Pick the flyout arrow on the Views panel, and select **SE Isometric** from the flyout menu.

A 3D model of the object appears on the screen.

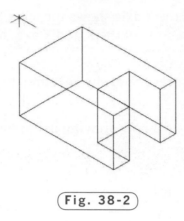

Fig. 38-2

3. Enter **Z** (for ZOOM) and **.9x**.

The object should look somewhat like the one in Fig. 38-2.

The HIDE command removes edges that would be hidden if you were viewing a real, solid 3D object. Hiding these edges produces a more realistic view of the object. This is called **hidden line removal**.

HIDE

4. Pick the **Hide** button on the Visual Styles panel.

The object should now look similar to the one in Fig. 38-3.

5. Return to the plan view of the object by selecting **Top** from the flyout menu on the Views panel of the Ribbon.

6. **ZOOM All**.

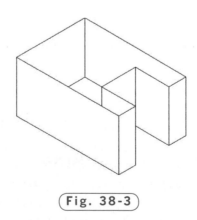

Fig. 38-3

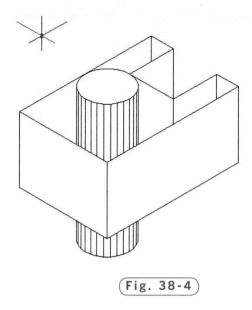

Fig. 38-4

Changing Elevation and Thickness

Let's add the steel rod to the 3D model at a new elevation and thickness.

1. Enter the **ELEV** command and set the elevation at **–1″** and the thickness at **6″**.
2. Draw a **2″**-diameter circle in the center of the model, as shown in Fig. 38-1 on page 512.
3. On the Views panel on the View tab of the Ribbon, select **SW Isometric** from the flyout menu.
4. Pick the **Hide** button to remove the hidden lines.

Your drawing should now look similar to the one in Fig. 38-4.

5. Save your work.

Viewing Options

AutoCAD offers several options for viewing models in 3D space. One method you've already used is the Views panel on the Ribbon. Another is 3D Orbit, an option that is both visual and easy to use. The Viewport Controls in the upper left corner of the drawing area are always available to change the view. The ViewCube is located in the upper right corner of the drawing area. Finally, you can use a tool called the SteeringWheel to manage the drawing view. The method you choose should be based on your personal preference. One method is only better than another if it's faster for you and if it accomplishes what you are trying to achieve.

3D Orbit

3D Orbit is a tool for manipulating the view of 3D models by clicking and dragging. It's especially useful because you can use it to manipulate 3D models while they are shaded or when hidden lines are removed. 3D Orbit's right-click shortcut menu offers options for removing hidden lines, rotating objects, panning, zooming, shading, perspective viewing, and clipping planes. You will explore these options in the following steps.

1. Turn **GRID** on.

NAVIGATE 2D

2. In the Navigate 2D panel on the View tab of the Ribbon, pick the arrow next to Orbit and select **Free Orbit** from the flyout menu.

This enters the 3DFORBIT command and activates 3D Free Orbit. When you enter 3D Free Orbit, AutoCAD displays an arcball. The **arcball** is the green circle divided into four quadrants by smaller circles. The UCS icon is the red, green, and blue icon made up of the X, Y, and Z axes. The icon serves as a visual aid to help you understand the direction from which you are viewing the model. You will learn more about user coordinate systems in later chapters.

The XY plane—the 2D plane on which all Z values are zero—is called the **ground plane**. The ground plane is the same plane on which you have created drawings throughout this book, but now you can view them in three dimensions.

When the 3DFORBIT command is active, the target of the view stays stationary and the point of view (also called the **camera**) moves around the target. The target point is the center of the arcball, not the center of the object(s) you are moving.

3. Position the cursor inside the arcball and click and drag upward to rotate the model.

The model rotates freely in three dimensions around the target. The grid, UCS icon, and ViewCube provide visual reference.

4. With the 3DFORBIT command active, right-click to display a shortcut menu.

The second line of the shortcut menu displays the current Orbit mode, which is Free Orbit.

5. Rest the cursor over **Other Navigation Modes** to view a cascading menu of options.

The Constrained Orbit mode limits the rotation of the model with respect to the Z axis and the grid. It allows rotation around the Z axis, but constrains the rotation around the XY plane to the top or bottom view.

Continuous Orbit mode allows you to animate the model in a continuous rotation. You can control the direction and speed of the continuous orbit by moving the cursor.

6. Experiment with the **Constrained Orbit** and **Continuous Orbit** modes on your own.

7. Reselect **Free Orbit** from the shortcut menu.

Rotating Objects in 3D Space

When you move the cursor over different parts of the arcball, the cursor icon changes. When you click and drag, the appearance of the cursor indicates the rotation of the view.

1. Study the cursor; then move it outside the arcball and then back inside it.

When the cursor is inside the arcball, it changes to a small sphere encircled by two lines, as shown in Fig. 38-5A. When this cursor is present, you can manipulate the view freely.

2. With the cursor inside the arcball, click and drag up and down, back and forth.

3. Move the cursor outside the arcball.

The cursor changes to a circular arrow around a small sphere, as shown in Fig. 38-5B.

When you click and drag outside the arcball, AutoCAD moves the view around an axis at the center of the arcball, perpendicular to the screen. This action is called a **roll**.

4. With the cursor outside the arcball, click and drag.

5. Move the cursor over one of the smaller circles on the left or right side of the arcball.

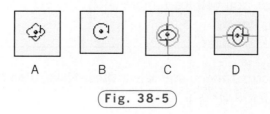

A   B   C   D

Fig. 38-5

The cursor becomes a horizontal ellipse around a small sphere as shown in Fig. 38-5C. When you click and drag from either of these points, AutoCAD rotates the view around the Y axis that extends through the center of the arcball.

6. With the cursor over one of the smaller circles at the left or right, click and drag to the left and right.

7. Move the cursor over one of the smaller circles on the top or bottom of the arcball.

The cursor becomes a vertical ellipse around a small sphere, as shown in Fig. 38-5D. When you click and drag from either of these two points, AutoCAD rotates the view around the horizontal or X axis that extends through the center of the arcball.

8. With the cursor over one of the smaller circles at the top or bottom, click and drag up and down.

Panning and Zooming

1. With the 3DFORBIT command active, right-click to display the shortcut menu and select **Other Navigation Modes** and **Pan**.

This causes AutoCAD to change to the pan mode within the 3DFORBIT command.

2. Click and drag to adjust the view of the model.

3. Right-click and select **Other Navigation Modes** and **Zoom**.

4. Zoom in and out on the model.

5. Display the shortcut menu and select **Reset View**.

Reset View resets the view back to how it was when you first started in a 3D Orbit mode. The Preset Views option displays a list of predefined views such as Top, Front, and NW Isometric.

 Notice that the background turns gray when you select Reset View. This background appears by default in the 3D workspace. You will learn more about the 3D workspace later in this chapter.

6. Display the shortcut menu and select **Other Navigation Modes** and **Free Orbit**.

The arcball returns.

7. Press **ESC** or **ENTER** to exit Free Orbit.

The arcball disappears and the three-colored UCS icon is replaced with the one-color coordinate system icon. Notice also that the background color in the drawing area is white.

Controlling Visual Aids

AutoCAD provides several visual aids to help orient you in 3D space as you work with objects in 3D Orbit.

1. Re-enter the **3DFORBIT** command.
2. From the shortcut menu, pick **Visual Aids** and **Compass**.

This displays a 3D sphere within the arcball composed of three ellipses representing the X, Y, and Z axes.

3. Rotate the model and pay attention to the 3D sphere.
4. From the shortcut menu, turn off **Compass**.

The Visual Aids shortcut menu also controls the display of the grid and the UCS icon. The grid represents the 0 elevation ground plane discussed earlier in this chapter. We will leave the grid and the USC icon displayed.

5. Rotate the model and pay attention to the movement of the ground plane.
6. Press **ESC** or **ENTER** to exit the Free Orbit mode.

The Steering Wheel

Notice the grayed-out, vertical bar on the right side of the drawing area. This is the Navigation Bar. This toolbar provides access to many useful navigation tools.

If the Navigation Bar is not displayed, select the [-] item on the Viewport Controls in the upper left corner of the drawing area and toggle the Navigation Bar on.

1. Pick the **Full Navigation Wheel** button on the Navigation Bar.

This enters the NAVSWHEEL command and attaches a large, multi-button icon to the pointer called the Full Navigation Wheel, as shown in Fig. 38-6.

NAVIGATION BAR

2. If the wheel on your screen does not match Fig. 38-6, right-click and select **Full Navigation Wheel** from the shortcut menu.

Study the buttons on the wheel. Several of the buttons should be familiar to you, such as Zoom, Pan, and Orbit. Notice also the close button and the down-facing arrow button, which displays the shortcut pull-down menu.

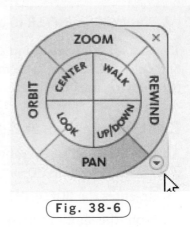

Fig. 38-6

3. Click and drag the **Zoom** button on the Full Navigation Wheel to use the Zoom Tool.

4. Experiment with the **Pan** and **Orbit** buttons.

Notice the green Pivot icon that appears on the screen while you are using the Zoom Tool and the Orbit Tool. This is the center point of the orbit and zoom actions.

5. Pick the **Center** button on the Full Navigation Wheel and drag the wheel to the lower left corner of the sheet-metal shroud. When the green Pivot icon appears, release the mouse button.

6. Pick the **Orbit** button, and rotate the drawing.

Notice that the pivot of the orbit has now changed to the center point you selected.

7. Click once on the **Rewind** button.

AutoCAD takes you to the previous view.

8. Click on the **Rewind** button and hold the mouse button down.

The Rewind Tool appears, along with a row of thumbnails that represent all the previous views.

9. Drag the cursor across the row of thumbnails and watch the view on the screen change. Release the mouse button on a previous view of your choice.

10. Pick the down-facing arrow in the lower right corner of the Full Navigation Wheel to display the wheel's shortcut menu.

Notice the different options for wheels, including options for tour building, object viewing, full object viewing—plus mini and basic versions of each.

11. Experiment on your own with the shortcut menu. Select each of the mini wheels, basic wheels, view the SteeringWheel Settings... dialog box, select **Go Home,** then close the wheel.

Using the 3D Modeling Workspace

Now that you are working in three dimensions, it is time to revisit the 3D Modeling workspace.

1. Pick the **Workspace** button on the Quick Access toolbar, select the **3D Modeling** workspace, and close any open palettes and floating toolbars.

Focus your attention on the View panel of the Home tab on the Ribbon.

2. Pick the down arrow to the right of the text box to display the options under Visual Styles.

AutoCAD displays graphic representations of the available visual styles, including the familiar 2D Wireframe style, as shown in Fig. 38-7.

3. Pick the **Wireframe** style and observe the changes in the drawing area.

4. View the model in the **Hidden, Realistic,** and **Conceptual** visual styles.

5. Experiment with all the other visual styles.

6. Pick the down arrow to the right of the 3D Navigation text box and pick the **Top** view.

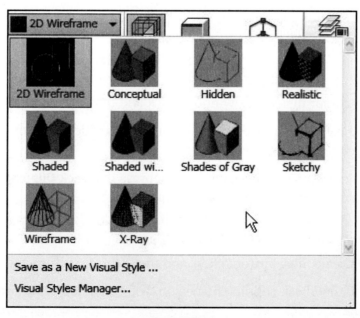

Fig. 38-7

AutoCAD displays the top view of the model.

7. Pick the down arrow again, and pick the **NW Isometric**, **Back**, and **NE Isometric** views.

8. Select the **View** tab of the Ribbon.

Notice the Navigate panel in the 3D Modeling workspace includes the Steering Wheels button. Now focus your attention on the Viewport Controls in the upper left corner of the drawing area.

9. Pick the **View Controls** command (it should currently read NE Isometric) and select **NW Isometric** from the pull-down menu that appears.

10. Pick the **Visual Styles Control** command (the third bracketed text item) and select **Conceptual**.

As you can see, the Viewport Controls are another easy way to control views and visual styles.

Clipping Planes

A **clipping plane** is a plane that slices through one or more 3D objects, causing part of the object on one side of the plane to be omitted. It is useful in viewing the interior of a part or assembly. You can use clipping planes to achieve views of an object that are similar to sectional views.

1. Enter the **3DCLIP** command at the keyboard.

The Adjust Clipping Planes window appears.

2. In the Adjust Clipping Planes window, depress the **Front Clipping On/Off** button. (It may already be depressed.)

3. With the cursor inside this window, click and drag up and down.

This moves the clipping plane up and down and provides a corresponding view of the model in the drawing area.

4. In the Adjust Clipping Planes window, depress the **Back Clipping On/Off** button, as well as the **Adjust Back Clipping** button.

5. With the cursor inside this window, click and slowly drag up and down to adjust the location of the back clipping plane.

6. Pick the **Create Slice** button, click, and move the cursor up and down.

With the front and back clipping planes on, you can adjust their location at the same time.

7. Depress the buttons to turn **Front Clipping** and **Back Clipping** off.

8. Close the Adjust Clipping Planes window.

The ViewCube

1. Pick the **2D Wireframe** visual style and **SW Isometric** view, both on the Viewport Controls in the upper left corner of the drawing area.

The visual style you have used thus far in this textbook is 2D Wireframe. The other visual styles are 3D visual styles.

2. Pick the **Wireframe** visual style.

Notice the changes in the drawing area. The background color is now dark gray, and the three-colored icon is present. These are the visual styles for all 3D visual styles.

Focus your attention on the ViewCube in the upper right corner of the drawing area, as shown in Fig. 38-8. You can toggle the ViewCube and Navigation Bar on and off with the Viewport Controls command in the upper left corner of the drawing area.

3. Pick the **FRONT** face of the ViewCube and observe the view change in the drawing area.

4. Pick the arrow above the cube to change to the TOP view.

5. Pick the **W** (for west) on the ViewCube compass to change to the LEFT view.

6. Pick the rotation arrows in the upper right corner of the ViewCube and observe the result in the drawing area.

7. Pick the corners and edges of the ViewCube and observe the results in the drawing area.

8. Pick the **Home** button in the upper left corner of the ViewCube.

As you can see, the ViewCube is a fast, powerful option for viewing your model in three dimensions.

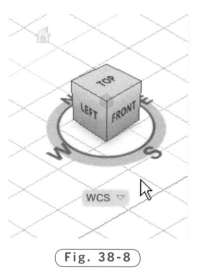

Fig. 38-8

Parallel and Perspective Views

Consider the buildings shown in Fig. 38-9. Both are shown from the same point in space. However, the one on the left is shown in **parallel projection.** The lines are parallel and do not converge toward one or more vanishing points. The building on the right is a **perspective projection.** Parts of the building that are farther away appear smaller, the same as in a real-life photograph.

1. Right-click on the **ViewCube** and pick **Parallel** from the shortcut menu.

Notice the lines in the model are parallel.

2. Right-click on the **ViewCube** and pick **Perspective** from the shortcut menu.

Notice the model looks more realistic, with the lines in the model converging toward a vanishing point.

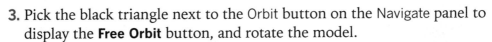

NAVIGATE

3. Pick the black triangle next to the Orbit button on the Navigate panel to display the **Free Orbit** button, and rotate the model.

As you rotate the model, notice the color difference between the "sky" and the "ground." The boundary between the two shades represents the horizon line. When perspective projection is on, AutoCAD provides one shade for the ground plane, with subtle differences near its horizon line, and another for the space above the ground plane.

4. Press **ESC** to exit 3D Orbit.

5. Save your work and exit AutoCAD.

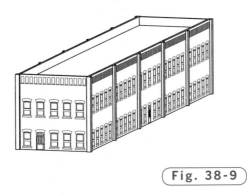

Fig. 38-9

Chapter 38 Review & Activities

REVIEW QUESTIONS

Answer the following questions on a separate sheet of paper.

1. The extrusion thickness of an object is specified with which command?
2. Briefly explain the process of creating objects (within the same model) at different elevations and thicknesses.
3. With what command are hidden lines removed from 3D objects?
4. Why is the 3D Free Orbit feature especially useful, compared to other 3D viewing options?
5. Briefly explain how the 3D Modeling workspace and the Ribbon can be useful for 3D modeling.
6. What is the difference between parallel projection and perspective projection?
7. What is a clipping plane?

CHALLENGE YOUR THINKING

These questions are designed to further your knowledge of AutoCAD by encouraging you to explore the concepts presented in this chapter. Answer each question on a separate sheet of paper.

1. Of what benefit are front and back clipping planes? Describe at least one situation in which clipping planes could save drawing time or reveal important information about a design.

Continued

2. Enter the *x*, *y*, and *z* coordinate values for the alphabetically identified points in Fig. 38-10. To get you started, point A has coordinate values of 0,0,0.

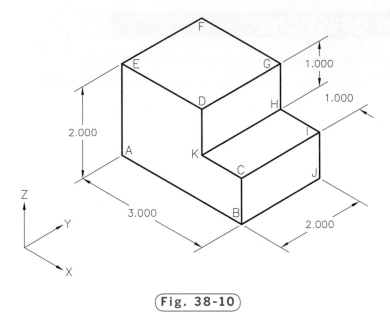

Fig. 38-10

● APPLYING AUTOCAD SKILLS

Work the following problems to practice the commands and skills you learned in this chapter. Draw the following objects and generate a 3D model of each.

1. Set the elevation at 1″ to create a 3D model of the guide shown in Fig. 38-11.

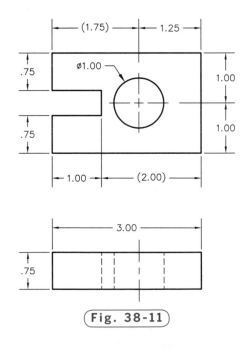

Fig. 38-11

2. To draw the shaft shown in Fig. 38-12, set the elevation for the inner cylinder at 0. Set the snap resolution before picking the center point of the first circle.

3. Draw the gear blank and hub shown in Fig. 38-13 as a basic 3D model. In the AutoCAD Classic workspace, view the model using each of the buttons on the View toolbar. Are all the views practical for this model? Which ones are not?

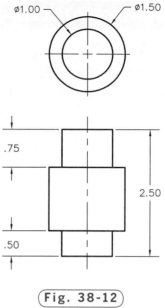

Fig. 38-12

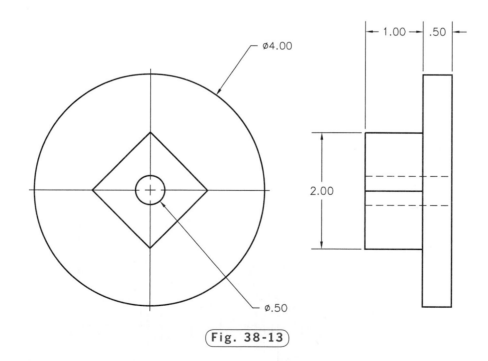

Fig. 38-13

Continued

● USING PROBLEM-SOLVING SKILLS

Complete the following activities using problem-solving skills and your knowledge of AutoCAD.

1. The gear manufacturer for whom you supply the gear blank (see "Applying AutoCAD Skills" problem 3) wants to see the blank in a realistic form. Provide a 3D view of the gear blank in the Realistic visual style. Save and plot the drawing. Did it plot the way you expected?

2. Use your knowledge of AutoCAD to create the shaft, gear, and cam shown in Fig. 38-14 as closely as possible.

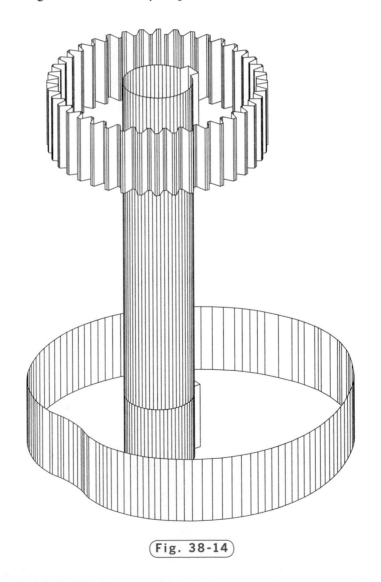

Fig. 38-14

Chapter **39**

Drawing Objects in 3D

Objectives

- Draw objects at different Z elevations
- Draw objects using coordinate entry
- Mirror objects in 3D
- Create 3D faces

Vocabulary

elevations
3D faces

In Chapter 38, you were introduced to the third axis of the Cartesian coordinate system, the Z axis. AutoCAD offers a variety of ways to create objects in 3D space. In this chapter, you will use familiar tools and new commands to create objects in three dimensions.

Drawing 2D Objects in 3D Space

The ELEVATION command allows you to create objects at different **elevations**, or Z heights.

1. Start AutoCAD and create a new drawing as follows.

   - Pick the **Use a Template** option.
   - Select the **acad3d.dwt** template.
   - Enter the **UNITS** command and pick **Feet** as the Insertion Scale.
   - Right-click on the **ViewCube** and pick **Parallel** projection from the shortcut menu.
   - Select the **Top** view and **ZOOM All**.

2. Set snap spacing to **.5** and grid spacing to **1.0**.

3. Create a new layer named **Objects**, assign the color red to it, and make it the current layer. (Enter **LAYER** at the keyboard or use the **Layer Properties** button on the Layers panel of the Home tab or the Palettes panel of the View tab.)

4. Draw the base of a storage shed as shown in Fig. 39-1 using the **LINE** command. Place the lower left corner at $x = 1$, $y = 1$. The shed's base is $8' \times 10'$.

As you draw the lines, look at the coordinate display on the status bar. Notice that the Z value of the lines is zero because you are drawing on the ground plane, where $z = 0$.

5. Enter the **ELEV** command and set the new default elevation to **8**. This is the elevation at the top of the shed's walls. For the new default thickness, enter **0**.

6. Move the cursor around in the drawing area and notice the Z coordinate value on the status bar.

As you can see, you are now on a work plane of $z = 8$. The objects you draw now will be 8 feet above the shed's base.

7. Use the **LINE** command to draw the outline of the shed at the top of the walls. The walls are vertical, so the outline will be identical to the outline of the base, $8' \times 10'$. Note that object snap must be off during this operation to avoid snapping to points at the 0 elevation.

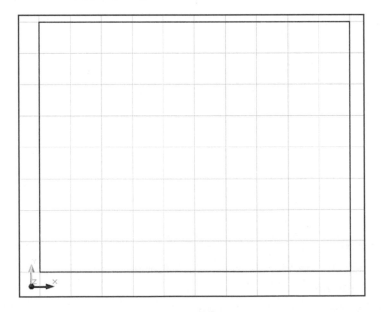

Fig. 39-1

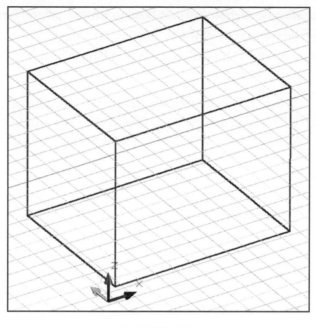

Fig. 39-2

8. Select the **SW Isometric** view using the ViewCube or the Viewpoint Controls. (You may need to turn the grid off to see the base outline clearly.)

9. Use the **LINE** command to draw the four vertical corners of the shed, as shown in Fig. 39-2.

HINT | Turn snap off and object snap on. Use the Orbit command to rotate the model slightly to make endpoint selection easier.

10. Save your work as **workshed.dwg**.

3D Drawing by Coordinate Entry

Let's add a roof to our storage shed.

1. In the status bar, pick the **Dynamic Input** button to deselect it.

This deactivates AutoCAD's dynamic input. Coordinate entry with dynamic input is preset in the Pointer Input Settings dialog box and can yield unpredictable results. Refer to Chapter 7 for more information on coordinate entry with dynamic input.

STATUS BAR

2. To the left of the the Dynamic Input button on the status bar, pick the **Dynamic UCS** button to deselect it.

3. Display the View tab of the Ribbon and pick the **World** button from the Coordinates panel.

Before entering endpoint coordinates for the new roofline, you must be sure that the dynamic UCS feature is turned off and the coordinate system is set to the standard World coordinate system. You will learn more about user coordinate systems in Chapter 40.

4. Enter the **LINE** command.

5. In response to Specify first point, type **0,0,8** and press **ENTER**.

6. In response to Specify next point, type **12,0,8** and press **ENTER**.

7. Enter **12,5,12** for the third point and **0,5,12** for the fourth point.

8. Type **C** to close the line segments and terminate the LINE command.

9. View the shed from the **Front** and **Left**; then return to the **SW Isometric** view.

Mirroring Objects in 3D

The ability to mirror objects is not limited to two dimensions. Let's use the 3D MIRROR command to create the other half of the roof.

1. Pick the **World** button on the Coordinates panel.

2. Pick the **3D Mirror** button on the Modify panel of the Home tab, select lines **1**, **2**, and **3**, as shown in Fig. 39-3, and press **ENTER**.

3. In reply to Specify first point of mirror plane, select the **ZX** option to specify a mirror plane that is parallel to the XZ plane.

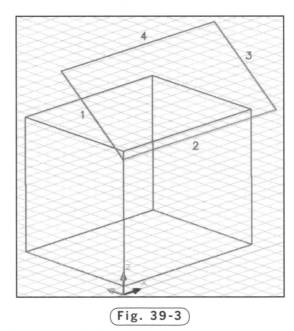

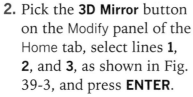

Fig. 39-3

4. For a point on the mirror plane, select either endpoint of line 4. Do not delete the source object.

5. View the shed from the **Front** and **Left**; then return to the **SW Isometric** view.

As you can see, you have now created the back half of the roof.

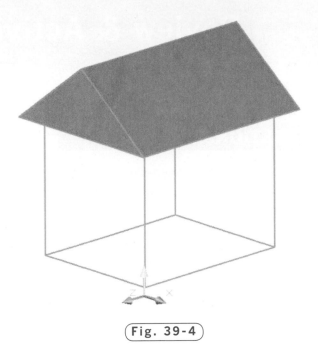

(Fig. 39-4)

Creating 3D Faces

Thus far, you have created only wireframes in three dimensions. In this context, a wireframe is a representation of an object that uses lines and curves to define its boundaries. The shed you have created should have solid walls and a solid roof, of course. The 3DFACE command creates **3D faces**, which are three- or four-sided planar meshes, to make the roof and walls appear solid.

1. On the View panel of the Home tab, select the **Conceptual** visual style.
2. Enter **3DFACE** at the keyboard.
3. Pick the four corner points of one side of the roof in a clockwise or counterclockwise order and press **ENTER**.

You have placed a 3D face on the front section of the roof so that it now appears solid.

4. Enter the **3DFACE** command, select the three points on the left end of the roof, and press **ENTER**.

This creates a face to define the triangular gable end of the roof on the left side of the shed. Your model should look similar to the one in Fig. 39-4.

5. Repeat this process to create 3D faces for the other two roof surfaces.
6. Save your work and exit AutoCAD.

Chapter 39 Review & Activities

● REVIEW QUESTIONS

Answer the following questions on a separate sheet of paper.

1. Can the ELEVATION command be used to create 2D objects on XZ and YZ planes? Explain.

2. In practical terms, what happens when you change the elevation from 0 to 5?

3. What UCS must be active before you create a 3D model by entering 3D coordinates?

4. In your own words, define *wireframe*.

5. Describe a 3D face object.

● CHALLENGE YOUR THINKING

These questions are designed to further your knowledge of AutoCAD by encouraging you to explore the concepts presented in this chapter. Answer each question on a separate sheet of paper.

1. Briefly explain the difference between the ELEVATION command and the THICKNESS system variable.

2. As you can tell, entering coordinates for object creation in 3D can be tedious and complex. Without looking ahead in the book, can you think of easier ways to create geometry in three dimensions?

● APPLYING AUTOCAD SKILLS

Work the following problems to practice the commands and skills you learned in the chapter.

1. Open worshed.dwg and save it as prb39-1.dwg. Create 3D faces for the four walls of the shed. View the shed from several views and with several different visual styles.

2. Open prb39-1.dwg and save it as prb39-2.dwg. Create a 3D face for a concrete pad under the shed. The pad measures 18′ × 20′ and extends beyond the shed walls evenly (5′) on all sides. When you are finished, the model should look similar to the one in Fig. 39-5.

Fig. 39-5

● USING PROBLEM-SOLVING SKILLS

Complete the following problems using problem-solving skills and your knowledge of AutoCAD.

1. Your proposed lamp shade design needs to be drawn in three dimensions before your supervisor will approve the design. Draw a 3D model of the lamp shade shown in Fig. 39-6. Omit the hanger. Create 3D faces where appropriate. Save the drawing as ch39-lampshade.dwg.

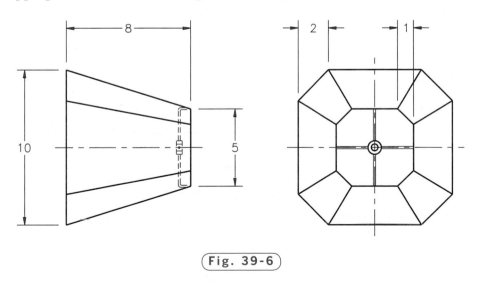

Fig. 39-6

Continued

2. Create the 3D representation of a sphere shown in Fig. 39-7. The center point of the sphere is the origin (0,0,0). Use the ELEVATION command and the data in Table 39-1 to create the series of 2D circles. The center of the circle at each elevation is located at 0,0. Save the drawing as ch39-globe.dwg.

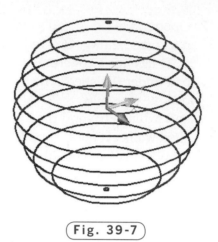

Fig. 39-7

Table 39-1. 2D Circle Data

Elevation (Z Height)	Center of Circle	Diameter of Circle
5.0	0,0	.25
4.0	0,0	6.00
3.0	0,0	8.00
2.0	0,0	9.00
1.0	0,0	9.85
0.0	0,0	10.00
–1.0	0,0	9.85
–2.0	0,0	9.00
–3.0	0,0	8.00
–4.0	0,0	6.00
–5.0	0,0	.25

Engineer

Most jobs provide opportunities for workers to fine-tune their skills. Young employees often start out in entry-level positions and, by showing their employers that they are hardworking, loyal, and motivated individuals, they are able to advance their status. Some workers move around within one company while others may work for several employers, holding jobs at different levels. Both of these career paths provide a learning platform for employees interested in advancing their status and earnings.

Growing on the Job

Drafters have a variety of opportunities to move upward in the workplace. Those who develop a particular area of interest, such as mechanical, electrical, architectural, or aeronautical, may eventually wish to work as an engineer. Enrolling in college programs and receiving a degree in a concentrated area is a requirement for any drafter interested in advancing to an engineering position.

© Ryan McVay/Getty Images/RF

Thinking About It

Engineers use their expertise to assist drafters, builders, and maintenance crews. They provide detailed working instructions and drawings for others involved in the production process, and they work closely with drafters to design systems and products. Engineers are problem-solvers; they are able to think clearly and logically in order to derive possible solutions to both existing and potential problems. The ability to work with others is also an important quality. Engineers often find themselves acting as project managers. Being able to prioritize and organize tasks helps them maintain an efficient working team.

▶ Career Activities

1. Write a report about the various types of engineers and their typical duties.

2. Are employment forecasts encouraging for engineers?

Chapter 40

User Coordinate Systems

Objectives

- Create user coordinate systems
- Move among UCSs as necessary to work on a 3D Model
- Create, name, and save UCSs

Vocabulary

UCS icon
**user coordinate systems
   (UCSs)**

AutoCAD's default coordinate system is the world coordinate system (WCS). In this system, the X axis is horizontal on the screen, the Y axis is vertical, and the Z axis is perpendicular to the XY plane (the plane defined by the X and Y axes).

You may create drawings in the world coordinate system, or you may define your own coordinate systems, called **user coordinate systems (UCSs)**. The advantage of a UCS is that its origin is not fixed. You can place it anywhere within the world coordinate system. You can also rotate or tilt the axes of a UCS in relation to the axes of the WCS. Many drawing aids (such as the grid, snap, coordinate display, and polar tracking features) move, rotate, or tilt to match the current UCS. These are useful features when you are creating a 3D model.

The **UCS icon** shows the orientation of the axes of the current coordinate system. Its default location is in the lower left corner of the drawing area. When you are viewing the plan view of the WCS in the 2D Wireframe visual style, the Z axis of the UCS icon is not visible, as shown in Fig. 40-1.

Fig. 40-1

Fig. 40-2

In the 3D visual styles, the UCS icon consists of three thicker, shaded arrows. The X axis is red, the Y axis is green, and the Z axis is blue. The 3D UCS icon is shown in Fig. 40-2.

Creating UCSs

The UCS command enables you to create a new UCS.

1. Start AutoCAD and open **workshed.dwg**. Select the **3D Modeling** workspace and the **SW Isometric** view.

2. Display the View tab on the Ribbon.

3. On the Coordinates panel, select the **World** button and then the **Origin** button.

COORDINATES

4. Use the endpoint object snap to pick the lower left corner of the shed, as shown in Fig. 40-3.

Notice that the UCS icon has moved to a new location, but it retains the same orientation as the WCS.

5. With snap on, draw a **3′**-diameter circle on the floor of the shed. Place the center of the circle **4′** from the front of the shed and **5′** from the left side of the shed, as shown in Fig. 40-3.

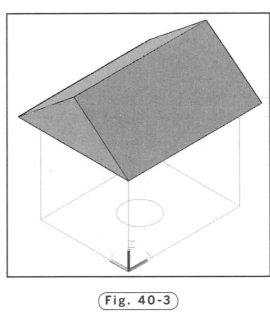

Fig. 40-3

Notice how easy it is to position the circle accurately when the coordinate display and snap are oriented to the new UCS.

Using Three Points

Next, let's create a door on the front of the shed. To do this, we will create a new UCS on the plane where the door should be. One of the most useful ways to define a new UCS is to select three points that lie on the axes. The first point becomes the origin and the other two lie on the X and Y axes.

COORDINATES

COORDINATES

1. Erase the circle you just created.

2. On the View panel of the Home tab, pick the **Wireframe** visual style.

3. Pick the **World** button on the Coordinates panel to reset the current UCS to the world coordinate system. Check to be sure dynamic input is off.

4. Pick the **3 Point** button on the Coordinates panel.

5. Once again, snap to the lower left corner of the shed for the new origin point.

6. In response to the next prompt, snap to the lower right corner of the shed.

This defines the positive orientation of the X axis.

7. In response to the next prompt, snap to the top of the wall at the front left corner of the shed, directly above the new origin.

Notice that the drawing does not change. However, the crosshairs, grid, and UCS icon shift to reflect the new UCS.

8. Use the **RECTANG** command to create a door. Type **3.5,0** to specify the first corner and **6.5,6.5** to specify the other corner.

As you can see, when you create a UCS oriented to the plane of the front wall of the shed, it greatly simplifies the process of drawing objects there. Now let's add some skylights to the roof.

9. From the View panel, switch to the **Hidden** visual style.

This will give you a clearer view of the roof surface.

COORDINATES

10. Pick the **3 Point** button on the Coordinates panel.

11. For the origin, snap to the front left corner of the roof.

12. Snap to the front right corner of the roof to define the orientation of the X axis.

13. Snap to the roof peak on the left side of the shed to define the orientation of the Y axis.

The new UCS is oriented to the front side of the roof. Notice also that the ViewCube is oriented to the new UCS.

14. Draw two 2′ × 3′ skylights on the front roof. Locate the skylights 2′ from the downhill corners of the front roof, as shown in Fig. 40-4.

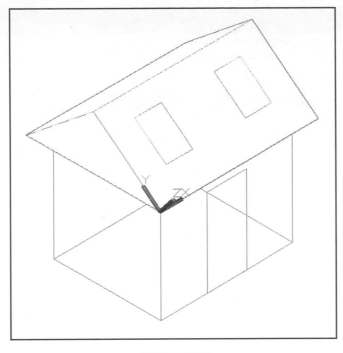

Fig. 40-4

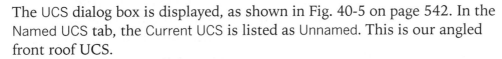

Managing UCSs

The purpose of creating user coordinate systems is to make it easier to work with 3D models. When you have created several UCSs, you need to be able to move from one to another quickly. It is helpful to save UCSs so you can do this without re-creating each UCS.

COORDINATES

1. Pick the **Named UCS...** button on the **Coordinates** panel.

The UCS dialog box is displayed, as shown in Fig. 40-5 on page 542. In the Named UCS tab, the Current UCS is listed as Unnamed. This is our angled front roof UCS.

2. Select **Unnamed** from the list of named UCSs, type **frontroof**, and press **ENTER**.

3. Select the **World** UCS, pick the **Set Current** button, and pick **OK**.

The world coordinate system becomes the current UCS.

4. In the Coordinates panel, pick the down arrow to the right of the Named UCS Combo Control text box to view the list of UCS options.

5. Select the **frontroof** UCS.

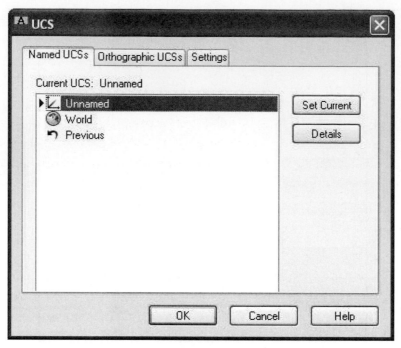

Fig. 40-5

The saved frontroof UCS becomes the current UCS again.

COORDINATES

6. Pick the **Named UCS...** button on the Coordinates panel to display the **UCS** dialog box.

Notice the choices of UCSs. In addition to the World UCS and the newly created frontroof UCS, AutoCAD provides orthographic UCSs oriented to the six viewpoints in three-dimensional space (Top, Bottom, Front, Back, Left, and Right). Previous makes the previous UCS the current UCS. AutoCAD saves the last ten UCSs, so you can back up with repeated Previous selections.

7. Pick **Cancel** to close the dialog box.

Now focus your attention on the other tools on the Coordinates panel (Fig. 40-6). The Show UCS button flyout allows you to show, show at the origin, or hide the UCS icon. The X, Y, and Z button flyout allows you to

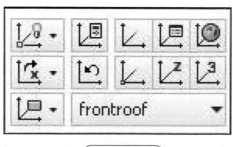

Fig. 40-6

rotate the UCS around each of the coordinate axes. The third flyout set allows you to align the UCS with the screen, an object, or a face on a 3D solid.

You have already used four of the other eight buttons: World, Named UCS..., Origin, and 3 Point. The Properties button allows you to control the visual appearance of the UCS icon itself. The Z Axis Vector button allows you to specify the positive direction of the Z axis. The Previous button restores the previous UCS. Finally, the UCS button allows you to specify any UCS with keyboard entry.

8. Create and name several new UCSs, then delete them in the **UCS** dialog box.

9. From the View panel of the Home tab, change the current view to **Left**.

Notice that the UCS icon has changed location and orientation, and that the current UCS, shown in the text box of the Coordinates panel, is now the Left orthographic UCS. When you change views, AutoCAD automatically changes the current UCS to match the view. If you want frontroof to remain the current UCS after selecting a new view, you must reselect it with the Named UCS button.

Dynamic UCS is a feature of AutoCAD that moves the UCS automatically. However, dynamic UCS only works on the faces of solid models, 3D objects you have yet to create. You will learn about dynamic UCS in Chapter 43.

10. Change the current view to **SE Isometric** and the current UCS to **World**.

11. Select the UCS icon.

Notice that four grips appear, one for the origin and one for each of the three axes.

12. Rest the pointing device over each of the grips and read the options available.

13. Move the origin and rotate two of the axes to become familiar with the capabilities of the UCS grips.

14. Save your work, and exit AutoCAD.

● REVIEW QUESTIONS

Answer the following questions on a separate sheet of paper.

1. What is the primary benefit of AutoCAD's user coordinate system?
2. What purpose does the UCS icon serve?
3. Describe the 3 Point option of the UCS command.
4. Briefly describe the function of the View button on the Coordinates panel.
5. In this chapter, you created a UCS for the front roof using the 3 Point button. Which other option could you have used to create this UCS?

● CHALLENGE YOUR THINKING

These questions are designed to further your knowledge of AutoCAD by encouraging you to explore the concepts presented in this chapter. Answer each question on a separate sheet of paper.

1. Explain the difference between a view and a UCS and describe how the two can be used together to create complex 3D models.
2. Look back at Chapter 39, in which you created the shed drawing. Now that you know how to create and change UCSs, would you have created the shed differently? Explain.

● APPLYING AUTOCAD SKILLS

Work the following problems to practice the commands and skills you learned in this chapter.

1. Open workshed.dwg. Use the 3 Point button to create a UCS on the back wall of the shed. Save the UCS using a name of your choice. With this as the current UCS, add two windows to the back wall. Choose the shape, size, and location of the windows. Save the drawing as prb40-1.dwg.
2. Open the workshed.dwg and view it in an isometric view of your choice. Create a UCS with the View button. Then make a border and title block for the 3D model. Save your work as prb40-2.dwg.

• USING PROBLEM-SOLVING SKILLS

Complete the following problems using problem-solving skills and your knowledge of AutoCAD.

1. Make a 3D drawing of the track shown in Fig. 40-7. Your supervisor wants further design work on the 60° angular face. Create a UCS on the face and save it in preparation for the redesign. Save the drawing as prb40-track.dwg.

2. Open worshed.dwg and create the gable-end roof vent shown in Fig. 40-8A. First, view the shed from the left. Create a new UCS as shown by snapping to the midpoint of the horizontal line. The vent is six-sided and is located 1.5' above the bottom of the gable. The size of the vent is your choice. Add an identical vent to the opposite side of the shed. When you are finished, your drawing in the Wireframe style should look similar to the one in Fig. 40-8B. Save your drawing as prb40-shed.dwg.

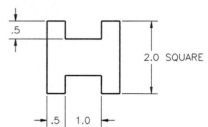

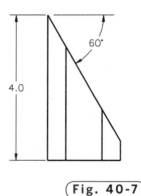

Fig. 40-7

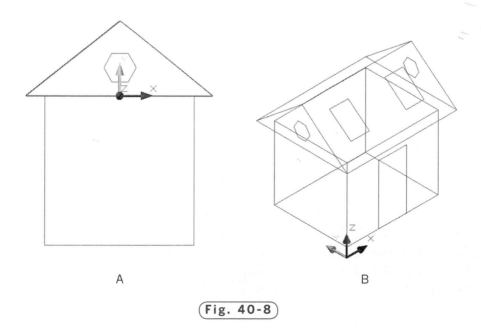

A   B

Fig. 40-8

3D Modeling

Sometimes, to represent an object accurately, you need an alternative to orthographic projections. Two such alternatives are isometric views and 3D models. AutoCAD provides special options to facilitate such drawings.

▶ Description

This project includes two problems:

1. Draw an isometric version of a die (one of two dice), as shown in Fig. P8-1. Include the spots, indicating values of one, two, and three on the three faces showing. Center the spots accurately, and make the length of each side 2″ long. Dimension the length, height, and depth of the die.

2. Create a 3D model of a die the size of a 3″ cube. Draw the spots (numbered one through six) on each of the six faces, symmetrically arranged upon each surface, as shown in Fig. P8-2A and B. Looking at an actual die may help you create an accurate model.

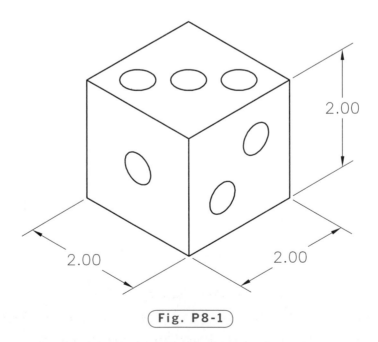

Fig. P8-1

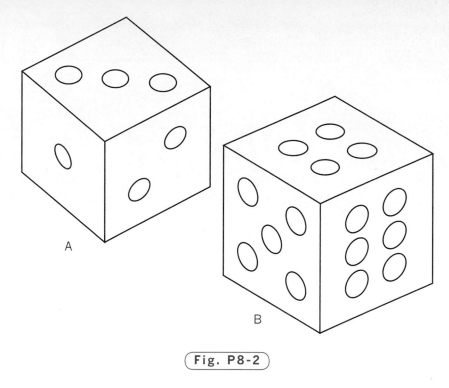

▶ Hints and Suggestions

1. Remember that circles appear elliptical in isometric drawings. Use the Isocircle option with the correct plane orientation.

2. Use the DIVIDE command to place the spots. Remember to set PDMODE to a value that allows you to see the divisions, and use REGEN to make the divisions appear.

3. Create two viewports of the 3D model to show all six surfaces. You should only need to create one model.

▶ Summary Questions

Your instructor may direct you to answer these questions orally or in writing. You may wish to compare and exchange ideas with other students to help increase efficiency and productivity.

1. What effect does toggling the isometric plane have upon the use of the ortho feature?

2. Distinguish between thickness and elevation.

3. Of what value is the 3D globe (compass) when you work on 3D models?

4. Describe one possible advantage of reorienting the UCS icon.

Chapter 41

Solid Primitives

Objectives

- Define solid modeling
- Create solid primitives
- Modify solid primitives

Vocabulary

gizmo
polysolid
primitives
solid modeling
subobject

Computer modeling of three-dimensional volumes of solid parts is called **solid modeling**. This type of modeling has become increasingly important because designs created with solid models more accurately reflect the actual physical properties of the part. The appearance of a solid model is somewhat similar to that of a 3D wireframe with faces, but the two are quite different. Solid models are closed volumes, like a real physical object, and they contain physical and material properties.

Solid **primitives** are the basic building blocks of 3D modeling. Examples of primitives include boxes, spheres, cones, and cylinders. Several different solid primitives are shown in Fig. 41-1. In this chapter, you will learn how to create solid primitives by specifying the necessary points and dimensions. You'll also learn to modify solid primitives: move, rotate, and scale, as well as using grips to modify a **subobject** on the primitive. (A subobject is a face, edge, or corner of a solid.) In later chapters, you will learn to combine primitives with other three-dimensional objects to create more complicated shapes.

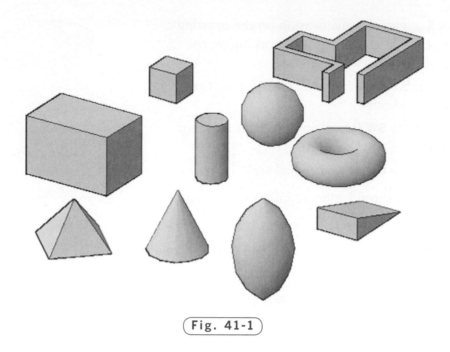

Fig. 41-1

Creating Solid Primitives

The CYLINDER command allows you to create a solid cylinder primitive of any size.

1. Start AutoCAD and use the **Acad3d.dwt** template file to create a new drawing.

2. Select the **3D Modeling** workspace.

3. Create a new layer named **Objects**, assign the color red to it, and make it the current layer.

4. Set grid to **1** and snap to **.25** and pick **Parallel** projection from the shortcut menu on the ViewCube.

5. Save your work in a template file named **solid.dwt**. For the template description, enter **For 3D models**, and pick **OK**.

6. Close **solid.dwt** and use it to create a new drawing file named **primit.dwg**.

7. Select the **Hidden** visual style from the View panel and **ZOOM All**.

8. Pick the **Cylinder** button from the Primitive panel of the Solid tab of the Ribbon.

This activates the CYLINDER command.

⬜ Cylinder

PRIMITIVE

9. Pick a point at any location on the drawing area. (Notice that you are working on the ground plane, where $z = 0$.)

10. Enter **.5** for the base radius and **2** for the height of the cylinder.

11. Select the **SW Isometric** viewpoint from the Viewport Controls.

12. Select the **Conceptual** visual style.

The cylinder should appear similar to the one in Fig. 41-2.

Fig. 41-2

The TORUS command lets you create donut-shaped solid primitives.

Torus
PRIMITIVE

13. Zoom out and pan if necessary to make space for the torus.

14. Pick the **Torus** button from the flyout menu on the Primitive panel.

This activates the TORUS command.

15. Pick a center point for the torus (anywhere) and enter **1** for the radius of the torus.

The radius of the torus specifies the distance from the center of the torus to the center of the tube.

16. Enter **.5** for the radius of the tube.

Fig. 41-3

A torus appears, as shown in Fig. 41-3.

If you enter a negative number for the radius of the torus, a football-shaped solid primitive appears, as shown in Fig. 41-4.

Torus
PRIMITIVE

17. Enter the **TORUS** command and pick a center point. Enter a radius value of **–1** and a tube radius value of **2**.

A **polysolid** is a solid primitive that is similar to a two-dimensional polyline, but is three-dimensional.

Polysolid
PRIMITIVE

18. Pick the **Polysolid** button from the flyout menu on the Primitive panel.

Fig. 41-4

19. Before picking a starting point, select the **Width** option and type **.25** for the width of the polysolid.

20. Select the **Height** option and type **1**.

21. Pick a point in an open area of the ground plane and move the crosshairs.

The new polysolid begins to take shape.

22. Pick a series of points to create the polysolid. When you are finished, press **ENTER**.

An example of a polysolid is shown in Fig. 41-5. The wall width and height of your polysolid should be similar, but the shape will probably be quite different. Note that you can create curved polysolid segments with the Arc option.

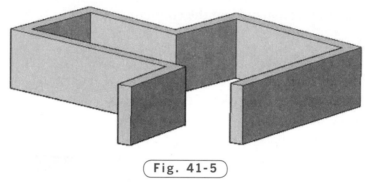

Fig. 41-5

23. On your own, create a cone, a wedge, a box, a sphere, and a pyramid using the solid primitive buttons on the Primitive panel.

24. Save your work.

Modifying Solid Primitives

Let's modify solid primitives using 3D commands, grips, and subobject selection.

1. Use the **solid.dwt** template file to create a new drawing named **3Dgrips.dwg**.

2. Select the **NE Isometric** viewpoint and the **Wireframe** visual style.

3. Use the **BOX** command to create a box primitive of any size and shape on the ground plane.

PRIMITIVE

Moving Objects in 3D

MODIFY

The 3DMOVE command allows you to move solids in 3D space.

1. Pick the **3D Move** button on the Modify panel of the Home tab.

2. Select the box primitive and press **ENTER**.

The Move gizmo appears at the center of the selection set. A **gizmo** is an icon that simplifies the selection process for 3D commands. The Move gizmo consists of three axis handles and a center box.

3. Move the crosshairs to the middle of the gizmo and right click to display the shortcut menu.

4. Verify that the **Move** gizmo is checked, then review the other commands on the shortcut menu.

5. Select the endpoint on the front lower corner of the box, as shown in Fig. 41-6.

6. Pick a second point in the drawing area to move the solid.

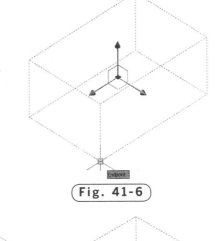

Fig. 41-6

You can use the Move gizmo to constrain the direction of a move.

7. Pick the **3D Move** button, select the box primitive, and press **ENTER**.

8. Move the crosshairs over the green axis handle on the Move gizmo until it is highlighted and a green vector appears, as shown in Fig. 41-7, and pick the handle.

9. Move the crosshairs in the drawing area to move the box and notice that its movement is now constrained along the selected axis vector.

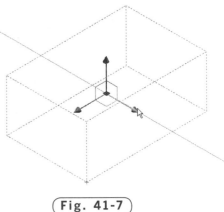

Fig. 41-7

10. Pick a point to complete the 3D move.

Rotating Objects in 3D

MODIFY

The 3DROTATE command allows you to rotate objects in 3D space.

1. Pick the **3D Rotate** button on the Modify panel.

2. Select the box primitive and press **ENTER**.

The Rotate gizmo appears. Like the Move grip tool, the Rotate gizmo consists of three colored handles. However, the handles for this gizmo are circles, not axes.

3. Pick the front lower corner of the box as the base point.

The Rotate gizmo moves to the selected point.

4. In reply to Pick a rotation axis, move the crosshairs over the circular blue axis handle to highlight it. (It becomes yellow when it is highlighted.)

A blue vector appears, as shown in Fig. 41-8.

5. Pick the axis while the blue vector is present.

6. Pick an endpoint on the box primitive as the angle start point; then rotate the box dynamically by moving the crosshairs in the drawing area.

7. Pick anywhere in the drawing area to complete the 3D rotation.

8. Experiment with the **3DSCALE** command and the Scale gizmo on your own.

9. Erase the box primitive.

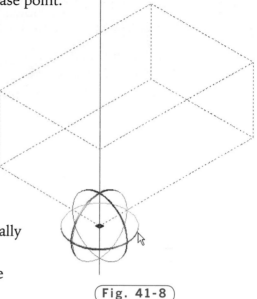

Fig. 41-8

Using Object Grips

In the previous exercise, you first entered the 3DMOVE and 3DROTATE commands and then selected the object. In this section, you will select the object first, before applying a command.

1. Create a new box primitive on the ground plane.

2. Pick the box with a single click and observe the grips and the Move gizmo that appear. (The Move gizmo should be the default gizmo, as shown in the Selection panel of the Solid tab.)

3. Rest the crosshairs on the rectangular grip on the bottom (down-facing) face of the box.

The Move gizmo appears on the grip. You can move the box primitive without selecting the 3D Move button or typing the 3DMOVE command.

4. Press **ESC** to deselect the grip tool location and the box primitive.

5. Select the box again.

6. Select the triangular grip on the top face of the box, as shown in Fig. 41-9, move the crosshairs upward, and pick a point to stretch the box (you may need to rotate the model using **Orbit** to see the grip clearly).

7. Select the square grip at the front lower corner of the box and stretch the box outward, as shown in Fig. 41-10.

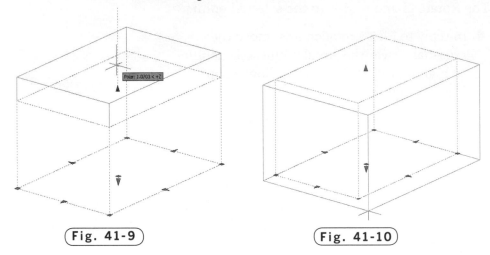

Fig. 41-9   Fig. 41-10

The corner grip allows you to stretch the solid in two directions at the same time.

Using Subobjects

The moving, rotating, and stretching operations you have just learned can be applied to any subobject—any face, edge, or corner of a solid.

SELECTION

1. Select the **Hidden** visual style.
2. Select the **Face** subobject filter button from the flyout menu below the No Filter button on the Selection panel of the Solid tab.

Notice that the face filter icon is now attached to the crosshairs.

3. Move the pick box over the top face of the box primitive.

The top face of the box becomes highlighted.

4. Pick the top face.

SELECTION

A face grip appears in the center of the top face, along with the Move (default) gizmo, as shown in Fig. 41-11.

5. Select the **Edge** subobject filter button from the flyout menu on the Selection panel.
6. Pick the front right edge of the top face.

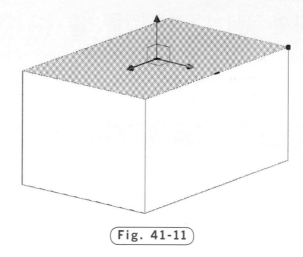

Fig. 41-11

An edge grip appears, as shown in Fig. 41-11.

7. Select the **Vertex** subobject filter button and pick the upper right corner of the box.

A vertex grip (corner grip) appears, as shown in Fig. 41-11.

Notice that the grip for each type of subobject has a distinct shape. As you can see, you can select one subobject or a set of subobjects.

We will now use subobjects to modify the shape of the solid.

8. Press **ESC** to clear the selection set.

9. Select the **Edge** subobject filter button and pick the front right edge of the top face.

10. Pick the edge grip and move the crosshairs in the drawing area.

Notice the change in the solid.

11. Press the **CTRL** key once to toggle to the next modification option.

Again, notice the change in the solid as you move the crosshairs.

12. Press the **CTRL** key again, observing the next option for modification.

As you can see, AutoCAD offers three different options for modifying the solid as you drag the edge.

13. Experiment on your own with subobject selection, selection sets, and modification options. Apply what you have learned to different primitives.

14. Save your work and close all drawing files.

SELECTION

● REVIEW QUESTIONS

Answer the following questions on a separate sheet of paper.

1. Describe the function of each of the following commands.
   a. CYLINDER
   b. TORUS
   c. CONE
   d. WEDGE
   e. BOX
   f. SPHERE
   g. PYRAMID
   h. POLYSOLID

2. With what command can you create a football-shaped primitive?

3. With what command can you rotate a solid primitive?

4. With what command can you create solid walls?

● CHALLENGE YOUR THINKING

These questions are designed to further your knowledge of AutoCAD by encouraging you to explore the concepts presented in this chapter. Answer each question on a separate sheet of paper.

1. If you could slice open a solid primitive, what would you find in the interior?

2. Experiment with the CYLINDER and CONE commands. Is it possible to create a solid primitive cylinder or cone at other than a 90° (right) angle to its base? Explain.

3. When a primitive solid is modified using a subobject selection set, the subsequent selection of the solid reveals fewer grip options. Why do you think this occurs?

• APPLYING AUTOCAD SKILLS

Work the following problems to practice the commands and skills you learned in this chapter.

1. Create a solid shaft that measures .5125″ in diameter by 10″ in length. Rotate the shaft so that its length is parallel to the WCS.

2. Create a solid ball bearing that measures .3625″ in diameter. Use a polar array to create a ring of 16 balls using a 1″ radius. View the bearings in an isometric view, and apply the Realistic visual style, as shown in Fig. 41-12. Save the drawing as 3dbearing.dwg.

Fig. 41-12

3. Create an inflated bicycle inner-tube that measures 24″ in diameter. Make the tube diameter 2″.

4. Create a football-shaped primitive of any size on top of a 6″ cube.

• USING PROBLEM-SOLVING SKILLS

Complete the following activities using problem-solving skills and your knowledge of AutoCAD.

1. Your company's graphics department is designing ornaments for various holidays. They need forms for the art decals. Create the solid primitives for them using the sizes shown in Fig. 41-13. Shade the primitives with the Conceptual visual style. Save the drawing as ornament.dwg.

2. You need forms around which to design a laptop computer carrying case. Do research to find the most common dimensions for laptops today. Then create two or three solid boxes to represent the most common sizes. Create a UCS that is parallel to the screen and add text to give the basic computer sizes. Shade the boxes in the Realistic style. Save the drawing as case.dwg.

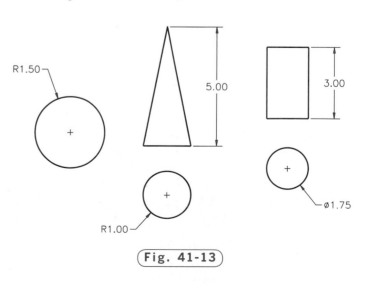

R1.50

5.00

3.00

R1.00

Ø1.75

Fig. 41-13

HVAC Designer

Have you read stories about the days before electricity, when a single fireplace was used to warm an entire household during the coldest months of winter? Fortunately, we live during a time when heating and air-conditioning systems are fairly common-place. Certainly you have a furnace in your home, and almost any store you visit in summertime will have its air condition-ing going full blast. HVAC (heating, ven-tilation, and air condi-tioning) designers are the people we have to thank for design-ing the heating and air-conditioning units that make our lives more comfortable.

© James Hardy/PhotoAlto/RF

Combining Skills

People who work with HVAC systems should be familiar with standard codes for heating and cooling systems. Math skills are important to determine heat loss and air flow calculations. HVAC designers may move into positions focusing on heating and drafting analy-sis and procedures. Computer skills are helpful for all indi-viduals involved in this field, including knowledge of CAD-related programs and word processing pro-grams.

Understanding Efficiency

HVAC designers create plans for heating and cooling systems for residential, com-mercial, industrial, and institutional uses. The most efficiently designed systems use techniques to improve the quality of air indoors. They can also reduce the amount of energy used to heat or cool a building. This can help businesses and homeowners save money. Energy-efficient systems are also beneficial to the environment.

▶ Career Activities

1. Interview someone who can tell you about employment of HVAC design-ers in your area. What education is required? Are apprenticeships available?

2. Has employment in this field been steady over the last few years? What are forecasts for future employment?

Mesh Modeling

Objectives

- Define mesh modeling
- Create mesh primitives
- Control mesh divisions
- Modify meshes and convert meshes to solids

Vocabulary

extrude
mesh
tessellations

3D modeling is not limited to solid primitives. Another type of 3D object is a **mesh**. Many of today's products, from cell phone housings and handheld remote controls to the exteriors of automobiles, are designed with curved, flowing styles. 3D mesh modeling allows you to create these smooth, organic shapes that are difficult to model with the solid modeling tools. Unlike solid primitives, mesh models are not closed volumes. However, mesh primitives can be modified more easily and more extensively than solid primitives. Mesh models can also be converted to solids at any time.

In this chapter, you will learn to create mesh primitives, control the appearance of meshes, modify meshes, and convert meshes to solids.

Creating Mesh Primitives

1. Start AutoCAD and use the **solid.dwt** template file to create a new drawing named **meshes.dwg**.
2. Select the **3D Modeling** workspace.

3. Pick the **Realistic** visual style and the **NE Isometric** view from the Viewport Controls.

The CONE command allows you to create a cone-shaped mesh primitive.

PRIMITIVES

4. Pick the **Mesh** tab, then pick the **Mesh Cone** button from the flyout menu below the Mesh Box button on the Primitives panel.

This activates the MESH_CONE command.

5. Pick a center point at any location on the drawing area. (Notice that you are working on the ground plane, where z = 0.)

6. Enter **1** for the base radius and **2** for the height of the cone.

A cone appears, as shown in Fig. 42-1.

Fig. 42-1

Now we will create a wedge mesh primitive using the MESH WEDGE command.

PRIMITIVES

7. Pick the **Mesh Wedge** button from the flyout menu on the Primitives panel of the Mesh tab.

8. Pick two diagonally opposite corner points to define the size of the wedge's base.

9. Enter **.75** for the height.

A wedge-shaped mesh primitive appears, as shown in Fig. 42-2.

The MESH BOX command enables you to create a cube-shaped mesh primitive.

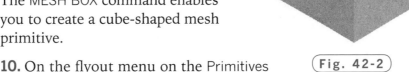

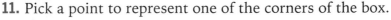

Fig. 42-2

10. On the flyout menu on the Primitives panel, pick the **Mesh Box** button.

This activates the MESH BOX command.

PRIMITIVES

11. Pick a point to represent one of the corners of the box.

The Center option permits you to create a box by specifying the center point of the box first.

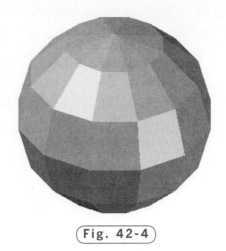

Fig. 42-3

Fig. 42-4

12. Pick the diagonally opposite corner of the rectangle's base.

13. Enter **2** for the height.

A box appears, as shown in Fig. 42-3.

 To create a cube, you can select the MESH BOX command's Cube option after selecting the first point.

The MESH SPHERE command creates a mesh sphere primitive.

14. Pick the **Mesh Sphere** button from the flyout menu on the Primitives panel.

This activates the MESH SPHERE command.

15. Pick a point to define the center of the sphere.

16. Enter **1** for the radius of the sphere.

The sphere appears, as shown in Fig. 42-4.

The MESH PYRAMID command creates a mesh pyramid primitive.

17. Pick the **Mesh Pyramid** button from the flyout menu.

18. Pick a point in an open area to define the center of the pyramid's base.

19. Enter **1** for the base radius and **1.5** for the height of the pyramid.

PRIMITIVES

PRIMITIVES

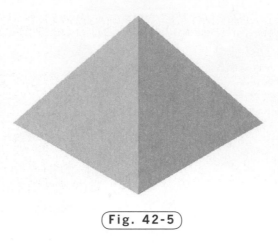

Fig. 42-5

The pyramid appears, as shown in Fig. 42-5.

> **NOTE**
> The Sides option specifies the number of sides on the pyramid. The Edge option allows you to define the base by specifying an edge instead of the center point.

20. On your own, create a cylinder and a torus using the mesh primitive buttons on the **Primitives** panel.

21. Save your work.

Controlling Mesh Divisions

When resting the crosshairs over any mesh primitive, dashed lines subdivide the surfaces on the mesh primitive. These planar subdivisions are called **tessellations**. Tessellations on curved surfaces are a collection of planar facets that approximate the actual curvature. You can control the amount of tessellation in meshes.

1. Pick the arrow in the lower right corner of the Primitives panel.

The Mesh Primitives Options dialog box appears, as shown in Fig. 42-6.

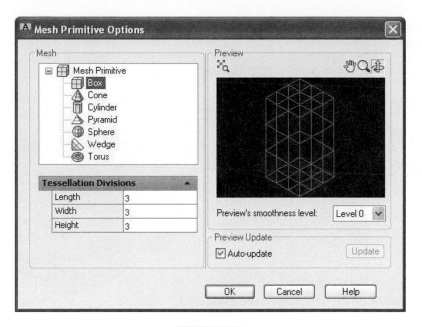

Fig. 42-6

2. Select each of the mesh primitives in the dialog box, adjust the tessellation divisions (all values), and notice how the model in the Preview pane changes. (Make sure the Auto-update box is checked.)

Note that increasing tessellation produces smoother models, but also requires more computing power and may affect the speed at which your system displays the 3D models.

3. Pick **Cancel** to close the dialog box and save your work.

Modifying Mesh Objects

3D meshes can be modified in a variety of ways that are impossible with solids. You can also convert a primitive solid to a mesh or convert a mesh to a solid.

1. Use the **solid.dwt** template file to create a new file named **meshmod.dwg**.

2. On the Viewport Controls in the upper left corner of the drawing area, select the **Shaded with edges** visual style and the **NE Isometric** view.

3. Select the **Solid** tab of the Ribbon and use the **BOX** command to create a solid box primitive of any size and shape on the ground plane.

4. Select the **Mesh** tab of the Ribbon.

PRIMITIVE

Smooth Object

MESH

5. Select the **Smooth Object** button from the Mesh panel, select the box, and press **ENTER**.

AutoCAD converts the solid primitive to a mesh.

6. Select the **Smooth More** button, select the mesh object, and press **ENTER**.

MESH

The edges and corners of the mesh are now smoother. The mesh object should look similar to the one in Fig. 42-7.

You can **extrude**, or add depth to, an individual face or group of faces on a mesh.

Fig. 42-7

7. Select the **Extrude Face** button from the Mesh Edit panel, select the center face on the top of the mesh, and press **ENTER**. (Make sure the default subobject filter is set to Face.)

Extrude Face

MESH EDIT

8. Move the crosshairs in the drawing area and watch the dynamic changes to the mesh, then select a point above the mesh to create a protrusion above the object.

Your mesh object should look somewhat like the one in Fig. 42-8.

9. Set the default subobject filter to **Edge**.

Fig. 42-8

Add Crease

MESH

10. Select the **Add Crease** button from the Mesh panel, select the four edges of the top face of the mesh object, as shown in Fig. 42-9. (You will need to rotate the view of the mesh object to select all four edges.)

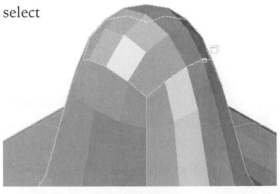

Fig. 42-9

HINT Use the Orbit button on the Navigation Bar to rotate the view dynamically.

11. Press **ENTER** twice.

As you can see, the edges of the top face are now creased, making the top face perfectly flat.

12. Set the default subobject filter to **No Filter**.

13. Select the **Refine Mesh** button on the Mesh panel, select the mesh object, and press **ENTER**.

MESH

The tessellation of the mesh object has increased. You can also refine individual faces on a mesh, rather than the entire object.

14. Select the **Convert to Solid** button on the Convert Mesh panel, select the mesh object, and press **ENTER**. (It may take a few seconds for the mesh to convert to a solid.)

CONVERT MESH

15. Select the **Realistic** visual style.

The modified mesh object is now a solid. Your mesh object should look similar to the one in Fig. 42-10.

As this exercise shows, mesh modeling allows you to create complex shapes that would be difficult to model with AutoCAD's solid modeling tools.

16. Save your work and exit AutoCAD.

Fig. 42-10

Chapter 42 Review & Activities

● REVIEW QUESTIONS

Answer the following questions on a separate sheet of paper.

1. With which command can you create a mesh cube?
2. With which command can you create a mesh pyramid?

● CHALLENGE YOUR THINKING

These questions are designed to further your knowledge of AutoCAD by encouraging you to explore the concepts presented in this chapter. Answer each question on a separate sheet of paper.

1. If you could slice open a mesh primitive, what would you find in the interior?
2. What are the relative advantages and disadvantages of using solid primitives versus using mesh primitives?
3. Can you think of products in your everyday life that have the clearly defined shape of solid primitives? Can you think of products in your everyday life that have the smoothed shapes of mesh objects?

● APPLYING AUTOCAD SKILLS

Work the following problems to practice the commands and skills you learned in the chapter.

1. Create a mesh torus with a 12″ radius and a ring diameter of 2″. At its center, create a mesh sphere with a 4″ diameter. View the model in an isometric view, and apply the X-Ray visual style. Your model should look like Fig. 42-11.

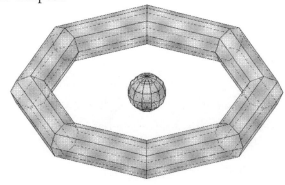

Fig. 42-11

2. Create a solid primitive cylinder with a 6″ diameter and a 6″ height. Convert it to a mesh using the Smooth Object command. Then smooth the mesh repeatedly until it cannot be smoothed further. Next, extrude the six pie-piece-shaped faces on the top center of the cylinder 6″. View the model with the Shaded with edges visual style. Your model should look like Fig. 42-12.

Fig. 42-12

● USING PROBLEM-SOLVING SKILLS

Complete the following problems using problem-solving skills and your knowledge of AutoCAD.

1. Let's revisit Chapter 41's Using Problem-Solving Skill, Problem 1. Open the Mesh Primitives Options dialog box and double the tessellation divisions for all primitives. Now create the mesh primitives for the ornaments as shown in Fig. 42-13. Shade the ornaments in the visual style. Save the drawing as mesh ornament.dwg.

2. The computer cases you created in Chapter 41 have been approved for further development. Open case.dwg and convert the solid primitives to meshes. Smooth the meshes to a degree you consider to be aesthetically pleasing, and then convert the meshes back to solids. Save the drawing as smoothcase.dwg.

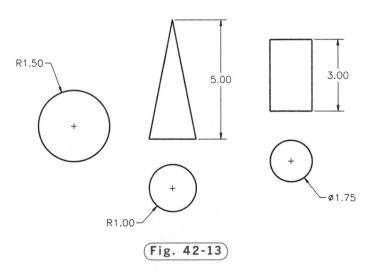

Fig. 42-13

Chapter 43

Basic Modeling

Objectives

- Explain the difference between a solid model, a mesh molel, and a surface model
- Create models from 2D objects with the **REVOLVE**, **EXTRUDE**, **SWEEP**, and **LOFT** commands
- Learn to use the **PRESSPULL** command

Vocabulary

dynamic UCS
helix
lofting
mesh model
revolve
solid model
surface model
sweeping

Many models are produced by first creating a 2D object or objects. For example, it is possible to **revolve** objects (turn them about an axis) to create a 3D model. In this chapter, we will do this. We will also create models by extruding, sweeping, and lofting 2D objects, and learn to modify solids using the PRESSPULL command.

The modeling techniques in this chapter produce either a solid model, a mesh model, or a surface model, depending on the defining geometry and the modeling technique used. A **mesh model** is produced from a mesh modeling operation. A **solid model** is produced when a solid modeling operation results in a closed (also called *watertight*) volume. A **surface model** is produced when the operation results in an open 3D surface. By definition, a surface has no thickness and no internal volume.

For example, when a circle is revolved around an axis to form a torus (donut), a solid model or mesh model is created. When a half circle is revolved around an axis, the result is a surface model. Fig. 43-1 demonstrates this example.

Fig. 43-1

Revolution

We will use the REVSURF command to create a mesh model of the shaft shown in Fig. 43-2.

1. Start AutoCAD and use the **solid.dwt** template file to create a new drawing named **shaft.dwg**.

2. Select the **3D Modeling** workspace.

3. Use a polyline to create the profile of the shaft, as shown in Fig. 43-3. Position point 1 on the ground plane at $x = 1$, $y = 1$. Make sure the polyline is closed. Omit the dimensions.

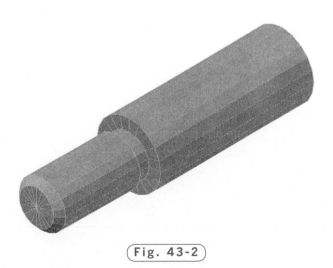

Fig. 43-2

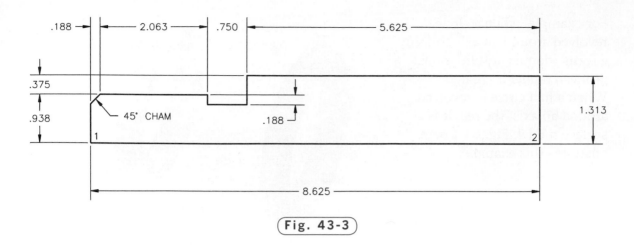

Fig. 43-3

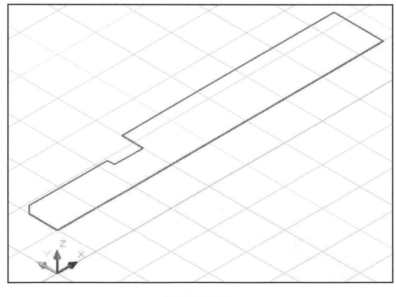

HINT Begin at point 1, work counterclockwise, and take advantage of polar tracking. Set the polar tracking angle to 45°, which will help you create the 45° chamfer.

4. Create a line that passes through points 1 and 2.

5. View the profile from the **SW Isometric** viewpoint and zoom out slightly.

6. Enter **ZOOM** and **.7x**. (You can also use the wheel button on the mouse to zoom out.)

You should now see a view similar to the one in Fig. 43-4.

Fig. 43-4

7. Enter **DELOBJ** and **0**.

DELOBJ is the Delete Object system variable. With the default value of 3, the 2D profiles used to create solids are deleted. The DELOBJ value of 0 retains the profile, which is what we wish to do.

8. Enter **SURFTAB1** and **20**.

SURFTAB1 is a system variable that controls the number of tesselations on a mesh model in the direction of revolution. The default value is 6, which would produce a revolved mesh with a hexagonal cross section.

9. Pick the **Revolved Surface Mesh** button on the Primitives panel on the Mesh tab.

This activates the REVSURF command.

PRIMITIVES

10. Select the polyline.

11. Pick the line as the axis of rotation.

12. Press **ENTER** twice to select the default 0- to 360-degree angle of revolution.

The mesh model of the shaft generates on the screen, as shown in Fig. 43-2 on page 569. Note that you could have used the REVOLVE command to produce a solid model, instead of a mesh model, from the polyline.

13. Save your work and close the drawing.

Extrusion

The solid model of an aluminum casting shown in Fig. 43-5 was created using the EXTRUDE command. This command allows you to **extrude**, or add depth to, a closed area. The EXTRUDE command works on closed 2D objects such as planar 3D faces, closed polylines, polygons, circles, ellipses, closed splines, and donuts. It is not possible to extrude objects contained within a block or a polyline that has a crossing or self-intersecting segment.

1. Begin a new drawing using the **solid.dwt** template file.

2. Using the **PLINE** command's **Arc** and **Line** options, approximate the size and shape of the casting shown in Fig. 43-6.

Fig. 43-5

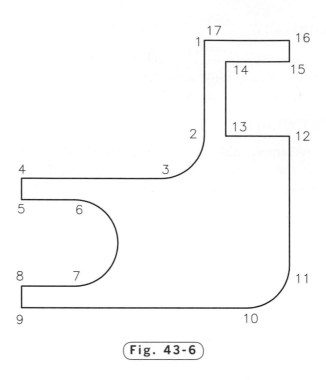

Fig. 43-6

The casting's overall size is 6.25″ × 6.25″. Pick the points in the order shown. Do not place the numbers on the drawing.

3. Save your work in a file named **extrude.dwg**.

4. Select the **SW Isometric** viewpoint.

MODELING

5. Pick the **Extrude** button on the Modeling panel of the Home tab.

6. Select the polyline and press **ENTER**.

7. Move the crosshairs up and down in the drawing area and notice that AutoCAD displays the height of the extrusion dynamically.

8. Select the **Taper angle** option and enter **5** (for 5 degrees of taper).

9. Enter **.75** for the height of the extrusion.

An extruded solid model appears, with a height of .75 and sides tapered at 5°, as shown in Fig. 43-5 on page 571.

10. In an open area of the screen, draw an arc and a circle of arbitrary size on the ground plane.

MODELING

11. Use the **EXTRUDE** command to extrude both objects **1.5** inches. Do not taper the extrusions.

Your drawing should look similar to the one in Fig. 43-7 on page 573.

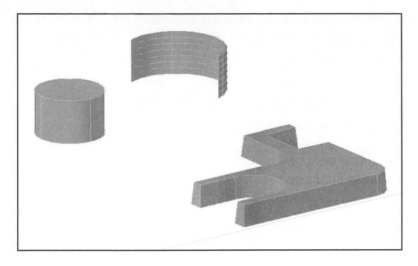

Fig. 43-7

12. Open the **Properties** palette and select each of the two new extrusions individually.

Notice that the extrusion created from the circle is a 3D solid, while the extrusion created from the arc is a surface. This is because the circle is a closed profile, like a closed polyline, while the arc is an open profile.

If we had used a series of lines instead of a polyline to create the profile of the aluminum casting, the EXTRUDE command would have produced a series of surfaces. All of the tapered wall areas would have appeared, but the model would have no top or bottom—and no closed volume.

13. Close the **Properties** palette, save your work, and close the drawing.

Sweeping

The SWEEP command creates a new solid or surface by **sweeping**, or moving a 2D profile along a path curve. The sweep curve can be open or closed, and 2D or 3D. We will demonstrate the SWEEP command by sweeping a circle along the path of a helix.

DRAW

1. Begin a new drawing using the **solid.dwt** template file.
2. Draw a **1″** line and a **1″**-diameter circle anywhere on the ground plane.
3. Pick the **Helix** button on the expanded Draw panel of the Home tab to enter the **HELIX** command.

This command creates a **helix**, or open 3D spiral, given the radius of the spiral at the top and bottom and the height of the spiral. It is often used to create springs and coils. You can also create a 2D spiral by specifying a height of 0.

4. Pick a point on the ground plane and enter **1.5** for the base radius.

5. For the top radius, enter **3.5**.

6. For the height, enter **10**.

The helix forms. Note that we accepted the default number of turns (3) and twist direction (counterclockwise).

MODELING

7. Pick the **Sweep** button from the flyout menu under the Extrude button on the Modeling panel of the Home tab.

8. Pick the **1″** line as the object to sweep and press **ENTER**.

9. Select the helix as the sweep path.

The 3D surface model—a helical ribbon—appears on the screen, as shown in Fig. 43-8.

MODELING

10. Undo the sweep and pick the **Sweep** button once again.

11. This time, pick the circle as the object to sweep and press **ENTER**.

12. Again select the helix as the sweep path.

A 3D solid model of the tapered spring appears, as shown in Fig. 43-9.

13. Save your work in a file named **sweep.dwg**.

Fig. 43-8

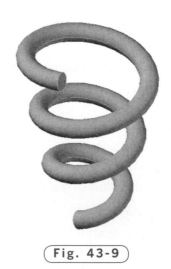

Fig. 43-9

Lofting

Creating a new solid model or surface by specifying a series of 2D cross sections is known as **lofting**.

1. Begin a new drawing using the **solid.dwt** template file.

2. Draw a **6″**-diameter circle on the ground plane. Center the circle at $x = $ **5**, $y = $ **5**.

3. Enter the **ELEV** command, specify a new elevation of **5** ($z = 5$), and a thickness of **0**.

4. Draw a **2″**-diameter circle at the $z = 5$ elevation. Center the circle at $x = $ **5**, $y = $ **5**.

5. Enter the **ELEV** command and specify a new elevation of **10** ($z = 10$) and a thickness of **0**.

6. Use the **RECTANG** command to draw a square at the $z = 10$ elevation. Locate the first corner at $x = $ **3**, $y = $ **3**. Locate the other corner at $x = $ **7**, $y = $ **7**. (If you are entering coordinates at the keyboard, you may need to turn dynamic input off on the status bar to get accurate results.)

7. View the drawing area from the **SE Isometric** viewpoint.

8. Pick the **Loft** button from the flyout menu below the Sweep button on the Modeling panel.

9. Select, in order, the large circle, the small circle, and the square; then press **ENTER**.

10. Select the **Settings** option and press **ENTER**.

MODELING

A 3D solid model is created, with a lofted contour that passes through the three cross sections. The Loft Settings dialog box is displayed as well. The lofted solid and the Loft Settings dialog box are shown in Fig. 43-10 on page 576.

AutoCAD offers several options to control the contour of the lofted solid. Initially, the Smooth Fit option is selected.

11. Select the **Ruled** option and notice that the lofting between the cross sections changes to straight lines.

12. Select the **Normal to** option and explore the different options in the drop-down menu.

The Draft angles option allows you to specify the angle of the lofted contour at the first and last cross section. By default, they are set to 90°.

13. Reselect **Smooth Fit** and pick the **Close surface or solid** box.

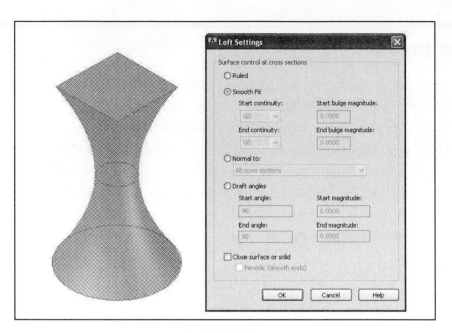

Fig. 43-10

This option creates a torus-like tube, with the lofted cross sections defining the inside surface of the tube.

14. Uncheck the **Close surface or solid** box and pick **OK**.

15. Save your work in a file named **loft.dwg**.

Pressing and Pulling Solids

The PRESSPULL command allows you to "press" or "pull" any closed 2D profile on the face of a solid.

1. Use the **solid.dwt** template to create a new drawing named **presspull.dwg**.

2. Select the **NE Isometric** viewpoint and the **Hidden** visual style.

3. Set snap and grid to **10**.

4. Make sure the Allow/Disallow Dynamic UCS button on the status bar is toggled on so that dynamic UCS is active.

STATUS BAR

Dynamic UCS works only on solid faces, and only from within active commands. It automatically and temporarily moves the current UCS to the selected face. This saves you from constantly managing and changing UCSs as you work on a model.

5. Use the **BOX** command to create a box primitive with a base measuring **40 × 20** units and a height of **40** units.

6. Enter the **CIRCLE** command and move the crosshairs over the large side face of the box solid.

Notice that the face becomes highlighted.

7. With snap off, select a point near the center of the face as the center of the circle, as shown in Fig. 43-11.

8. Enter a radius of **10**.

9. Select the **Presspull** button on the Modeling panel of the Home tab.

10. Pick inside the area bounded by the circle.

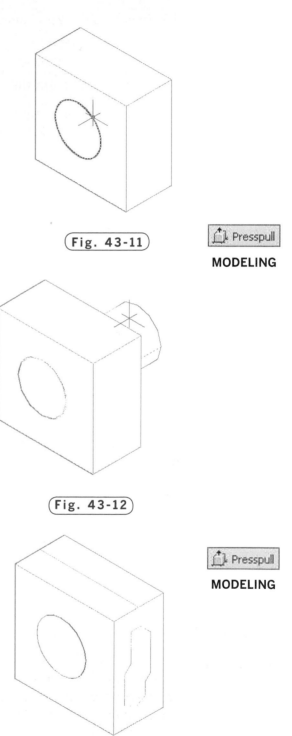

Fig. 43-11

Presspull

MODELING

11. Push the circle through the box solid until you see the extruded cylinder on the far side of the solid, as shown in Fig. 43-12, and click the mouse button.

A hole is created in the solid.

12. On the top face, use the **LINE** command and the **Midpoint** object snap to split the face into two long, equal rectangles, as shown in Fig. 43-13.

13. On the side face, create a closed 2D profile of arbitrary shape and size with the **PLINE** command, as shown in Fig. 43-13.

Fig. 43-12

14. Select the **Presspull** button and pick inside the back rectangle on the top face of the solid.

15. Type **10** and press **ENTER** twice.

Presspull

MODELING

Fig. 43-13

A raised step is created on top of the solid. As you can see, the PRESSPULL command works with keyboard entry.

16. Switch to the **Wireframe** visual style.

17. Pick the **Presspull** button, pick the area bounded by the closed polyline, and press inward a short distance.

18. View the model in the **Conceptual** visual style.

Your model should look similar to the one in Fig. 43-14. PRESSPULL is a powerful command for quick modifications to the faces of solids. This PRESSPULL command works with 2D and 3D curves as well. Also, when you hold the CTRL key down while using the command, the offset planar face will follow the taper angles of adjacent sides.

19. Save your work and close the **presspull.dwg** file.

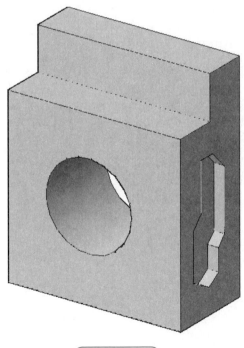

Fig. 43-14

● REVIEW QUESTIONS

Answer the following questions on a separate sheet of paper.

1. Describe each of the following commands.

   a. REVOLVE

   b. EXTRUDE

   c. SWEEP

   d. LOFT

   e. PRESSPULL

2. What command(s) would you use to create a 3D model of a quart oil container?

3. What effect does the Ruled surface control have on a lofted model?

● CHALLENGE YOUR THINKING

These questions are designed to further your knowledge of AutoCAD by encouraging you to explore the concepts presented in this chapter. Answer each question on a separate sheet of paper.

1. What command should be used to create a model of a toothpaste tube? . . . a model of a Frisbee®? Explain your answers.

2. Is it possible to extrude a wide polyline (one that has a defined width)? Explain.

Continued

• APPLYING AUTOCAD SKILLS

Work the following problems to practice the commands and skills you learned in this chapter.

1. Identify a simple object in the room. Create a profile of the object and then extrude it to create a solid model. Note that only certain objects—those that can be extruded—can be used for this problem.

2. Open shaft.dwg and save it as ch43-revolve.dwg. Erase the revolved mesh. Create two new revolved models using the X axis and Y axis options. Make one a solid model and the second a mesh model. Move one of the models so they do not overlap (use the Move button on the Modify panel of the Ribbon). Your completed drawing should look similar to Fig. 43-15. Save your work.

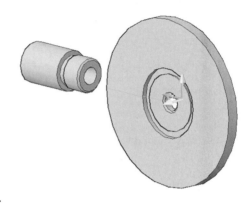

Fig. 43-15

3. Use the following four methods to create a solid cylinder with a radius of .4 and a height of 2.

   a. Draw a rectangle measuring .4 × 2 and revolve it 360°.

   b. Draw a circle with a radius of .4 and extrude it to a height of 2.

   c. Use the CYLINDER command to create a solid cylinder with a radius of .4 and a height of 2.

   d. Use Mesh Cylinder to create a mesh model with a radius of .4 and a height of 2.

4. Open sweep.dwg and save it with a new name of ch43-sweepprob.dwg. Erase the swept solid. Create a new profile by drawing a .5″ × 1.5″ rectangle on the ground plane. Create a new swept solid using the new rectangle as the profile, and use the Twist option to twist the swept profile 270°. Your drawing should look similar to the one in Fig. 43-16. Save your work.

Fig. 43-16

• USING PROBLEM-SOLVING SKILLS

Complete the following activities using problem-solving skills and your knowledge of AutoCAD.

1. Create a solid model of the boat trailer roller. Define the shape of the roller with a polyline, as shown in Fig. 43-17.

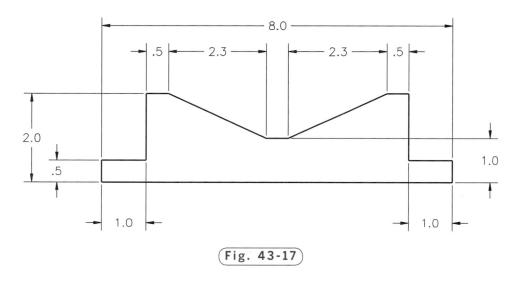

Fig. 43-17

Revolve the profile and view it from the SW Isometric viewpoint. Save the drawing as ch43roller.dwg. Your drawing should look similar to the one in Fig. 43-18.

Fig. 43-18

2. Create the compression spring shown in Fig. 43-19 using the appropriate 3D modeling commands. Make the spring 7″ long, with an overall diameter of 2.5″ and a wire diameter of .25″. Save your work as compression spring.dwg.

3. Use the LOFT command to create a lamp similar to the one shown in Fig. 43-20. The base has a diameter of 7″, and the lamp is 13.7″ tall. Determine how many cross-sectional profiles you need and what settings to use in the Loft Settings dialog box. Save your work as lamp base.dwg.

4. The bumper assembly for a bumper pool table is shown in Fig. 43-21. Use the dimensions of the three parts shown in Fig. 43-22 and use the REVSURF command to create three mesh models. Note that the bumper support has a square hole. Assume the hole to be circular. After you have finished modeling the three parts, place them in an isometric view. View the models in every 3D visual style. Save your work as bumper.dwg.

Fig. 43-19

Fig. 43-20

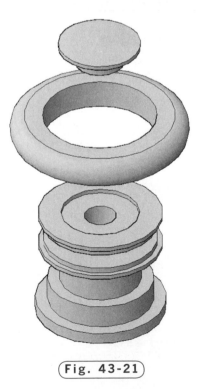

Fig. 43-21

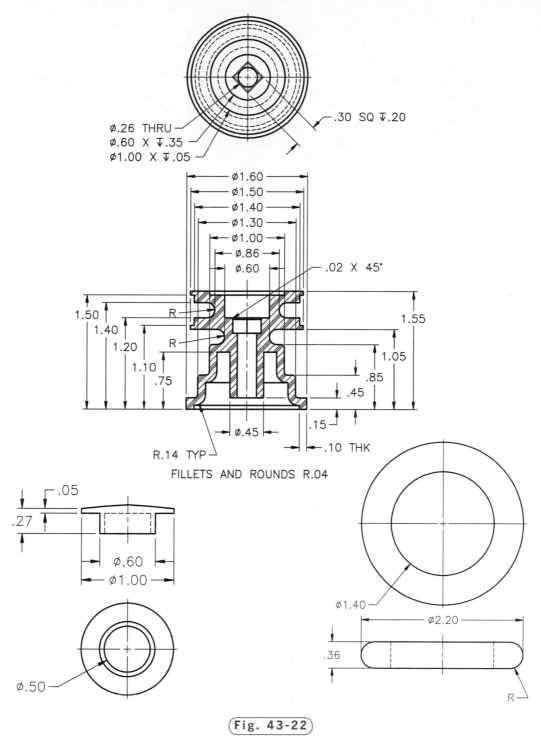

Ø.26 THRU
Ø.60 X ↧.35
Ø1.00 X ↧.05

.30 SQ ↧.20

Ø1.60
Ø1.50
Ø1.40
Ø1.30
Ø1.00
Ø.86
Ø.60

.02 X 45°

1.50
1.40
1.20
1.10
.75

R

R

1.55
1.05
.85
.45

.15
.10 THK

Ø.45

R.14 TYP

FILLETS AND ROUNDS R.04

.05
.27

Ø.60
Ø1.00

Ø.50

Ø1.40
Ø2.20

.36

R

Fig. 43-22

Problem 4 courtesy of Gary J. Hordemann, Gonzaga University

Chapter 44

Surface Modeling

Objectives

- Define surface modeling
- Create planar surfaces
- Create procedural surfaces
- Create NURBS surfaces

Vocabulary

NURBS surfaces
procedural surfaces

In addition to 3D solid and mesh objects, AutoCAD offers surface models. A surface is a thin shell that does not have mass or volume. There are two distinct types of surfaces in AutoCAD: procedural and NURBS. **Procedural surfaces** can be associative so that they are modified when the objects used to construct them are modified. **NURBS surfaces** (Non-Uniform Rational B-Splines) use settings such as degree, control vertices, and fit points that allow users to modify the surfaces by manipulating the settings.

In this chapter, we'll create several surfaces using new methods and some techniques that were introduced for the creation of solid models.

Planar Surfaces

1. Start AutoCAD and use the **solid.dwt** template file to create a new drawing named **planesurf.dwg**.

2. Select the **3D Modeling** workspace.

3. At the top of the Visual Styles panel of the View tab, pick the **Shades of Gray** visual style from the flyout menu.

4. On the Views panel of the View tab, pick the **SE Isometric** view from the flyout menu.

The PLANESURF command allows you to create a planar surface either by specifying a rectangular area or by selecting objects that form an enclosed area.

5. Draw a circle, a six-sided polygon, and a closed polyline of any shape and size anywhere on the ground plane.

6. Pick the **Planar** button on the Create panel of the Surface tab to enter the **PLANESURF** command.

CREATE

7. In response to Specify first corner or [Object], press **ENTER** to select the Object mode.

8. Select the circle, polygon, and polyline and press **ENTER**.

Three trimmed surfaces are created, defined by the three enclosed areas. The PLANESURF command works for any enclosed 2D area, including a series of line segments.

9. Pick the **Planar** button on the Create panel.

CREATE

10. In response to Specify first corner or [Object], pick a point in an open area of the ground plane.

11. Move the crosshairs and preview the planar surface in the drawing area.

12. Pick a second point to form a planar surface.

The PLANESURF command is not limited to the 2D ground plane. Let's create a surface on an angled plane.

13. On the Coordinates panel of the Home tab, select the triangle to the right of the **X UCS** button, then pick the **X UCS** button from the flyout menu.

COORDINATES

14. Enter **45** to rotate the current UCS 45° about the X axis.

15. Pick the **Planar** button again.

16. Pick two points to define the diagonally opposite corners of the planar surface.

17. Pick the **World UCS** button on the Coordinates panel on the Home tab.

Converting Planar Surfaces to Solids

The THICKEN command allows you to create a solid from a surface.

1. Pick the **Thicken** button on the Solid Editing panel of the Home tab.

2. Select the hex-shaped surface and the angled planar surface and press **ENTER**.

3. Specify a thickness of **4**.

Two solids are created. Your drawing should look similar to the one in Fig. 44-1. (Your polygonal surface will be different.)

4. Save your work.

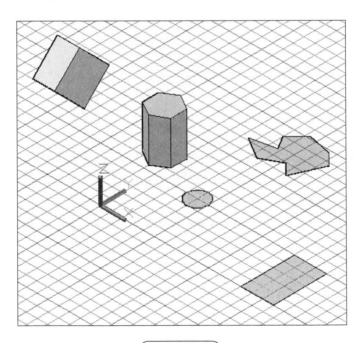

Fig. 44-1

SOLID EDITING

Working with Procedural Surfaces

1. Pick the **Save As...** button on the Quick Access toolbar and save the drawing as **procsurf.dwg**.

2. Delete all the objects in the drawing area.

3. Pick the **Shades with edges** visual style and the **NE Isometric** view.

4. Draw an arc on the ground plane of arbitrary shape and size.

5. On the Create panel of the Surface tab, toggle the **Surface Associativity** button on (the button is blue when it's on) and the **NURBS Creation** button off (the button is white when it's off).

CREATE

This will result in the creation of a procedural surface.

6. Pick the **Extrude** button on the Create panel, select the arc, and press **ENTER**.

CREATE

7. Move the crosshairs upward so that a surface begins to form dynamically and enter **6** for the height of the extrusion.

A procedural surface is created.

Now we will create another surface, offset from the first surface.

CREATE

8. Pick the **Offset** button on the Create panel, select the surface, and press **ENTER**.

AutoCAD displays a group of arrows indicating the direction of offset.

9. If the arrows point to the outside diameter of the arc, select the **Flip direction** option.

CREATE

10. Specify an offset distance of **1**.

A new surface is created, as shown in Fig. 44-2 on page 588.

11. Create a circle on the plane Y = 0, as shown in Fig. 44-2.

 Rotate the UCS so that the Y=0 plane becomes the ground plane, use the ViewCube to change the view to show the two surfaces, create the circle, then return to the World UCS.

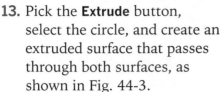

CREATE

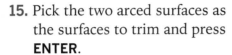

EDIT

12. Create a new layer named **Trim**, select the color blue, and make it the active layer.

13. Pick the **Extrude** button, select the circle, and create an extruded surface that passes through both surfaces, as shown in Fig. 44-3.

14. Pick the **Trim** button on the Edit panel of the Surface tab.

15. Pick the two arced surfaces as the surfaces to trim and press **ENTER**.

16. Pick the tubular surface as the cutting surface and press **ENTER**.

17. In response to Select area to trim, select the portion of the first two surfaces that lie inside the trim surface. (You may need to use the **Orbit** feature on the Navigation Bar to rotate the model so you are viewing from the end of the "tube" of the cutting surface to make these selections.)

18. Make **Objects** the current layer and freeze the Trim layer.

As you can see, the trim surface has trimmed a hole through each of the two surfaces.

19. Thaw the Trim layer.

20. Use the **Move** gizmo to move the tubular surface upward and clear of the other surfaces, as shown in Fig. 44-4 on page 589. If a warning dialog box appears, select **Continue**.

As you can see, the trim holes in the two surfaces have disappeared. This is because these procedural surfaces are associated: when one is modified, the others are affected by association.

21. **Undo** the 3D Move and save your work.

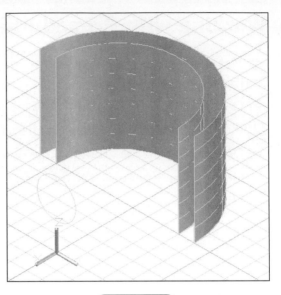

Fig. 44-2

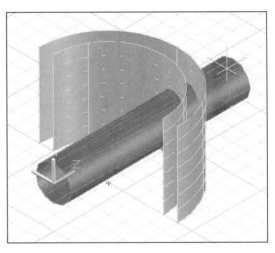

Fig. 44-3

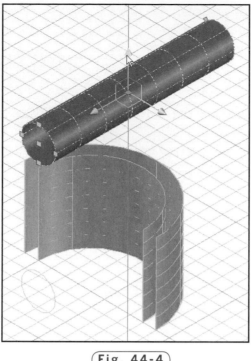

(Fig. 44-4)

Working with NURBS Surfaces

1. Start a new drawing using the **solid.dwt** template and save the drawing as **NURBSsurf.dwg**.

2. Use the **SPLINE** command to draw a spline on the ground plane of arbitrary shape and size.

3. On the Create panel of the Surface tab, toggle the **Surface Associativity** button off and the **NURBS Creation** button on.

4. Pick the **Revolve** button on the Create panel, select the spline, and press **ENTER**.

5. In response to Specify axis start point or define axis, select the **Y** option for the axis of revolution, and enter **180** for the angle of revolution.

6. Pick the **Show CV** button on the Control Vertices panel of the Surface tab, pick the surface, and press **ENTER**.

The NURBS control vertices are displayed, as shown in Fig. 44-5 on page 590.

Surface
Associativity

CREATE

NURBS
Creation

CREATE

Revolve

CREATE

Show
CV

**CONTROL
VERTICES**

Before editing control vertices, AutoCAD requires the surface to be rebuilt to a specific range of control vertices.

CONTROL VERTICES

7. Pick the **Rebuild** button on the Control Vertices panel of the Surface tab, and pick the surface.

8. In the Rebuild Surface dialog box, pick **OK** to accept AutoCAD's parameters to rebuild the surface.

9. Select and move any of the control vertices, and notice

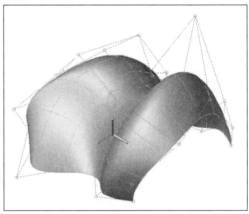

Fig. 44-5

the effect the movement has on the surface. Experiment on your own with the manipulation of several control vertices.

The CV Edit Bar button permits you to add and edit control vertices on a surface or a spline and to change tangency at specific points.

10. Pick the **CV Edit Bar** button on the Control Vertices panel, pick any point on the surface, and experiment with the effect your edits have on the surface.

11. Experiment on your own with the CV Edit Bar tool. When you are finished, your surface may look something like the surface in Fig. 44-6 or it may look quite different.

CV Edit Bar

CONTROL VERTICES

Fig. 44-6

> Note that solids and meshes can be converted to NURBS surfaces. Also, any surface can be turned into a solid using the Thicken command. In this way, all of the 3D modeling tools can be employed to produce the best solution for any given modeling problem.

12. Save your work and exit AutoCAD.

REVIEW QUESTIONS

Answer the following questions on a separate sheet of paper.

1. In your own words, describe the differences between procedural surfaces and NURBS surfaces.

2. Explain the purpose of the following surface modeling commands.

    a. Planar Surface

    b. Extrude Surface

    c. Revolve Surface

    d. Offset Surface

CHALLENGE YOUR THINKING

These questions are designed to further your knowledge of AutoCAD by encouraging you to explore the concepts presented in this chapter. Answer each question on a separate sheet of paper.

1. Explain surface associativity.

2. Use AutoCAD Help to learn more about Surface Blend, Surface Patch, and Surface Fillet commands.

Continued

• APPLYING AUTOCAD SKILLS

Work the following problems to practice the commands and skills you learned in this chapter.

1. Create the 3D model shown in Fig. 44-7 using the following steps.

   a. Create a solid primitive wedge on the ground plane with a 50″ × 50″ base and a 25″ height.

   b. Use the Surface Extrude command to extrude three surfaces 25″ from the three upper edges of the angled face.

   c. Use the Surface Fillet command to fillet the common edges of the new surfaces. Use a fillet radius of 5″.

   d. Use the Surface Blend command to create a surface on the fourth side of the angled area. Use a Bulge Magnitude of 0.0.

   e. View the model using an isometric viewpoint with the X-Ray visual style. The model should look similar to Fig. 44-7.

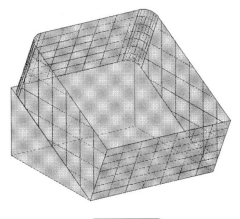

Fig. 44-7

2. Create three spheres, each with a radius of 15″: one mesh, one solid, and one NURBS surface. When you are finished, your 3D model space should look like Fig. 44-8. (To create a surface, convert a solid primitive sphere to a surface.)

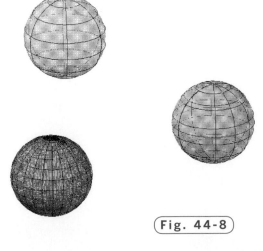

Fig. 44-8

• USING PROBLEM-SOLVING SKILLS

Complete the following activities using problem-solving skills and your knowledge of AutoCAD.

1. Open sweep.dwg and save it using the name surfsweep.dwg. Erase the swept solid. Create a new profile by drawing a simple S-shaped spline on the ground plane. Create a new swept surface using the spline as the profile and a twist angle of 0°. Move the first surface to the side, and create a second swept surface with a twist of 720°. In the Shaded with edges visual style, your drawing should look similar to the one in Fig. 44-9. Save your work.

Fig. 44-9

2. Open loft.dwg and save it using the name surfloft.dwg. Erase the lofted solid. With the NURBS Creation button toggled on, create a new lofted surface using Surface Loft. Select the Settings option and experiment with the settings in the Loft Settings dialog box. Finally, use the SURFACE_PLANAR command to create a surface to close the top of the model. In the Wireframe visual style, your drawing should look similar to the one in Fig. 44-10. Save your work.

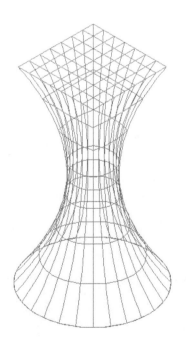

Fig. 44-10

Chapter 45

Boolean Operations

Objectives

- Prepare solid primitives for Boolean operations
- Use the Boolean subtraction operation to remove portions of a solid model
- Use the Boolean union operation to combine composite models

Vocabulary

Boolean logic
Boolean mathematics
Boolean operations
composite solids

Much of AutoCAD solid modeling is based on the principles of **Boolean mathematics**. Boolean math, also called **Boolean logic**, is a system created by mathematician George Boole for use with logic formulas and operations. AutoCAD **Boolean operations**, such as union, subtraction, and intersection, are used to create composite solids. **Composite solids** are composed of two or more solid objects. Using Boolean operations, let's create a composite solid of the steel support shown in Fig. 45-1.

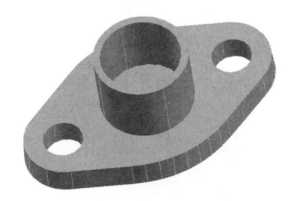

Fig. 45-1

Preparing the Base Solid

The first step in creating a composite solid is to create the base solid—the solid to which other solids will be joined.

1. Start AutoCAD and begin a new drawing using the **solid.dwt** template file.

2. Using the **PLINE Arc** and **Line** options, create the shape shown in Fig. 45-2 using the information below.

   - Use snap and grid settings of **.25** to produce the polyline accurately.

   - Pick the four points in the order shown. Each point falls on the snap grid.

   - In the PLINE command, use the **Arc's Radius** option.

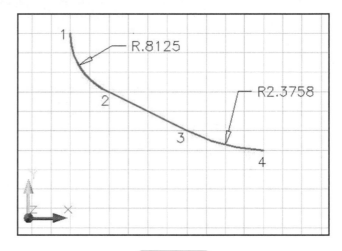

Fig. 45-2

You have just completed one-quarter of the object shown in Fig. 45-3.

3. Mirror the shape to create one-half of the object, and mirror that to create the entire object.

4. Use the **JOIN** command to join the four polylines into one.

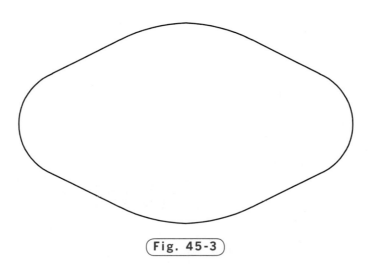

Fig. 45-3

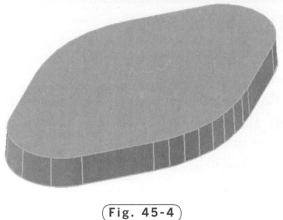

Fig. 45-4

MODELING

5. Select the **SW Isometric** viewpoint and the **Shaded with edges** visual style.

6. Save your work in a file named **composite.dwg**.

7. Pick the **Extrude** button on the Modeling panel of the Home tab.

8. Select the polyline and press **ENTER**.

9. Specify a taper angle of **3** degrees, and enter **.5** for the height.

A view similar to the one in Fig. 45-4 appears.

10. Pick the **Top** viewpoint, switch to the **Hidden** visual style, and **ZOOM All**.

The top view allows you to see the taper, as shown in Fig. 45-5, but without the two holes and the dimensions.

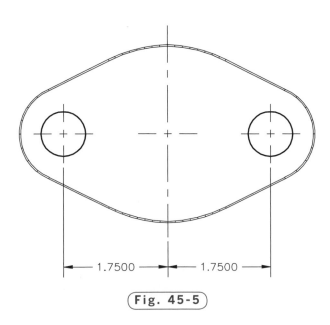

1.7500   1.7500

Fig. 45-5

11. On the Coordinates panel, pick the **World** button. (You can also select **WCS** at the flyout menu below the ViewCube.)

12. On the status bar, toggle **Allow/Disallow Dynamic UCS** off.

COORDINATES

13. Pick the **Cylinder** button on the Modeling panel and place two cylinders as shown in Fig. 45-5 to represent the holes. Make their radius **.375** and their height **.5**.

 HINT Create one cylinder and use the COPY or MIRROR command to create the second.

You have just created two solid cylinders within the original extruded solid.

14. Select the **SW Isometric** viewpoint and the **Conceptual** display.

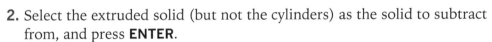

Subtracting the Holes

The SUBTRACT command performs a Boolean operation that creates a composite solid by subtracting one solid object from another.

SOLID EDITING

1. Pick the **Subtract** button from the Solid Editing panel to enter the **SUBTRACT** command.

2. Select the extruded solid (but not the cylinders) as the solid to subtract from, and press **ENTER**.

3. Select the two cylinders as the solids to subtract, and press **ENTER**.

AutoCAD subtracts the volume of the cylinders from the volume of the extruded solid.

4. Rotate the model to see that the cylinders have been removed, creating two holes in the solid.

The composite solid should look similar to the one in Fig. 45-6.

(Fig. 45-6)

Adding the Support Cylinder

The UNION command permits you to join two solid objects to form a new composite solid.

1. Select the **Top** view and the **World** UCS.

2. Place a solid cylinder at the center of the model. Make the diameter **1.75″**, and make it **1.75″** in height.

 Use the snap grid to snap to the center of the model.

3. Using the same center point, place a second cylinder inside the first. Specify a diameter of **1.5″** and a height of **1.75″**.

4. Select the **SW Isometric** viewpoint.

5. Subtract the smaller cylinder from the larger cylinder.

SOLID EDITING

The model is currently made up of two separate solid objects. Let's use the UNION command to join them to form a single composite solid.

SOLID EDITING

6. Pick the **Union** button on the Solid Editing panel.

This enters the UNION command.

7. Select the two solid objects and press **ENTER**.

The two objects become a single composite solid.

8. Pick the **Conceptual** visual style.

Your model should look similar to the one in Fig. 45-1 on page 594.

You have learned to create composite solids by adding and subtracting solid objects. You can also create them by overlapping two or more solid objects and calculating the solid volume that is common to each of the objects—the intersecting volume. The INTERSECT command performs this Boolean operation. Experiment on your own with the INTERSECT command.

9. Save your work and exit AutoCAD.

REVIEW QUESTIONS

Answer the following questions on a separate sheet of paper.

1. Describe the procedure used to create a hole in a solid object.
2. What is a composite solid?
3. Explain the purpose of the following solid modeling commands.
   a. SUBTRACT
   b. UNION
   c. INTERSECT

CHALLENGE YOUR THINKING

These questions are designed to further your knowledge of AutoCAD by encouraging you to explore further the concepts presented in this chapter. Answer each question on a separate sheet of paper.

1–2. Describe how you would create the solid models shown in Figs. 45-7 and 45-8. Be specific.

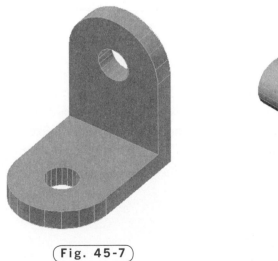

Fig. 45-7

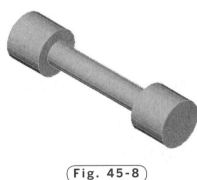

Fig. 45-8

Continued

● APPLYING AUTOCAD SKILLS

Work the following problems to practice the commands and skills you learned in this chapter.

1. Create the composite solid shown in Fig. 45-9.

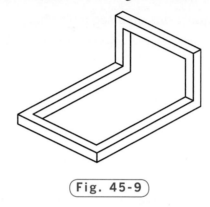

Fig. 45-9

2. Create the composite solid shown in Fig. 45-10.

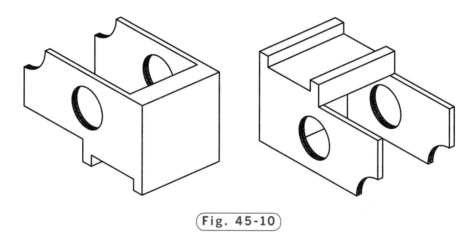

Fig. 45-10

3. Identify a machine part, such as a gear on a shaft or a bracket. Sketch the part in your mind or on paper. Using AutoCAD's solid primitives and Boolean operations, shape the part by adding and subtracting material until it is complete.

4. Draw a 3D model of the height gage shown in Fig. 45-11 using the EXTRUDE command and Boolean operations. Can you use a box or a wedge to create the V-groove? Why or why not? Save the drawing as ch45htgage.dwg.

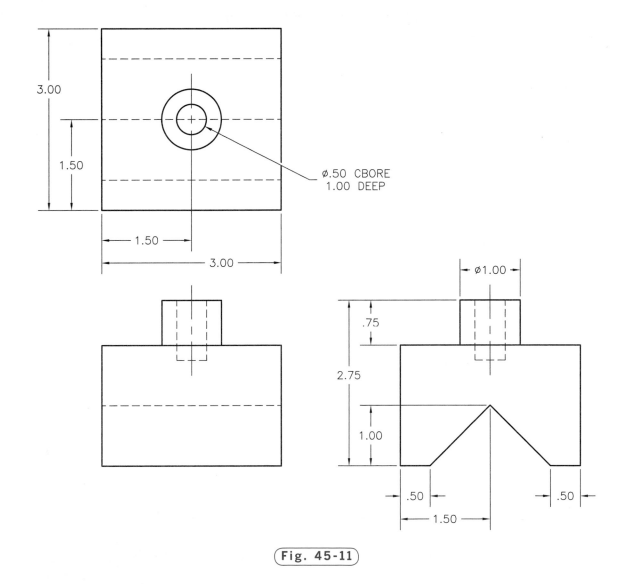

Ø.50 CBORE
1.00 DEEP

(Fig. 45-11)

Continued

5. Your customer wants to see a 3D view of the base plate shown in Fig. 45-12. Create the base, extrude it, and then use Boolean operations to create the 3D model. View the base plate in the Conceptual visual style, as shown in Fig. 45-13. Save the drawing as ch45baseplate.dwg.

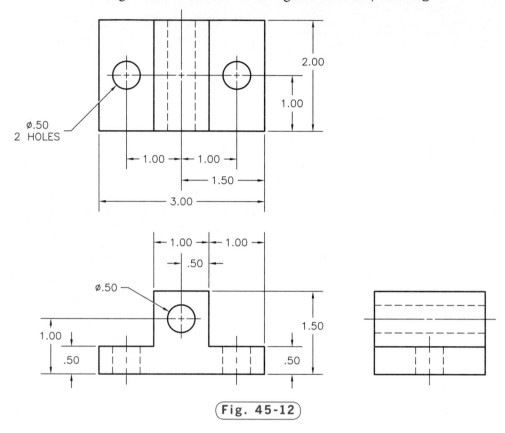

Fig. 45-12

Figure 45-12 prepared by Gary J. Hordemann, Gonzaga University

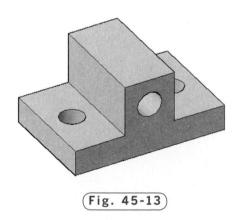

Fig. 45-13

● USING PROBLEM-SOLVING SKILLS

Complete the following activities using problem-solving skills and your knowledge of AutoCAD.

1. Create the fluid coupling end cap shown in Fig. 45-14 using the REVOLVE command. Begin by creating a layer named Endcap, and set the color to cyan. The object's profile is shown in Fig. 45-15.

Create the revolved model of the endcap on layer Objects and freeze the Endcap layer. Enter ISOLINES and enter a value of 25. (The ISOLINES system variable controls the number of tessellation lines used in the Wireframe and Realistic visual styles.) View the endcap in all four of the 3D visual styles at various isometric viewpoints. Save your work as endcap.dwg.

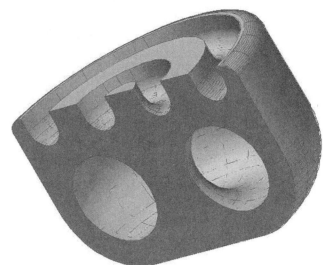

(Fig. 45-14)

Problem 1 courtesy of Gary J. Hordemann, Gonzaga University

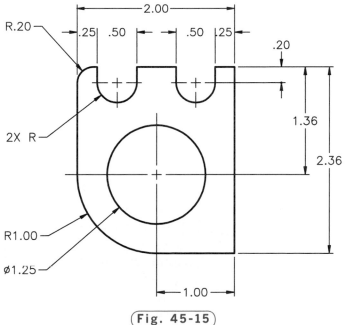

(Fig. 45-15)

Continued

2. Create the shaft clamp shown in Fig. 45-16 using the SUBTRACT and EXTRUDE commands. Create the .40-diameter hole by inserting and subtracting a cylinder. Display the model in the Conceptual visual style.

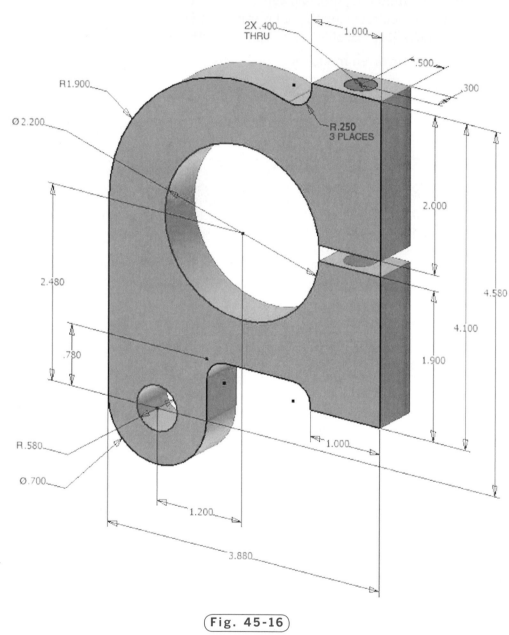

2X .400
THRU

1.000

.500

.300

R1.900

Ø 2.200

R.250
3 PLACES

2.000

2.480

4.580

.780

4.100

1.900

R.580

Ø .700

1.000

1.200

3.880

Fig. 45-16

Courtesy of Gary J. Hordemann, Gonzaga University

3. Create a solid model of the adjustable pulley half shown in Fig. 45-17. If you allow for the counterbored hole when drawing the profile, you will not have to bring in and subtract cylinders. After revolving the profile, go to the front view and insert and subtract the cylinders for the two smaller holes.

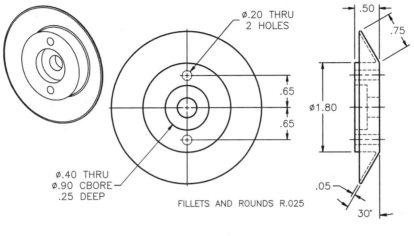

Fig. 45-17

Chapter **46**

Modifying Solid Models

Objectives

- Add chamfers and fillets to solids
- Adjust solids using the **SLICE** and **SHELL** commands
- Test for interference between solids

Vocabulary

interference

In the last five chapters, you have been introduced to many 3D–modeling methods. When these methods are used in combination, AutoCAD becomes a powerful tool for modeling solids and surfaces. In this chapter, you will be introduced to several techniques to adjust and modify existing solids.

Beveling and Rounding Solid Edges

Let's create the table shown in Fig. 46-1 on page 607 using the methods we've used in previous chapters. We will apply the CHAMFER and FILLET commands to make the filleted and beveled edges.

1. Start AutoCAD and use the **solid.dwt** template file to create a new file named **table.dwg**.

2. Select the **3D Modeling** workspace.

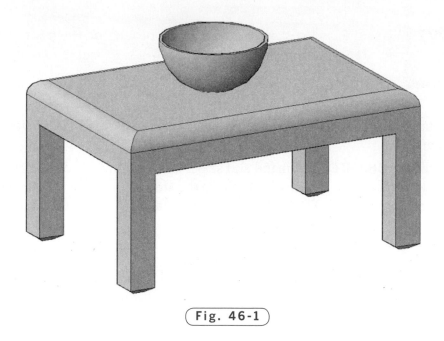

Fig. 46-1

3. Pick the **Box** button on the Modeling panel. Start at X = 2, Y = 2 and create a solid box **6** units in the X direction and **4** units in the Y direction. Make it **3** units in height and place it in the center of the screen.

4. Select the **SW Isometric** viewpoint and the **Hidden** visual style.

5. Use the **PRESSPULL** command to remove two rectangular boxes from the first one to form the basic table shown in Fig. 46-2. The legs are **.5** unit wide and the table top is **.75** thick.

MODELING

MODELING

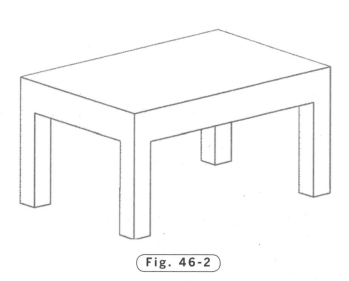

Fig. 46-2

HINT Use the LINE command to create 2D profiles on the front and left faces of the box primitive.

MODIFY

6. Rotate the table so that you can see the bottom edge of all four legs clearly.

7. Enter the **CHAMFER** command and select the bottom edge of one of the four table legs. If AutoCAD selects the side of the table rather than the bottom of the leg, select the **Next** option and press **ENTER**. When AutoCAD selects the bottom, press **ENTER**.

8. Enter a base surface chamfer distance of **.15**.

9. At the Specify other surface chamfer distance prompt, enter **.1**.

10. At the next prompt, pick all four edges that make up the bottom of the leg and press **ENTER**.

AutoCAD chamfers the edges.

11. Repeat steps 7 through 10 to chamfer the remaining three legs.

Notice that the lines you used to press the boxes to create the table legs now extend beyond the chamfered legs. It is safe to erase these lines now—doing so will have no effect on the model.

12. Erase the lines that extend beyond the chamfered legs.

The bottom of the legs should now look similar to those in Fig. 46-1 on page 607.

MODIFY

13. Enter the **FILLET** command, select any edge of the table top, and enter **.3** for the radius.

14. Select the **Chain** option, select the remaining three edges of the table top, and press **ENTER**.

15. Display the table in the **Conceptual** visual style.

16. Save your work.

The table should look very similar to the one in Fig. 46-1.

Slicing and Shelling a Solid Object

We will make a sphere and use the SLICE and SHELL commands to create the bowl shown on the table in Fig. 46-1.

1. Select the **World** button or enter the **UCS** command and type **W** to make sure you are in the world coordinate system. (The current UCS is also displayed at the bottom of the ViewCube.)

2. Select the **Sphere** button on the Modeling panel of the Home tab.

3. On the ground plane, create a sphere with a radius of **1**.

4. Select the **Top** viewpoint.

5. Pick the **Planar** button on the Create panel of the Surface tab.

6. Create a planar surface on the ground plane that overlaps the sphere on all sides.

7. Return to the **SW Isometric** view.

8. Select the **Slice** button on the Solid Editing panel of the Home tab.

9. Pick the sphere as the object to slice and press **ENTER**.

10. Select the **Surface** option plane.

11. Pick the planar surface you just created.

12. Pick the lower half of the sphere as the solid to keep.

13. Erase the planar surface.

MODELING

SOLID EDITING

We want to make the hemisphere into a bowl. The fastest way to do this is to shell it out.

14. Pick the **Shell** button from the last flyout menu on the Solid Editing panel.

This enters the Shell option of the SOLIDEDIT command.

15. Select the hemisphere.

SOLID EDITING

Next, we need to remove the top face from the shelling. If we don't, AutoCAD will hollow out the hemisphere, but it will not have an opening.

16. In response to Remove faces, pick the edge of the hemisphere.

The top face and the main body face share the same edge, so AutoCAD displays 2 faces found, 2 removed. We don't want to remove the main body face, so we need to add it back.

17. Select the **Add** option, select the rounded part of the hemisphere (do not select the top edge), and press **ENTER**.

18. For the shell offset distance, enter **.05** and press **ENTER** twice to terminate the command.

AutoCAD produces a shell, or bowl, as shown in Fig. 46-3.

Fig. 46-3

Testing for Interference

Let's move the bowl to the top of the table.

1. Select the **Top** viewpoint.

2. Use the **3DMOVE** command to move the bowl to the approximate center of the table.

3. Select the **Front** viewpoint.

4. Move the bowl again so that it appears to be resting on the table. Select the Z axis handle on the Move gizmo to be sure movement is constrained to the up-and-down direction.

The INTERFERE command enables you to check for **interference**, or overlap, between two or more solid objects. It is particularly useful when you are fitting solids together into an assembly.

SOLID EDITING

5. Select the **INTERFERE** button on the Solid Editing panel of the Home tab.

6. Pick the table and press **ENTER**.

7. Pick the bowl and press **ENTER**.

8. If the message Objects do not interfere appears on the Command line, move the bowl down 0.25 units and repeat steps 5 through 7 until the objects interfere with one another.

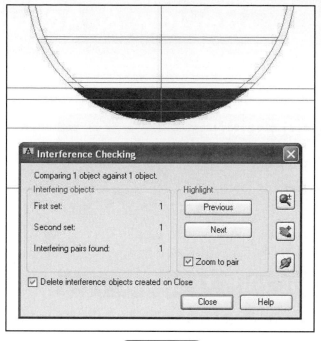

Fig. 46-4

When an interference occurs, the intersecting volume becomes shaded and the Interference Checking dialog box appears, as shown in Fig. 46-4. The dialog box provides Zoom Realtime, Pan Realtime, and 3D Orbit buttons and supports the number of interfering objects between the first and second selection sets.

9. Close the dialog box to exit the INTERFERE command.

10. Move the bowl so that it sits on the table, and switch to the **NE Isometric** view and the **Conceptual** visual style.

11. Save your work and exit AutoCAD.

● REVIEW QUESTIONS

Answer the following questions on a separate sheet of paper.

1. Explain how shelling is useful.
2. Describe the purpose of the INTERFERE command.
3. Explain in your own words how to modify the edges of a solid object.

● CHALLENGE YOUR THINKING

These questions are designed to further your knowledge of AutoCAD by encouraging you to explore the concepts presented in this chapter. Answer each question on a separate sheet of paper.

1. How do the CHAMFER and FILLET commands affect the appearance of solid objects containing outside corners?
2. Is it possible to shell a mesh object? Why or why not?
3. Is it possible to perform interference checking on mesh objects or surfaces? Explain your answer.

● APPLYING AUTOCAD SKILLS

Work the following problems to practice the commands and skills you learned in this chapter.

1. Open loft.dwg, a model you created in Chapter 43. We will use it to create a model of a candlestick. Rotate the model 180° to position the square end at the bottom. Draw two circles on the round top face using dynamic UCS and object snap. Make the larger circle the same size as the outer edge of the top face. Then use the PRESSPULL command to create a circular rim. Your candlestick should look similar to the one in Fig. 46-5 on page 613. Save your model as candlestick.dwg.

(Fig. 46-5)

2. Open ch45baseplate.dwg. Chamfer the two edges where the upper horizontal and vertical surfaces meet. Use a .25″ chamfer at 45°. Create fillets with a radius of .15 where the upper vertical surfaces meet the plate's horizontal surfaces. Create fillets with a radius of .15 at the two outer corners. Refer to Fig. 46-6 for specific locations.

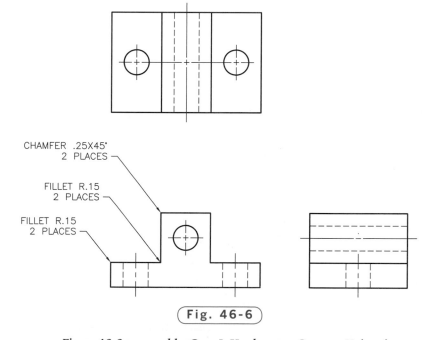

CHAMFER .25X45°
2 PLACES

FILLET R.15
2 PLACES

FILLET R.15
2 PLACES

(Fig. 46-6)

Figure 46-6 prepared by Gary J. Hordemann, Gonzaga University

Continued

• USING PROBLEM-SOLVING SKILLS

Complete the following activities using problem-solving skills and your knowledge of AutoCAD.

1. Use the PRESSPULL command to model the plate shown in Fig. 46-7. The dimensions are given in the orthographic projections in Fig. 46-8.

Use the FILLET command to create a fillet where the boss joins the main part of the plate. After you have finished the model, produce an isometric view in the Conceptual visual style. The drawing should look like the one in Fig. 46-7.

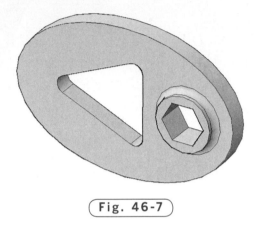

Fig. 46-7

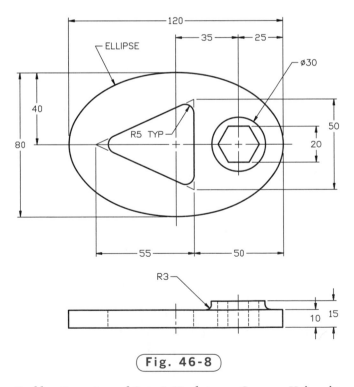

Fig. 46-8

Problem 1 courtesy of Gary J. Hordemann, Gonzaga University

2. Using the SUBTRACT and EXTRUDE commands, model the tube bundle support shown in Fig. 46-9. The dimensions are given in the orthographic projections shown in Fig. 46-10.

The rounded top can be produced in several ways. One way is to use the REVOLVE command to create a piece to be removed from the original extrusion using the SUBTRACT command. A second way is to use the REVOLVE command to create a rounded piece without holes; then use the INTERSECT command to obtain the common geometry. Try both ways. Display the model using an isometric viewpoint and the Conceptual visual style.

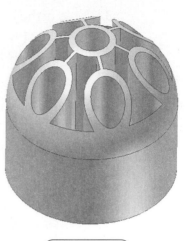

Fig. 46-9

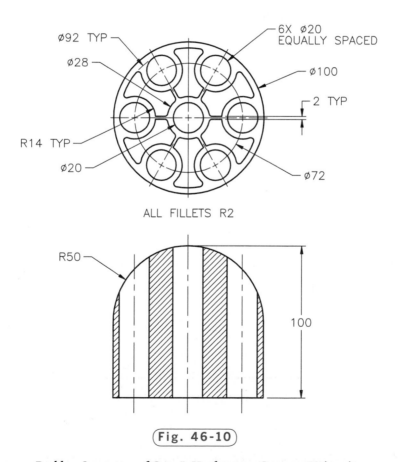

Fig. 46-10

Problem 2 courtesy of Gary J. Hordemann, Gonzaga University

Continued

3. Three orthographic views of a T-swivel support are shown in Fig. 46-11. Model the piece as a solid. Use the FILLET command to fillet all of the indicated edges.

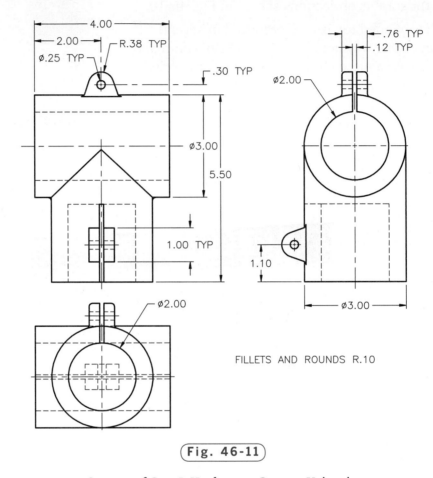

FILLETS AND ROUNDS R.10

$(\text{Fig. }46\text{-}11)$

Courtesy of Gary J. Hordemann, Gonzaga University

CAD Product Manager

Not too long ago, CAD drawings and models were created and used only by CAD departments. However, companies today share CAD files among multiple departments and locations. Improved drawing collaboration tools allow many people to work on these files. With CAD files being used for everything from design to marketing to computerized manufacturing, controlling these digital assets can quickly become a nightmare. This is especially true in large companies, which may need to track hundreds of thousands of CAD files.

© Ryan McVay/Getty Images/RF

Software Solutions

So how do they do it? The CAD product manager's job is to keep track of everything from who needs which files to revision records. But how can one person—or even an entire department—track thousands of digital files accurately? Many managers are turning to product lifecycle management (PLM) software. This software consists of a huge relational database that can track digital models, assemblies, development costs, rights, creation dates, usage, associated stock numbers, and just about any other information a company needs to track.

A Different Skill Set

In addition to being familiar with the CAD software being used in the company, a CAD product manager must be familiar with the needs of every department that uses the software and all of the products the company produces. To use PLM software, the manager should also have a working knowledge of databases, including relational databases. A logical mind and an excellent sense of humor may also come in handy!

▶ Career Activities

1. Make a list of the items you think would need to be tracked for each CAD file created by the company.

2. Research PLM software on the Internet. What options are available for companies of various sizes?

Documenting Solid Models

Objectives

- Create a full section from a solid model
- Create 2D flatshot views of a solid model
- Create an orthographic projection drawing sheet of a solid model

Vocabulary

live sectioning
section plane indicator

In this chapter, you will learn to create orthogonal, isometric, and section views from a solid model of a pulley. The SECTIONPLANE and FLATSHOT commands make this task quite simple. You will then create a multiview drawing of the pulley.

Creating the Solid Model

1. Start AutoCAD and use the **solid.dwt** template file to create a new drawing named **pulley.dwg**.

2. Select the **Top** viewpoint.

3. Set snap to **.25** and grid to **.25**.

4. Set the **ISOLINES** system variable to **60**, and select the **Realistic** visual style.

5. Create the polyline shown in Fig. 47-1 using the following information.

   - The lines on the grid are spaced .25 apart. All polyline endpoints lie on the grid, so use it to produce the object accurately.

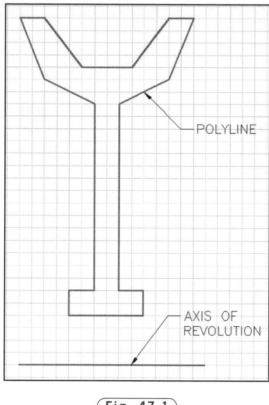

POLYLINE

AXIS OF
REVOLUTION

Fig. 47-1

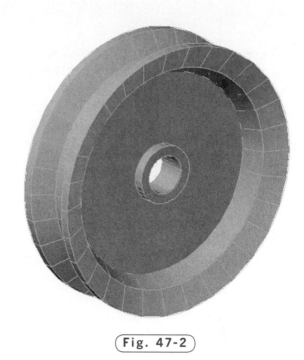

Fig. 47-2

- Produce the polyline in the top right quadrant of the drawing area.
- The axis of revolution is a line of arbitrary length.

6. Using the **REVOLVE** command, revolve the polyline to produce the pulley.

7. Select the **SE Isometric** viewpoint. Your view should be similar to the one in Fig. 47-2.

8. Erase the line used for the axis of revolution and save your work.

Creating a Full Section

A full section of a part is a view that slices all the way through the part, showing the interior of the part as it appears at that slice, or cross section. The SECTIONPLANE command creates a full cross section of a solid model.

1. Create a new layer named **Views**, assign the color blue to it, and make it the current layer.

2. Pick the **Section Plane** button on the Section panel.

3. Select the **Orthographic** option.

Section
Plane

SECTION

4. In response to Align section to, select the **Front** option.

A transparent plane, called a **section plane indicator**, appears and a full section of the pulley is created, as shown in Fig. 47-3. The SECTIONPLANE command automatically places the section plane at the center of all the 3D objects in the drawing area. This is why the section plane cuts through the exact center of the pulley.

5. Pick the section plane indicator.

Notice that the section plane indicator has grips, as shown in Fig. 47-3. Using the grips, you can move, rotate, and stretch the section plane, as well as specify which half of the pulley is hidden.

6. Experiment with moving and selecting the section plane grips; then undo all the changes you have made until the section plane returns to the original plane, as shown in Fig. 47-3.

7. With the section plane indicator selected, right-click to view the shortcut menu.

8. Pick the **Live section settings...** option to view the Section Settings dialog box.

Fig. 47-3

The Section Settings dialog box allows you to specify many settings for sectioning, including **live sectioning** (which provides realtime views while manipulating the grips) and the settings for the 2D or 3D sections you create.

9. Explore the settings in the Section Settings dialog box. Note that you can change the hatch pattern in a 2D or 3D section.

10. Pick **Cancel** to close the Section Settings dialog box.

11. Right-click to view the shortcut menu and select the **Generate 2D/3D section...** option.

12. In the **Generate Section/Elevation** dialog box, select **2D Section/Elevation** and pick the **Create** button.

13. Pick an insertion point to the left of the pulley and accept the default scale and rotation angle.

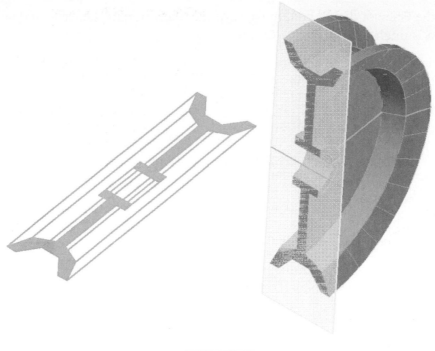

Fig. 47-4

A 2D section of the pulley at the section plane is created on the ground plane, as shown in Fig. 47-4.

14. Erase the section plane indicator.

15. Save your work.

Creating a 2D Representation

The FLATSHOT command creates a 2D, or "flattened," representation of all the 3D objects in the current view. The view can be an orthogonal view, predefined isometric view, or any other view that you create using 3D orbit. The 2D representations are created as blocks (as was the 2D section you have already created). The 2D objects are placed on the XY plane of the current UCS.

1. Switch to the **Top** view.

2. Select **WCS** in the menu below the ViewCube to make WCS the current UCS.

3. Pick the **Flatshot** button on the expanded Section panel.

SECTION

The Flatshot dialog box appears. It provides options for the destination and the representation of the lines in the 2D flatshot.

4. In the Flatshot dialog box, make sure the **Show** box in the Obscured lines area is unchecked and pick **Create**.

A 2D view of the top of the pulley appears, just as it would look when viewed from above, with hidden lines removed.

5. Pick a location to the left of the full section view and accept the default values for scale and rotation angle.

6. Switch to the **Right** view.

7. Make sure **WCS** is the current UCS.

SECTION

8. Pick the **Flatshot** button.

9. In the Flatshot dialog box, make sure the **Show** box in the Obscured lines area is unchecked and pick **Create**.

Because this view is perpendicular to the ground plane, we will need to specify the Z location of the new flatshot.

10. In response to Specify insertion point, snap to the center point of the pulley. Accept the default values for scale and rotation angle.

11. Pick the **NE Isometric** view and repeat the steps to create an isometric flatshot positioned on the ground plane.

12. Pick the **Top** view and use the **MOVE** command to arrange the 2D objects so they do not overlap. When you are finished, your drawing should look similar to the one in Fig. 47-5 on page 623.

13. Save your work.

Creating a Drawing

As you may recall, orthographic projection is a method of creating views of an object that are projected at right angles onto adjacent planes. With the SECTIONPLANE and FLATSHOT commands, we have already created 2D orthogonal and isometric projections of the pulley, as well as a full section view. Now we will create a drawing of those views. This process is similar to the multiple viewport drawing you created in Chapter 25, but since we have all the views projected onto the ground plane, we only need a single viewport.

1. Freeze layer **Objects** so that only the four 2D objects are visible.

2. Pick the **Layout1** tab at the bottom of the drawing area.

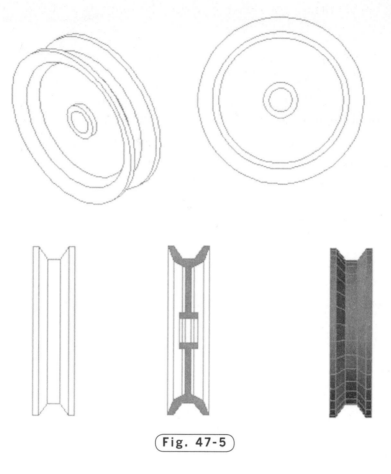

Fig. 47-5

AutoCAD switches to paper space, and the paper space icon replaces the coordinate system icon. The four 2D objects representing the views are present, but they are not positioned or aligned in the conventional style for orthogonal projection.

3. Pick the **Model** tab at the bottom of the drawing area to return to model space.

4. Use the **MOVE** command to arrange the four 2D objects so that the front view is in the lower left, the section view is in the lower right, and the isometric view is in the upper right.

5. Rotate the 2D top view 90° and place it above the front view.

6. Use object snap and ortho to align the top view and the section view to the front view precisely.

When you are finished, your drawing area should look similar to the one in Fig. 47-6 on page 624.

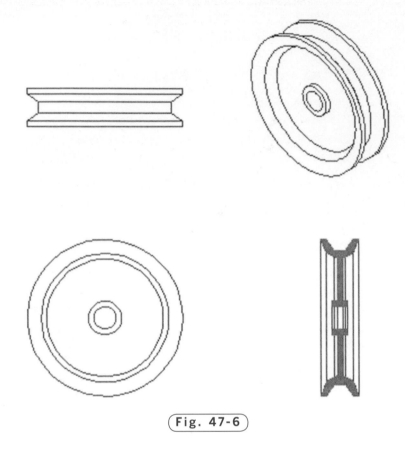

Fig. 47-6

7. Pick the **Layout1** tab to switch to paper space.

The drawing sheet now looks correct, with views properly positioned, oriented, and aligned. As you can see, the SECTIONPLANE and FLATSHOT commands make the production of multiview drawings fast and simple.

8. If necessary, resize and move the viewport to center the objects on the drawing sheet.

9. Plot the drawing.

10. Save your work.

Creating a Drawing Automatically

The VIEWBASE command allows you to create a single viewport drawing comprised of several orthogonal projections automatically.

1. Open the drawing **composite.dwg** that you created in Chapter 45.
2. Select the **Layout1** tab at the bottom of the drawing area to switch to paper space.
3. Pick the border of the default viewport to select it, and erase it.
4. On the Layout tab of the Ribbon, pick the **From Model Space** button from the flyout menu under the Base button.

**CREATE
VIEW**

This enters to VIEWBASE command. Notice that AutoCAD locks the front view of the solid model to the crosshairs and displays a contextual Ribbon tab named Drawing View Creation.

5. Move the crosshairs to the lower left quadrant of the drawing sheet and pick the position of the base view.

Now turn your attention to the options available on the Ribbon. You can change the position, orientation, scale, and visual style of views here.

6. Select the default values by selecting the **OK** button on the Create panel.
7. Move the crosshairs to a position above the base view and notice that the top view of the model is now projected and locked onto the crosshairs.
8. Pick a point in the upper left quadrant of the drawing sheet to position the top view.
9. Create a side view to the right of the base view and an isometric view in the upper right quadrant, then press **ENTER**.

Now we will create a section view through the center of the fitting using the VIEWSECTION command, as shown on Fig. 47-7 on page 626.

10. On the Create View panel, pick the **Section** button.
11. Pick the top view as the parent view, select two points to define the section line, and press **ENTER**.

CREATE VIEW

12. Pick a point in the top left center of the drawing layout to position the section view and press **ENTER** to exit the VIEWSECTION command.

AutoCAD allows you to edit the views you have created. Also, if you modify the solid or surface in model space, AutoCAD updates the views in paper space.

13. Pick the **Edit View** button on the **Modify View** panel, select the isometric view, change its view style to **Shaded with visible lines** on the Appearance panel, and pick the **OK** button.

14. Select the **Model** tab to enter model space.

15. Pick the **3D Scale** button on the Modify panel of the Home tab, and make the steel support solid larger by a scale factor of **1.5**.

16. Return to paper space by picking the **Layout1** tab.

Notice the message 5 view(s) updated successfully in the Command line window. We will need to reposition the views and resize the section line, however.

17. Move the views as needed to avoid overlapping.

18. Select the section line in the top view and drag its grips to the left and right beyond the edge of the part.

19. Move the section view identifier text up into the space between the views.

Your drawing should now look similar to Fig. 47-7.

20. Save the drawing as **viewbase.dwg** and exit AutoCAD.

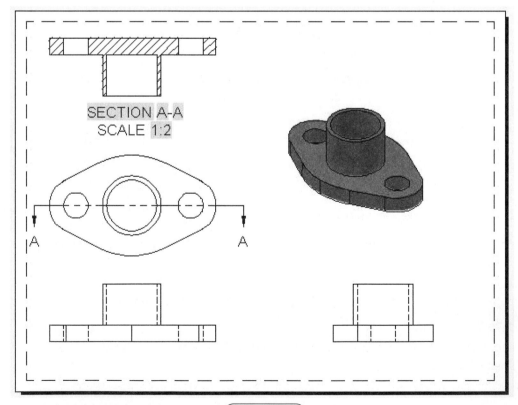

Fig. 47-7

Chapter 47 Review & Activities

● REVIEW QUESTIONS

Answer the following questions on a separate sheet of paper.

1. Explain the purpose of the following solid modeling commands.
   a. SECTIONPLANE
   b. FLATSHOT
   c. VIEWBASE
   d. VIEWSECTION
2. What is a section plane indicator?
3. Describe each of the grips on the section plane indicator and explain the function of each.
4. What type of object is the 2D representation that results from the FLATSHOT command?

● CHALLENGE YOUR THINKING

These questions are designed to further your knowledge of AutoCAD by encouraging you to explore the concepts presented in this chapter. Answer each question on a separate sheet of paper.

1. Investigate the other options in the SECTIONPLANE command. What option would you use to create a jogged (nonplanar) section of an object?
2. Consider the advantages and disadvantages of creating all views in model space and using one viewport in paper space. Can you think of a situation in which it would be preferable to use multiple viewports?
3. How many projected views of a base view can be created? Consider all orthogonal and isometric views.

● APPLYING AUTOCAD SKILLS

Work the following problems to practice the commands and skills you learned in this chapter.

1. Select any of the solid models you created in the previous chapters, or create a new one of your own. Then create front, top, right, isometric, and full section views of the model.

Continued

2. Draw a solid model of the bushing and insert shown in Fig. 47-8. Then create a full sectional view of each to describe them completely. Change the hatch pattern for the insert to ANSI31. Save the drawing as ch47bushing.dwg.

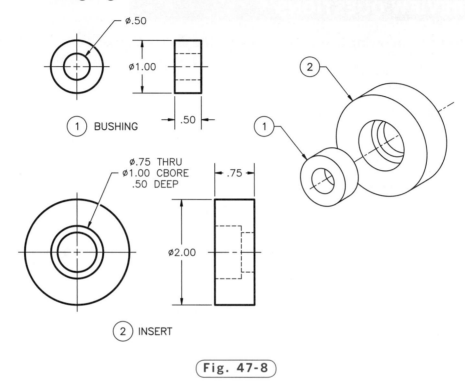

Fig. 47-8

Adapted from a drawing by Gary J. Hordemann, Gonzaga University

3. You are given a file containing the solid model and sectional view of the bushing and insert (problem 2 above). The manufacturing department needs a set of 2D working drawings. Create 2D front and side views of the bushing and insert. Save the drawing as ch47bushing2.dwg.

• USING PROBLEM-SOLVING SKILLS

Complete the following activities using problem-solving skills and your knowledge of AutoCAD.

1. Open bumper.dwg, which you created in Chapter 43. Use the FLATSHOT and SECTIONPLANE commands to create a drawing in paper space that is similar to Fig. 43-22 on page 583. Omit the dimensions and notes. Save your drawing as ch47bumper.dwg.

2. A solid model of the rocker arm shown in Fig. 47-9 is needed, along with a full section and a complete set of 2D drawings, for the design and manufacturing departments. Provide the necessary drawings. Save the drawings using names of your choice.

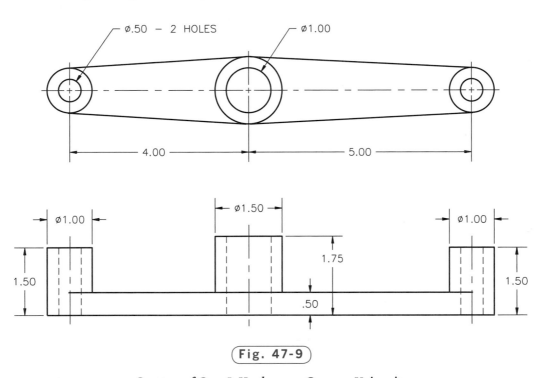

Fig. 47-9

Courtesy of Gary J. Hordemann, Gonzaga University

Chapter **48**

Visualization and Navigation

Objectives

- Apply AutoCAD's visual styles to a solid model
- Render a model with user-defined lighting and materials
- Place a camera in a model and use it to define a viewport
- Navigate a model using AutoCAD's walk and fly modes

Vocabulary

glyph
opacity
rendering

AutoCAD provides many tools to enhance the appearance of a model. These visualization tools help to produce a photorealistic representation, complete with realistic lighting, shading, and shadows. You can also assign a material specification to objects, giving them the patterns and textures of actual materials such as wood, concrete, or plastic. In this chapter, you will learn to use these visualization tools.

You are already familiar with many of AutoCAD's tools for navigating a model. You have moved around the drawing area by zooming, panning, and orbiting. In this chapter, you will learn to use an additional navigation tool: the camera.

Visualization Tools

Let's start by taking a look at the visualization tools AutoCAD provides.

1. Start AutoCAD and open the drawing named **table.dwg**, which you created in Chapter 46. Save the file as **visual.dwg**.

2. Select the **3D Modeling** workspace.

3. Create two new layers named **Table** and **Bowl**. Make the Table layer gray and the Bowl layer blue.

4. Move the table to layer Table and the bowl to layer Bowl. Freeze all other layers so that only the two solids are visible.

5. Select the **SW Isometric** viewpoint and the **Conceptual** visual style.

6. Select the **View** tab on the Ribbon, then pick the **X-Ray Effect** button on the Visual Styles panel.

VISUAL STYLES

X-ray mode allows you to view faces on objects transparently. The value of transparency, or **opacity**, is controlled by the VSFACEOPACITY system variable. A value of 100 makes objects opaque; a value of 0 makes them transparent. The default value is 60.

7. Pick the **X-Ray Effect** button again to turn X-ray mode off.

8. On the Visual Styles panel, click the arrow next to the **Warm-Cool Face Style** button to see the additional choices of face styles.

Warm-Cool Face Style is another name for the Conceptual visual style. The other choices are No Face Style (the Hidden visual style) and Realistic Face Style (the Realistic visual style).

9. View the table and bowl in all face styles and then return to the **Warm-Cool Face Style**.

Modifying Edge Settings

1. Click the arrow next to the Facet Edges button and select the **No Edges** button from the flyout menu. Observe the effect on the model.

VISUAL STYLES

2. Select the **Isolines** button from the flyout menu.

VISUAL STYLES

In Isolines mode, AutoCAD displays edges on curved objects. The system variable ISOLINES controls the number of edges. After you change the ISOLINES value, you need to regenerate the screen to display the change.

3. Enter the **ISOLINES** system variable at the keyboard and enter a new value of **50**.

4. Type **REGEN** and press **ENTER** to regenerate the screen.

Your model should now look similar to the one in Fig. 48-1.

5. View the model in all combinations of visual styles and edge modes.

6. Return to the **Warm-Cool Face Style** and **Isolines** edge mode.

7. Set the **ISOLINES** system variable to **10** and regenerate the screen.

8. Click on the arrow in the lower right corner of the Visual Styles panel to open the **Visual Styles Manager** palette.

9. Double-click on the **Realistic** visual style graphic at the top of the Visual Styles Manager palette to switch to the Realistic visual style.

Now focus on the Face Settings area of the Visual Styles Manager palette.

10. Click on the field to the right of Lighting Quality, and view the model in **Faceted**, **Smooth**, and **Smoothest** lighting. Observe the appearance of the bowl and the fillets on the rounded table edge.

11. Select the **Sketchy** visual style and focus on the entire area below Edge Settings.

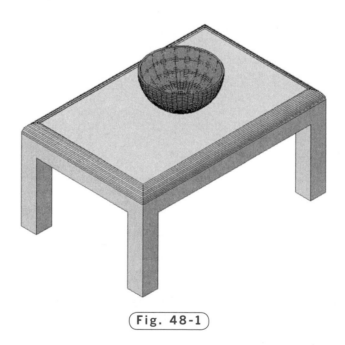

Fig. 48-1

Edge Modifiers control the appearance of edges. Line Extensions causes edge lines to overhang their endpoints. Jitter increases the number of edge lines and makes them irregular. When edges are overhung and jittered, it gives the drawing a hand-sketched, pencil-drawn appearance. Silhouette Edges thicken the edges that define the silhouette of the objects. Occluded Edges and Intersection Edges are active only in the Facet edges mode. As their names imply, they control the display of obscured (hidden) and intersection edges.

12. Experiment on your own with all the **Edge Settings** controls. Attempt to duplicate the hand-drawn look of Fig. 48-2.

13. Select the **Conceptual** visual style and close the **Visual Styles Manager** palette.

Fig. 48-2

Lighting Options

AutoCAD applies default lighting to models when they are shaded. Default lighting illuminates all faces on objects evenly and follows your viewpoint as you move around in the model. You can turn default lighting off and apply different lighting options to the model. These options include user-defined lights and lighting from the sun.

1. On the expanded Lights panel of the Render tab, pick the **Default Lighting** button to toggle default lighting off.

LIGHTS

LIGHTS

**SUN &
LOCATION**

2. Select the **Ground Shadows** button from the flyout menu below the No Shadows button.

Shadows appear below the table on the ground plane.

3. Pick the **Sun Status** button on the Sun & Location panel to toggle the sun lighting mode.

4. Move the **Date** and **Time** sliders on the Sun & Location panel left and right and notice the effect on the model.

The properties of sun lighting are controlled in the Sun Properties palette. You can open the Sun Properties palette by picking the arrow in the lower right corner of the Sun & Location panel. The geographic location of the model's sun lighting can be changed from the Sun Properties palette or by picking the Set Location button on the Ribbon.

5. Pick the arrow in the lower right corner of the Sun & Location panel to display the **Sun Properties** palette.

**SUN &
LOCATION**

6. Change the date to **July 4, 2013**, and change the time to **2:00 PM**.

7. Pick the **Set Location** button, select **Enter the location values,** enter a Latitude of **37.7950 North** and a Longitude of **122.3940 West** (San Francisco), and pick **OK**.

Your model should look similar to the one in Fig. 48-3.

**SUN &
LOCATION**

8. Pick the **Sun Status** button to toggle the sun lighting mode off, pick the **No Shadows** button to turn ground shadows off, and close the Sun Properties palette.

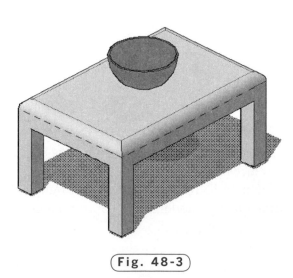

(Fig. 48-3)

Another option to default lighting and sunlight is user-created lights. We will now create a spotlight.

9. Select the **Wireframe** visual style.

10. Make sure dynamic input is off and the UCS is the world coordinate system.

11. Pick the **Spot** button from the flyout menu below the Create Light button on the Lights panel of the Render tab.

LIGHTS

This enters the SPOTLIGHT command.

12. For the spotlight's source location, enter **1, 7, 6**.

13. For the target location, pick the front right corner of the table top, as shown in Fig. 48-4.

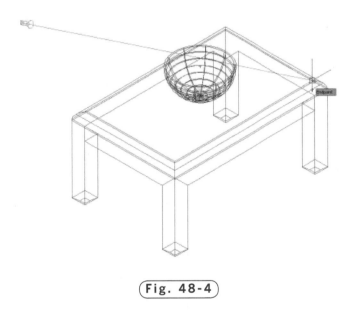

(Fig. 48-4)

14. Enter **X** to exit the **SPOTLIGHT** command.

15. Select the **Realistic** visual style.

A spotlight glyph appears on the screen and the model is now shaded by the spotlight. A **glyph** is an icon that has a specified location in the drawing area and represents one of AutoCAD's tools, such as a spotlight or camera.

15. Pick the light glyph and use the 3D grips to position the light so that the bowl is highlighted on the tabletop.

16. Save your work.

Selecting Materials and Rendering a Model

Now we will assign materials to the objects in the model. **Rendering** combines visual styles, lighting, materials, and shadows to create a photorealistic image of the model.

MATERIALS

1. On the Materials panel of the Render tab, select the **Materials Browser** button.

The Materials Browser palette appears. Material choices are displayed in a grid in the lower right of the palette.

2. Expand the size of the Materials Browser palette, then expand the width of the Name, Type, and Category columns.

3. Click on the column heading **Type** and scroll down the list to review the many choices of wood materials.

4. Select the **Maple - Solid Natural Polished** material and move the cursor into the drawing area.

The image of the material is locked to the cursor.

5. Pick the table solid.

6. Scroll through the materials list and review the many choices of metal materials.

7. Select the **Copper - Polished** material, move the cursor into the drawing area, and pick the solid model of the bowl.

RENDER

The appearance of the solids has not changed much, but when the model is rendered, you will see a difference.

8. On the Render tab, pick the **Render** button on the Render panel.

If a dialog box appears to inform you that the Autodesk Medium Image Library is not installed, select the third option, Work without using the Medium Images Library.

A new window, called the Render window, appears on the screen in front of the AutoCAD window. The rendered image of the table and bowl are displayed, as shown in Fig. 48-5 on page 637, complete with your user-defined spotlighting. Notice the realistic shadows and the reflection of light from inside the bowl.

The Render window displays an information pane on the right and a file management pane on the bottom. You can save or delete a rendered image in this pane.

Fig. 48-5

 Your rendering may look different, depending on the location of your spotlight and the amount you modified the grips on the light glyph.

9. Right-click Output File Name at the bottom of the Render window (the default name for the file will be **visual-Temp 000** or a similar name) and pick **Save...** from the shortcut menu.

10. Save the rendered image in your drawing folder as a JPEG file named **Render.jpg**, using the default JPEG image options.

11. Experiment on your own with rendering the model.

   • Use different lighting sources, materials, and viewpoints.

   • Select the spotlight glyph and move its grips to change the spotlight's location and cone size.

   • Render the model using the **Presentation** preset, which you can select at the top of the Render panel.

12. Close the Render window and the Materials Browser palette and save **visual.dwg**.

Navigation Tools

We will now take a look at AutoCAD's other navigation tools.

1. If the **Sun Status** is currently off, turn it on.
2. Select the **SteeringWheel.**
3. Use the **SteeringWheel** to zoom, pan, and orbit in the drawing area.

 You can also use shortcuts to navigate. Zoom by rolling the center (wheel) mouse button. Pan by holding the center mouse button down and dragging the cursor in the drawing area. Enable 3D orbit by holding the SHIFT key and center mouse button down simultaneously and dragging the cursor in the drawing area.

4. Close the **SteeringWheel** and select the **SE Isometric** viewpoint.
5. Reopen the **SteeringWheel** and click on the **Look** button.
6. Click and hold the left mouse button and move the cursor in the drawing area.

The Look navigation mode is different than the Pan mode, as you can see. Panning changes the viewpoint on a flat plane, maintaining a constant distance and angle from the objects in the drawing. Look swivels the viewpoint around a point in 3D space, similar to moving a camera around on a fixed tripod.

7. Close the **Steering Wheel**.

Creating a Camera

We will now create a custom camera, which is a saved viewpoint in a 3D drawing.

1. Select the **SW Isometric** viewpoint.
2. Make sure dynamic input is off and the current UCS is the world coordinate system.
3. Enter **CAMERA** at the keyboard.

This enters the CAMERA command. A camera glyph is attached to the crosshairs in the drawing area.

4. For the camera location, enter **3, -3, 2.5**.

5. For the target location, pick the right rear corner of the tabletop, as shown in Fig. 48-6.

6. Select the **Name** option and name the new camera **MyCamera**.

7. Enter **X** to exit the camera command.

8. Select the camera glyph.

When you select the camera glyph, dotted lines appear that define the camera's lens length and field of view. Grips are also displayed. The grips allow you to change the camera location, target location, and lens length (or field of view).

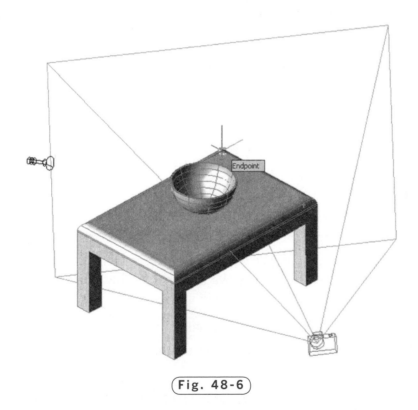

Fig. 48-6

Fig. 48-7

The Camera Preview window also appears. The preview pane allows you to see the model through the lens of the camera.

9. In the Camera Preview window, select the **Conceptual** visual style.

10. Click and drag one of the grips to increase the camera's field of view until the entire table is displayed in the preview pane, as shown in Fig. 48-7.

11. Pick the **Four: Equal** option from the **Viewport Configuration** List drop-down menu in the Model Viewports panel on the View tab.

Four viewports are displayed. We will keep an isometric, and make one a front view and another a top view. We will change the fourth to the MyCamera viewpoint.

12. Click in the lower right viewport to make it active.

13. Pick the down-facing arrow in the Views panel to display the available views in the flyout menu.

Notice that MyCamera has now been added to the list of view options.

14. Pick the **MyCamera** view for the lower right viewport.

The lower right viewport now displays the view from MyCamera.

15. Create the other views and improve the appearance of all four views.

- Create a front view and a top view in the two views on the left side.
- Center and zoom the objects in all four viewports.
- Align the top view with the front view.

Your drawing area should now look similar to the one shown in Fig. 48-8.

16. Render the MyCamera viewport.

17. Save and close **visual.dwg** and exit AutoCAD.

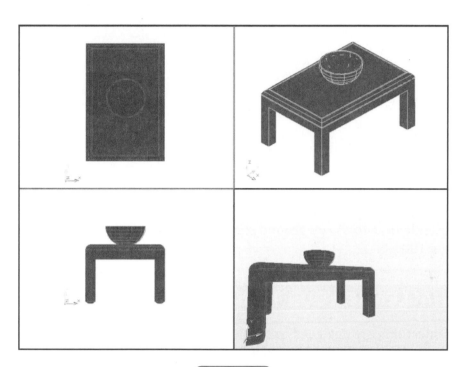

Fig. 48-8

● REVIEW QUESTIONS

Answer the following questions on a separate sheet of paper.

1. Describe each of the following AutoCAD visualization modes.

   a. X-ray Effect

   b. Line Extensions

   c. Edge Jitter

   d. Silhouette Edges

   e. Ground Shadows

   f. Sun Status

2. Identify the two glyphs you created in this chapter and explain the function of each.

3. List the nine navigation modes available in AutoCAD. (*Hint:* They are listed in the shortcut menu available during any navigation command.)

4. Describe how to set up ground shadows for January 1 and high noon in New York, NY.

● CHALLENGE YOUR THINKING

These questions are designed to further your knowledge of AutoCAD by encouraging you to explore the concepts presented in this chapter. Answer each question on a separate sheet of paper.

1. What is the difference between the Orbit and Look modes of the SteeringWheel? Can you think of an instance in which Look would be more useful than Orbit and vice versa?

2. Review all the options available in the Materials Browser palette. If you were designing the interior of a house, what material would you select for a tile floor? for a leather couch? for a copper sculpture?

● APPLYING AUTOCAD SKILLS

Work the following problems to practice the commands and skills you learned in this chapter.

1. Open the model presspull.dwg, which you created in Chapter 43. Light the model with sunlight for January 1, 2012, at 2:00 PM for Tulsa, OK. With the grid on, your model should look similar to the one in Fig. 48-9.

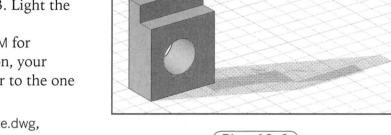

Fig. 48-9

2. Open the model composite.dwg, which you created in Chapter 45. Create a camera at the top center of the support cylinder, with a target directly down into the cylinder, as shown in Fig. 48-10. (Snap to circle centers to locate the camera and the target.) Name the camera CenterCam.

Use the camera's grips to move the camera up several inches above the solid. (Constrain the movement to the Z axis only with the Move grip tool.) Then expand the camera's field of view to encompass all of the model, as shown in Fig. 48-11 on page 644.

Finally, view the solid from camera CenterCam. Select the Realistic visual style, turn Facets on, and apply some edge extension and jitter. Your drawing area should look similar to that shown in Fig. 48-12 on page 644. Save the model as CompositeCam.dwg.

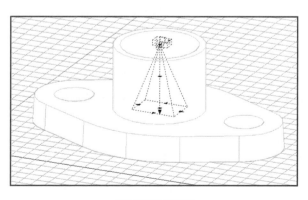

Fig. 48-10

Continued

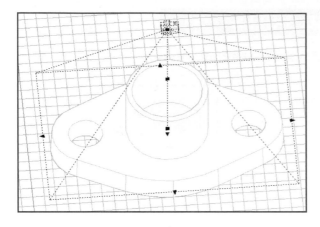

Fig. 48-11

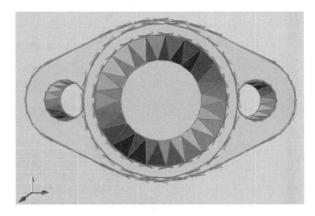

Fig. 48-12

● USING PROBLEM-SOLVING SKILLS

Complete the following activities using problem-solving skills and your knowledge of AutoCAD.

1. Start with the model table.dwg and create the simple hotel room shown in Fig. 48-13. The following information will help you create each solid in the model.

- Create the walls using the POLYSOLID command. They are 10′ high and .25′ wide. The room measures 18′ × 30′.

- Create the door and windows using the PRESSPULL command. The door is 3′ × 7′. The two matching windows are of arbitrary size.
- Create the floor using the PLANESURF and THICKEN commands. The value for thickness is −1.
- The bed is a box measuring 6′ × 7′ × 2.5′.
- If necessary, scale, rotate, and move the table and bowl to match their appearance in Fig. 48-13.

Put each object on its own layer and give each layer a unique color. Save your model as room.dwg.

2. Apply the following materials to the objects in room.dwg:

Walls: Stucco: Troweled - White

Floor: Flooring: Wood: Red Oak - Natural Classic

Bed: Fabric: Velvet - Red

Table: Finish: Wax - Varnish

Bowl: Glass: Glazing: Mirrored

Render the model with late afternoon sunlight, as shown in Fig. 48-14.

Finally, create a spotlight in the top corner of the room, to the right of the

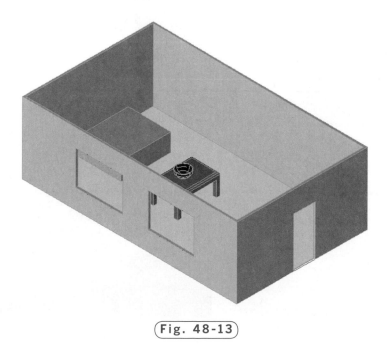

Fig. 48-13

Continued

door, and locate its target so that the light falls across the table and bowl onto the bed. Set the intensity of the spotlight to a value of 10. Adjust the spotlight's properties until the rendered view of the room looks similar to the view in Fig. 48-15. Notice that this rendering contains both late afternoon sunlight and light from the spotlight. Save your work.

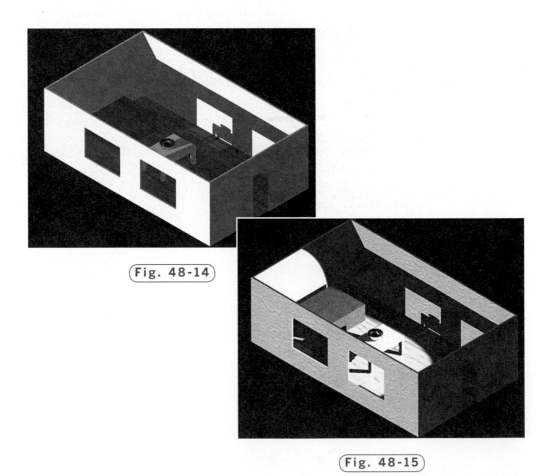

Fig. 48-14

Fig. 48-15

AutoCAD at Work

Pipeline Drafter

Although we don't always realize it, we sometimes take modern-day conveniences for granted. If we want a drink of water, all we have to do is go to the kitchen sink and help ourselves to a glass. If we feel like taking a bath, we can just walk into the bathroom and fill the tub. The water we use for these purposes doesn't just magically appear, of course. Someone had to plan a way to direct water from a main source to our homes. Thanks to people like pipeline drafters, we are able to enjoy one of our most important natural resources by merely turning a nozzle.

A Knack for Design

In addition to finding ways to channel water into homes, pipeline drafters design piping systems to move substances in refineries, oil and gas fields, sanitation systems, and chemical plants. The drawings they create are used for construction and layout of parts. An accurately designed system transports substances efficiently from place to place. It also resists leakages or damage

© 2007 Getty Images, Inc./RF

that may be potentially harmful to the environment.

Analyzing the Situation

Pipeline drafting requires knowledge of the materials used to build pipeline systems and the quality and safety standards for each. Analytical skills are necessary to determine which materials are best suited for a particular system in a particular environment. Drafters frequently consult with surveyors, scientists, and engineers to ensure proper design, so communication and "people" skills are an important part of the job, as well.

▶ Career Activities

1. Write a report about the work of a pipeline drafter. What challenges are involved in this type of drafting?

2. Why is it important for pipeline drafters to consult with other people when designing piping systems?

Benefits of Solid Modeling

Objectives

- Split a solid model in half
- Calculate mass properties of a solid model
- Create an **STL** file of a solid model for prototype fabrication

Vocabulary

additive manufacturing
bisect
XYZ octant

Solid modeling provides many benefits that are "downstream" from the solid model. As you saw in Chapter 47, the solid model of the pulley was used to create a production drawing. Other examples of benefits include mass properties generation, detail drafting, finite element analysis, and the fabrication of physical parts. Another benefit is the natural link to additive manufacturing. **Additive manufacturing** is a process of joining liquids, powders, filaments, or sheet materials to form parts from 3D model data. The parts are usually produced layer upon layer, as opposed to a subtractive manufacturing methodology, such as machining. Once you have a solid model, it becomes very easy to output the data needed to drive an additive manufacturing system.

In this chapter, you will split the pulley you created in Chapter 47, as shown in Fig. 49-1 on page 649. You will use the pulley half to calculate mass properties and create data that can be used to produce a prototype part.

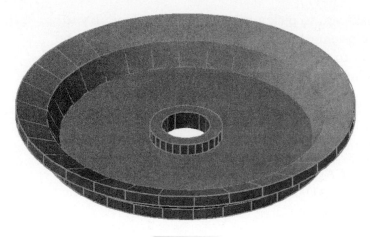

Fig. 49-1

Slicing the Pulley

The SLICE command cuts through a solid and retains either or both parts of it. The command is especially useful if you want to create a mold for half of a symmetrical part, such as a pulley.

1. Start AutoCAD and use the **pulley.dwg** drawing file to create a new file named **half.dwg**.

2. Freeze all layers except for layer **Objects**.

3. View the model from the **Top** view and switch to the **Wireframe** visual style.

4. Select the **World** UCS.

5. Pan and zoom the pulley to the center of the screen.

6. Turn snap on and draw a line that bisects the pulley.

When you **bisect** an object, you divide it into two equal halves. The view of the pulley should be similar to the one in Fig. 49-2.

7. Pick the **Slice** button on the Solid Editing panel of the Home tab to enter the **SLICE** command.

8. Select the pulley, press **ENTER**, and select the **YZ** option.

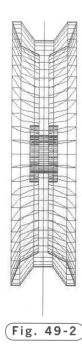

Fig. 49-2

SOLID EDITING

AutoCAD asks you to pick a point on the YZ plane. Try to visualize the orientation of this plane, which is vertical and perpendicular to the screen. "Vertical" is the Y direction, and "perpendicular to the screen" is the Z direction.

9. Pick the endpoints of the center line to define the slicing plane.

10. Pick a point anywhere to the right of the slicing plane.

This tells AutoCAD that you want to keep the right half of the pulley. AutoCAD cuts the pulley in half and leaves the right half on the screen.

11. Pick the polyline that you used to create the pulley in Chapter 47 and move it to layer **0**.

Because layer 0 is frozen, AutoCAD displays a dialog box to let you know that the object has been moved to a frozen layer.

12. Erase the center line.

13. Select the **Shaded with edges** visual style.

Your model should now look like the one in Fig. 49-3.

14. Save your work.

Fig. 49-3

Calculating Mass Properties

Because solid models contain volume information, it is possible to obtain meaningful details about the model's physical properties. For example, you can generate mass properties, including the center of gravity, mass, surface area, and moments of inertia. Properties such as mass and volume permit engineers to analyze the part for structural integrity.

1. Enter the **MASSPROP** command at the keyboard.

2. Select the solid model and press **ENTER**.

AutoCAD generates a report similar to the one shown in Fig. 49-4.

3. Press **ENTER** again to see the entire report.

4. Enter or select the **Yes** option in response to Write analysis to a file? and press **ENTER** or pick the **Save** button to use the default **half.mpr** file name.

AutoCAD creates an ASCII (text) file, assigning the MPR file extension to it automatically. The MPR file extension stands for "Mass Properties Report."

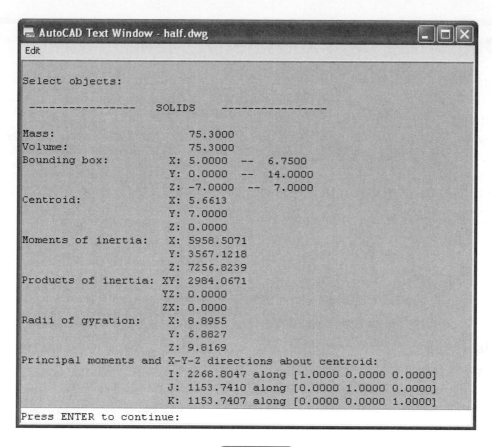

```
AutoCAD Text Window - half.dwg                    [_][□][X]
 Edit

Select objects:

---------------         SOLIDS       ---------------

Mass:                    75.3000
Volume:                  75.3000
Bounding box:      X: 5.0000  --   6.7500
                   Y: 0.0000  --  14.0000
                   Z: -7.0000 --   7.0000
Centroid:          X: 5.6613
                   Y: 7.0000
                   Z: 0.0000
Moments of inertia:  X: 5958.5071
                     Y: 3567.1218
                     Z: 7256.8239
Products of inertia: XY: 2984.0671
                     YZ: 0.0000
                     ZX: 0.0000
Radii of gyration:   X: 8.8955
                     Y: 6.8827
                     Z: 9.8169
Principal moments and X-Y-Z directions about centroid:
                     I: 2268.8047 along [1.0000 0.0000 0.0000]
                     J: 1153.7410 along [0.0000 1.0000 0.0000]
                     K: 1153.7407 along [0.0000 0.0000 1.0000]
Press ENTER to continue:
```

Fig. 49-4

Let's locate and review the file.

5. Close the AutoCAD Text window, if you haven't already, and minimize AutoCAD.

6. Start Notepad.

7. Select **Open...** from the File pull-down menu.

 Change the file name extension from txt to mpr to list only those files that have the MPR extension.

8. Locate and open the file named **half.mpr**.

The contents of the file should be identical to the information displayed by the MASSPROP command.

9. Exit Notepad and maximize AutoCAD.

Outputting Data for Additive Manufacturing

The STLOUT command creates an STL file from a solid model. STL is the file type required by most additive manufacturing systems, such as fused deposition modeling and laser sintering. These machines enable you to create a physical part from a solid model. The part can serve many purposes, such as a pattern for prototype tooling.

Positioning the Solid Model

Before creating the STL file, you must orient the part in the position that is best for the additive manufacturing system. In most cases, the part should be lying flat on the WCS. You can accomplish this in the current drawing by rotating the part around the Y axis.

Rotate Gizmo

SELECTION

1. Select the **SE Isometric** viewpoint.
2. Select the **Rotate Gizmo** from the Gizmo flyout on the Subobject panel.
3. Select the solid. Notice the Rotate Gizmo, with its three colored ellipses, appears.
4. Pick to highlight the green handle on the gizmo, because you want to rotate the part around the Y axis.
5. Dynamically rotate the solid so that the sliced side is facing down, then enter **90** for the rotation angle.

It is important that you rotate the part in the correct direction. We do not want to position the part upside-down. Also, if the part lies outside the positive XYZ octant, the STLOUT command will not work. The **XYZ octant** is the area in 3D space where the *x*, *y*, and *z* coordinates are greater than 0.

6. If necessary, move the part so that it lies entirely in positive 3D space.
7. Pick the **SW Isometric** view, as shown in Fig. 49-5 on page 653.

To make sure the part lies entirely in positive 3D space, turn on the grid and view the model from the Front viewpoint. Move it up using the 3DMOVE command, if necessary.

The part should appear to lie on the WCS. Let's check to be sure.

8. Using the **ID** command, pick a point at the bottom of the part.

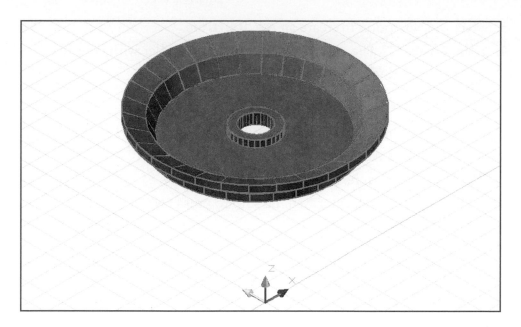

Fig. 49-5

HINT Use the Nearest object snap to snap to a point at the bottom of the part.

The value of the z coordinate should be greater than 0.0000. If it is 0.0000 or a value less than 0, you must move the solid upward into the positive XYZ octant.

9. If necessary, move the solid upward into the positive XYZ octant.

Creating the STL File

Now that the pulley is properly positioned, you can create the STL file.

1. Select the **World** UCS.

2. Enter the **STLOUT** command, pick the solid, and press **ENTER**.

AutoCAD wants to know whether you want to create a binary STL file. Normally, you would create a binary file because binary files are much smaller than ASCII files. However, you cannot view the contents of a binary STL file, and we want to view its contents.

3. Select the **No** option to create an ASCII **STL** file.

4. In the **Create STL File** dialog box, browse to the folder with your name, then pick the **Save** button to accept **half.stl** as the name of the file.

Reviewing the STL File Data

Let's review the contents of the STL file.

1. Minimize AutoCAD and start NotePad.

2. Select **Open...** from the File pull-down menu.

3. In the File name box, enter ***.stl**. Be sure to press **ENTER**.

4. Open the file named **half.stl**.

The contents of the file appear. The text listing in Fig. 49-6 shows the first part of the file. STL files consist mainly of groups of x, y, and z coordinates. Each group of three defines a triangle.

```
solid AutoCAD
facet normal 0.0000000e+000  0.0000000e+000  –1.0000000e+000
        outer loop
                vertex 8.4399540e+000  4.8596573e+000  1.0000000e–001
                vertex 8.5000000e+000  4.2500000e+000  1.0000000e–001
                vertex 5.8750000e+000  4.2500000e+000  1.0000000e–001
        endloop
endfacet
facet normal 0.0000000e+000  0.0000000e+000  –1.0000000e+000
        outer loop
                vertex 4.8750000e+000  4.2500000e+000  1.0000000e–001
                vertex 2.2500000e+000  4.2500000e+000  1.0000000e–001
                vertex 2.3100460e+000  4.8596573e+000  1.0000000e–001
        endloop
endfacet
facet normal 0.0000000e+000 0.0000000e+000  –1.0000000e+000
        outer loop
                vertex 4.9130602e+000  4.4613417e+000  1.0000000e–001
                vertex 4.8750000e+000  4.2500000e+000  1.0000000e–001
                vertex 2.3100460e+000  4.8596573e+000  1.0000000e–001
        endloop
endfacet
facet normal 0.0000000e+000  0.0000000e+000  –1.0000000e+000
```

Fig. 49-6

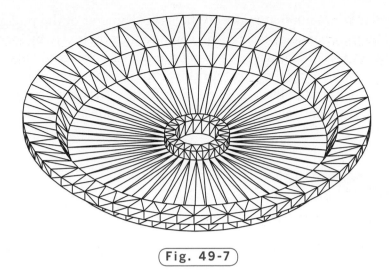

Fig. 49-7

5. After you have reviewed the file, close NotePad and maximize AutoCAD.

6. Save your work and exit AutoCAD.

The view shown in Fig. 49-7 was created using a special utility called stlview.lsp. This utility reads and displays the contents of an ASCII STL file.

As you can see, an STL file is indeed made up of triangular facets. You can adjust the size of the triangles using the FACETRES (short for "facet resolution") system variable by changing its value before you create the STL file. A high value creates a more accurate part with a smoother surface finish, but the file size and the time required to process the file increase. You should set it as high as necessary to produce an accurate part. Knowing what this value should be for a given part comes with experience and experimentation.

Part Creation

Hundreds of organizations around the world can create parts from STL files. If you were to send the half.stl file to one of them, that organization could create a part for you, typically for a fee. Autodesk (the maker of AutoCAD) has partnered with service providers to make 3D printing available. The Send to 3D Print Service button in the Output tab opens a dialog box that allows you to order one or more parts.

Figure 49-8 on page 656 shows pictures of a completed plastic prototype part. It was created from the half.stl file using a stereolithography additive manufacturing system.

Notice the flat segments that make up the large diameter of the pulley. They are caused by the triangular facets. You could reduce their size by increasing the value of FACETRES before creating the STL file. You could also reduce their visibility by sanding the part, but this would change the accuracy of the part. The part pictured in Fig. 49-8 was built from a version of half.stl with FACETRES set at a relatively low value to show the undesirable effect of large facets.

Fig. 49-8

Prototype part courtesy of Laser Prototypes, Inc. of Denville, New Jersey
(Photos courtesy of Terry Wohlers)

Chapter 49 Review & Activities

Answer the following questions on a separate sheet of paper.

1. Describe the SLICE command.

2. When would the SLICE command be useful?

3. AutoCAD can create files with an MPR file extension. How are these files created, and what information do they provide?

4. Name at least three mass properties for which AutoCAD makes information available about a solid model.

5. Explain the purpose of the STLOUT command.

6. What is the purpose of creating an STL file?

7. What AutoCAD system variable affects the size of the triangles in an STL file?

8. What is the primary advantage of creating a binary STL file? ...an ASCII STL file?

● CHALLENGE YOUR THINKING

These questions are designed to further your knowledge of AutoCAD by encouraging you to explore the concepts presented in this chapter. Answer each question on a separate sheet of paper.

1. Describe a situation in which you might need to slice an object using the 3 Points option.

2. Explore the different additive manufacturing systems that are commercially available. Write an essay that compares each of them.

Continued

● APPLYING AUTOCAD SKILLS

Work the following problems to practice the commands and skills you learned in this chapter.

1. Create ASCII and binary STL files from the solid model stored in the composite.dwg file, which you created in Chapter 45. Compare the sizes of the two files and then view the contents of the ASCII STL file.

2. Open the height gage you created in "Applying AutoCAD Skills" problem 4 in Chapter 45. If you have not yet created the model, create it now. Then make two slices. Make the first slice through the V groove and the second slice perpendicular to the first and in the same plane.

3. If you know of an additive manufacturing service provider in your area, request a demonstration of its system(s). Explore the possibility of having the company create a prototype part for you from an STL file.

4. Open the file named shaft.dwg, which you created in Chapter 43. Use the SLICE command to split the model down the middle to create a pattern. Use STLOUT to create a binary STL file.

5. Produce and save a list of the mass properties for shaft.dwg.

● USING PROBLEM-SOLVING SKILLS

Complete the following activities using problem-solving skills and your knowledge of AutoCAD.

1. The bushing and insert shown in Fig. 49-9 on page 659 are part of an aerospace project. Weight is critical. Slice the bushing and insert. Determine the volume of each part. Write each of the mass property reports to a file and print the files. Then select a material for each part, look up its density, and compute its mass. (To find density values, you may have to go to the reference section of a suitable library or search the Internet.)

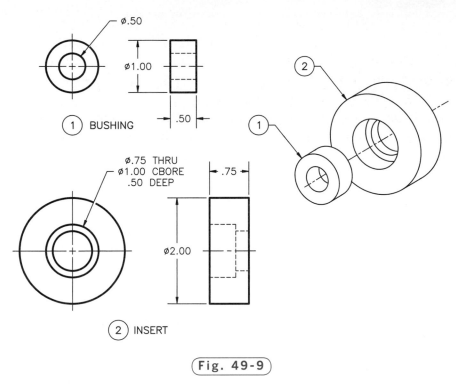

⌀.50

⌀1.00

.50

(1) BUSHING

⌀.75 THRU
⌀1.00 CBORE
.50 DEEP

.75

⌀2.00

(2) INSERT

(2)

(1)

Fig. 49-9

Courtesy of Gary J. Hordemann, Gonzaga University

2. Position the bushing and insert from problem 1. Then make an ASCII STL file of the bushing only. View the STL file, but do not print it.

Designing and Modeling a Desk Set

AutoCAD is capable of creating 2D drawings, 3D surface models, and 3D solid models. The differences are meaningful. Although all three types of drawings yield information, 3D models allow more realistic views. Solid modeling, in addition, allows determination of mass properties and the generation of production data such as that used in automated machine tool processes. Familiarity with solid modeling enhances the ability of the drafter/designer to create solutions; it is well worth the effort to learn.

▶ Description

Your company has sold the desk set shown in Fig. P9-1 for several years. Your supervisor has determined that it is time to modernize and streamline the desk set to meet current market demand.

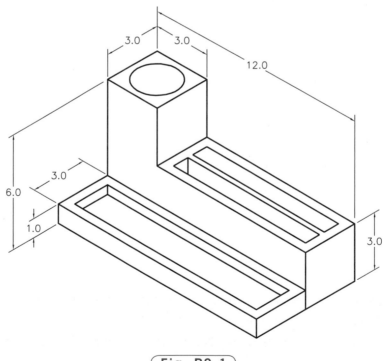

Fig. P9-1

1. Create a solid model of the desk set shown in Fig. P9-1. Include the round space to store pens, pencils, and the like; a large, flat, shallow area for items such as paper clips and rubber bands; and a medium-height section for a note pad, business cards, and similar papers. Wall thickness should range between .1″ and .185″. Assign a plastic material to the model. Save it as desk1.dwg.

2. Save desk1.dwg with a new name of desk2.dwg. Alter the design to modernize it or to make it more useful. For example, you may decide to provide an inclined surface on the two medium-height holes so that the one in back rises above the one in front of it. You may also want to round or bevel the edges to give it a streamlined look, or even add or subdivide the compartments.

3. Prepare a presentation for your supervisor. In your presentation, compare and contrast the new and old designs and explain why the new design is superior. Prepare rendered images of both desk sets to support your presentation.

▶ Hints and Suggestions

1. There is often more than one way to construct a solid model. If your approach does not seem to be working, try another.

2. Create the recesses for the compartments (such as the pencil holder) using the SUBTRACT command or the PRESSPULL command.

3. Use the UNION command as necessary so that the final product is a single composite solid.

▶ Summary Questions

Your instructor may direct you to answer these questions orally or in writing. You may wish to compare and exchange ideas with other students to help increase efficiency and productivity.

1. In your desk set design, how could the area actually being used to hold items be calculated?

2. Could the entire desk set design be created via extrusion? Explain how it could or why it could not.

3. What are some of the advantages of using predefined solid primitives in the creation of more complex solid models?

4. What is meant by "downstream benefits" of solid modeling?

Additional Problems

Introduction

The following problems provide additional practice with AutoCAD. These problems encompass a variety of disciplines. They will help you expand your knowledge and ability and will offer you new and challenging experiences.

The key to success is to **plan before you begin**. Review the options for setting up a new drawing and consider which commands and features you might use to solve the problem. As you discover new and easier methods of creating drawings, apply these methods to solving the problems. Since there is usually more than one way to complete a drawing, experiment with alternative methods. Discuss these alternatives with other users and create strategies for efficient completion of the problems.

Remember, there is no substitute for practice. The expertise you gain is proportionate to the time you spend on the system. Set aside blocks of time to work with AutoCAD, think through your approach, and enjoy this fascinating technology.

Each problem in this section is preceded by an icon that describes its level of difficulty, as shown below. The level of difficulty assigned to each problem assumes that you have learned the AutoCAD techniques and skills necessary to work the problem successfully. For some of the Level 3 problems, you may be required to expand your experience by combining your knowledge of AutoCAD with knowledge and skills that are not taught in this book.

L1 Uses basic AutoCAD skills

L2 Uses intermediate AutoCAD skills

L3 Uses advanced AutoCAD and problem-solving skills

L1 PROBLEM 1

Create the geometric design using the grid, snap, the LINE command, and (optionally) the MIRROR command. Set grid to 1 and snap to .25.

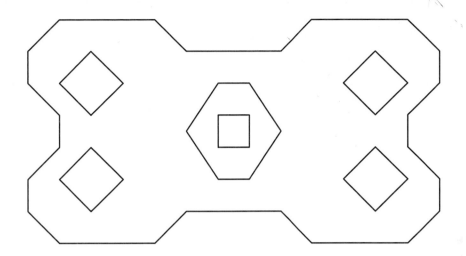

L1 PROBLEM 2

Create the geometric design using the grid, snap, and objects such as lines, arcs, circles, and fillets. Set grid to 1 and snap to .5.

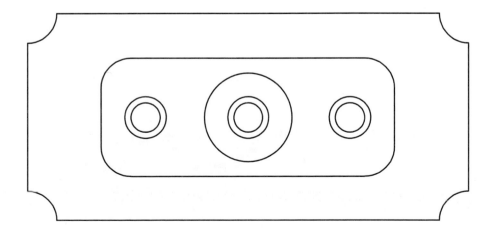

Problems 1 and 2 courtesy of Carole J. Williams, South Harrison High School

L1 PROBLEM 3

Create the first geometric figure using the POLYGON and ARRAY commands. Inscribe the polygon in a circle with a radius of 2. Then array the polygon three times to create the figure. Then copy the first geometric figure to a new location in the drawing area. Use the TRIM command to create the second figure.

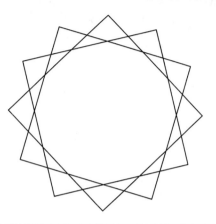 

L1 PROBLEM 4

Create the set of Olympic Rings using the CIRCLE, COPY, BREAK, and BHATCH commands. Create five layers using colors of blue, black, red, yellow, and green and create each ring on a separate layer. The first circle is blue, the second is yellow, the third is black, the fourth is green, and the fifth is red. Set snap to .25 and grid to .50. The circles that make up the outer part of each ring have a radius of 2.75, and the circles that define the inner part of the ring have a radius of 2.25. The centers of the rings are 6.00 units apart, and the centers of the two bottom rings are 5.50 units below the centers of the three top rings.

Problems 3 and 4 courtesy of Carole J. Williams, South Harrison High School

L1 PROBLEM 5

Use the CIRCLE and ARRAY commands to create each design variation shown. All of these variations were created by starting with a basic circle with a radius of 2.

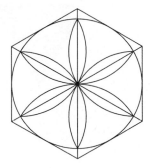

L1 PROBLEM 6

Use the CIRCLE, ARC, LINE, ERASE, and BHATCH commands to create the geometric design. Circumscribe a polygon about a circle with a radius of 3. Then construct lines from the center of the circle to each of the vertices. Use the Start, End, and Center options of the ARC command to create six arcs. Start each arc at one vertex and end it at the next alternate vertex, using the middle vertex as the center. Save the design and copy the entire figure to a new location in the drawing area. Alter the copy to make it look like the six-petal flower.

Problems 5 and 6 courtesy of Carole J. Williams, South Harrison High School

L1 PROBLEM 7

Use the LINE, MIRROR, and BHATCH commands with the necessary object snaps to create the two geometric figures. Set grid at 1 and snap at .25.

- Begin with an equilateral triangle with a side length of 3.
- Construct a second equilateral triangle with a vertex at the midpoint of the base of the first triangle and a side length of 1.5.
- Using LINE, extend the base of the second triangle by 1.25 on each side to create the top of the trapezoid.
- At each endpoint of the line created in the previous step, draw a segment of length .75 at an angle of 225° on the left side and an angle of –45° on the right side.
- Connect the inner endpoint of the left side to the endpoint of the inner right side.
- Mirror the top half of the design to complete the bottom half.
- Copy the entire figure to a new location in the drawing area.
- Use a solid hatch to fill in the areas shown.

Courtesy of Carole J. Williams, South Harrison High School

L1 PROBLEM 8

Create and fully dimension the drawing of a link.

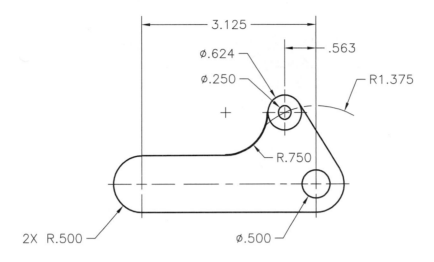

L1 PROBLEM 9

Create and fully dimension the drawing of a filler.

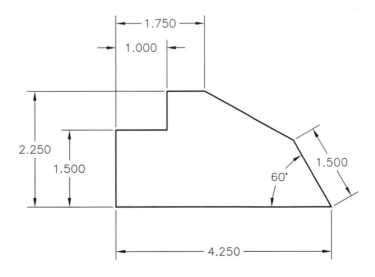

L1 PROBLEM 10

Create and fully dimension the drawing of a link.

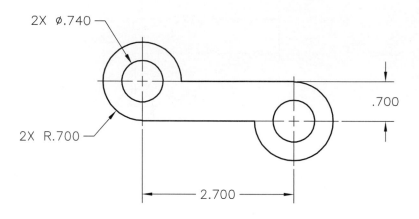

L1 PROBLEM 11

Create and fully dimension the orthographic views of the block.

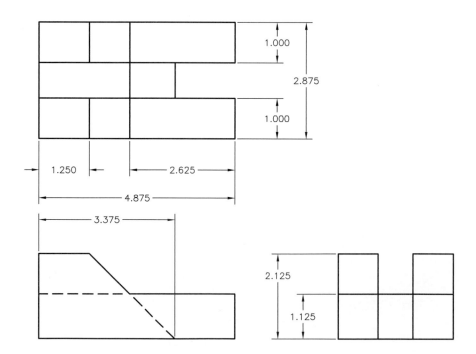

L1 PROBLEM 12

Create and fully dimension the drawing of a gage.

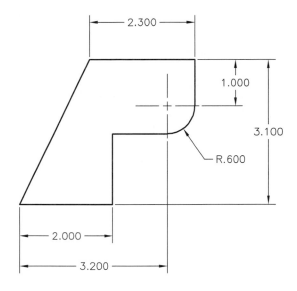

L1 PROBLEM 13

Create and fully dimension the drawing of a shaft bracket.

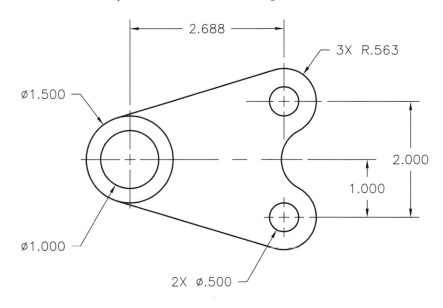

L1 PROBLEM 14

Create and dimension the drawing of an end block.

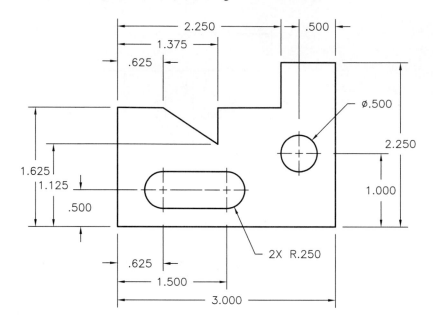

L1 PROBLEM 15

Create and dimension orthographic views of the angle block.

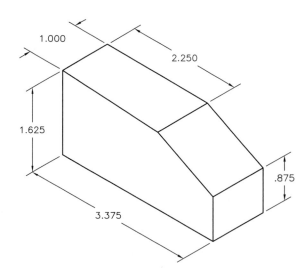

L1 PROBLEM 16

Create and fully dimension orthographic views of the nesting block.

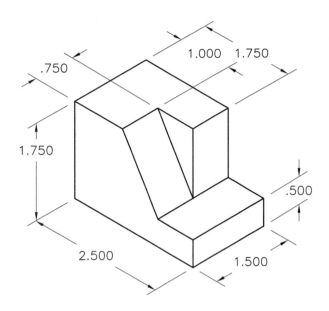

L1 PROBLEM 17

Create and fully dimension orthographic views of the locator.

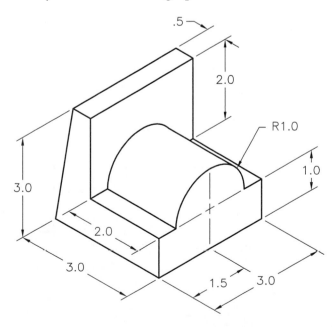

L1 PROBLEM 18

Create and fully dimension orthographic views of the step block.

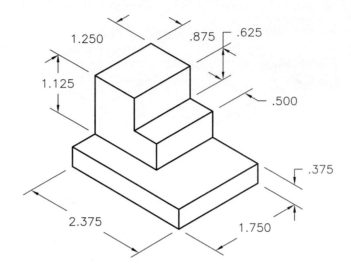

L1 PROBLEM 19

Create and fully dimension orthographic views of the block.

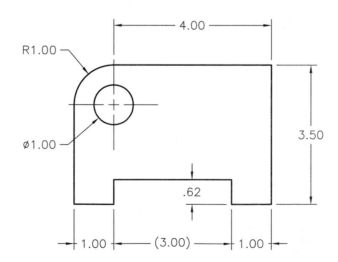

L1 PROBLEM 20

Create the drawing of an angle bracket. Use the dimensions shown, but dimension the drawing using decimal units.

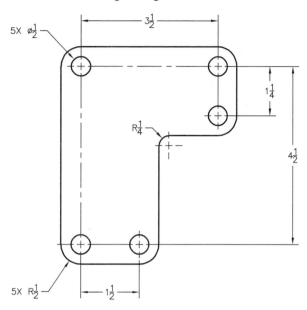

L1 PROBLEM 21

Create and dimension the drawing of a gasket.

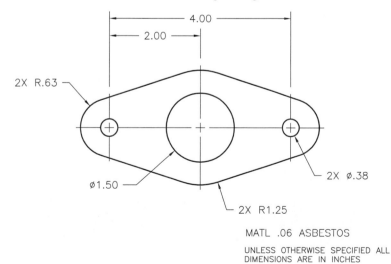

MATL .06 ASBESTOS

UNLESS OTHERWISE SPECIFIED ALL
DIMENSIONS ARE IN INCHES

Adapted from the textbook Drafting Fundamentals *by Scott, Foy, and Schwendau*

L1 PROBLEM 22

Create the electrical symbols as shown. Estimate their sizes. Produce
a block of each symbol to create a symbol library.

RESISTOR   THERMISTOR   VARIABLE RESISTOR

DIODE   ZENER   CONNECTION

CAPACITOR   CRYSTAL   TERMINAL

INDUCTOR   SWITCH   SPEAKER

TRANSISTORS   LOGICAL GATES

NPN   PNP   AND   NAND

JFET—N   JFET—P   OR   NOR

THYRISTORS   COMMON GROUND   CHASSIS GROUND

Courtesy of Robert Pruse, Fort Wayne Community Schools

L1 PROBLEM 23

Create the following international symbols.

Courtesy of Joseph K. Yabu, Ph.D., San Jose State University

L1 PROBLEM 24

Create the figure shown below using a pattern of circles of radius 2 and the ARRAY, BREAK, TRIM, and BHATCH commands.

Courtesy of Carole J. Williams, South Harrison High School

L1 PROBLEM 25

Create the floor plan drawing of a computer lab. Determine the drawing area according to space requirements. Estimate the dimensions that are not given.

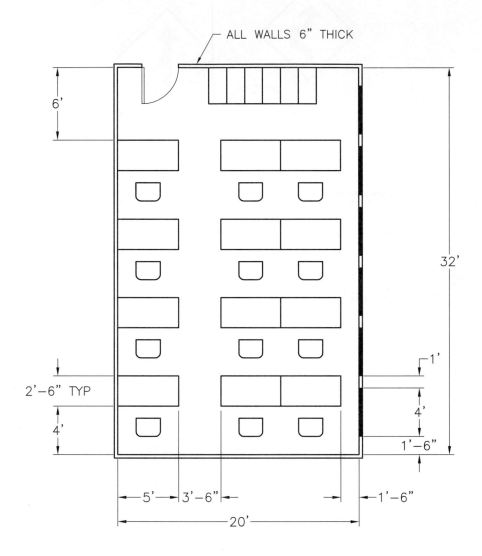

ALL WALLS 6" THICK

L2 PROBLEM 26

Create and fully dimension the drawing of an adjustable link.

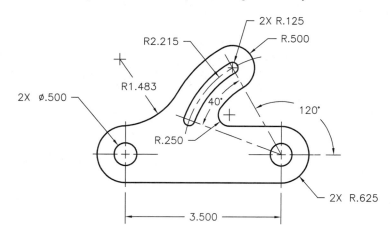

L2 PROBLEM 27

Create and fully dimension the drawing of an idler plate.

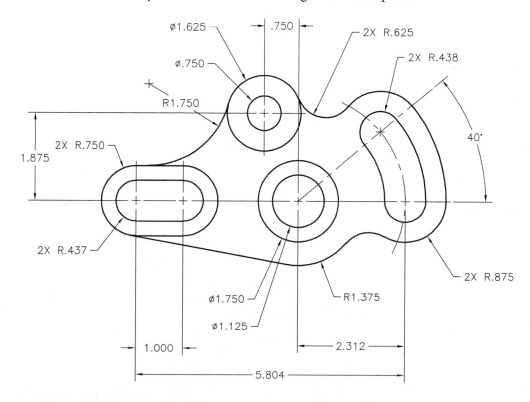

L2 PROBLEM 28

Create and dimension orthographic views of the support.

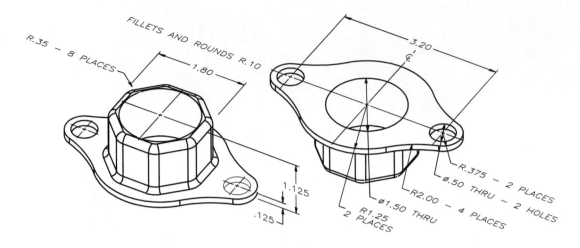

FILLETS AND ROUNDS R.10

R.35 – 8 PLACES

1.80

3.20

R.375 – 2 PLACES

ø.50 THRU – 2 HOLES

R2.00 – 4 PLACES

ø1.50 THRU

R1.25 2 PLACES

1.125

.125

Courtesy of Gary J. Hordemann, Gonzaga University

L2 PROBLEM 29

Create and dimension the drawing of a cover plate.

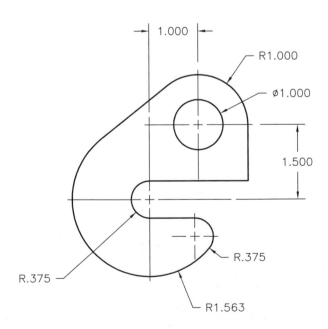

1.000

R1.000

ø1.000

1.500

R.375

R.375

R1.563

L2 PROBLEM 30

Create and dimension orthographic views of the base. Use a drawing
area of 17 × 11.

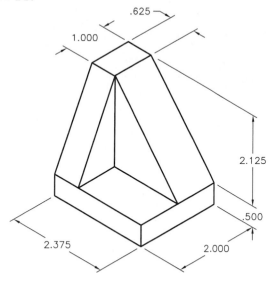

L2 PROBLEM 31

Create and dimension orthographic views of the base. Use a drawing
area of 17 × 11.

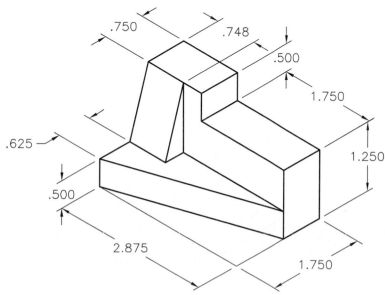

L2 PROBLEM 32

Create and fully dimension the orthographic views of the angle block.

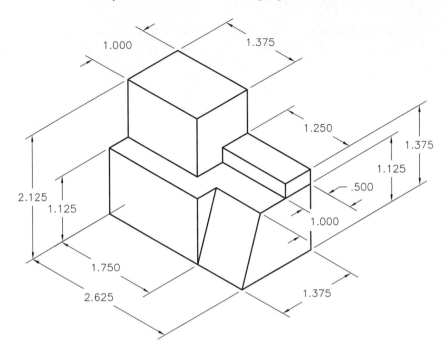

L2 PROBLEM 33

Create and fully dimension the orthographic views of the cradle.

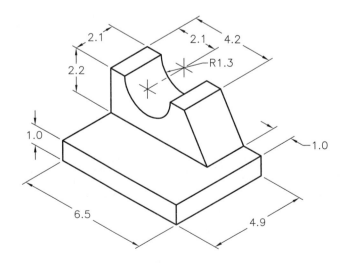

L2 PROBLEM 34

Draw the hex ratchet. Begin by creating the circles for the ends of the ratchet. Then create the lines and polygon. Use the BREAK command to remove parts of both circles as indicated to finish the drawing. Use a drawing area of 280 × 215. Dimension the drawing in millimeters.

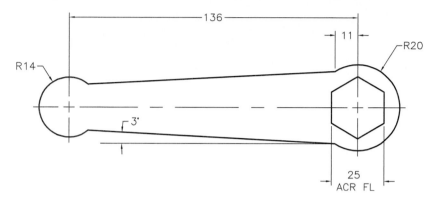

Adapted from the textbook Drafting Fundamentals *by Scott, Foy, and Schwendau*

L2 PROBLEM 35

Create and fully dimension the drawing of a rod support.

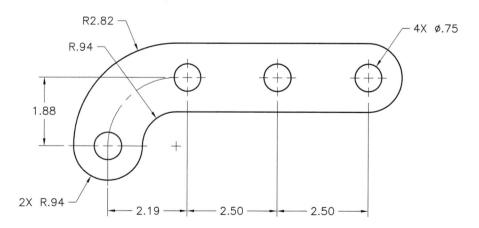

L2 PROBLEM 36

The mechanism below consists of three parts: a driver, a follower, and a link. Use the dimensions shown and the CIRCLE, OFFSET, and FILLET commands to draw the front view.

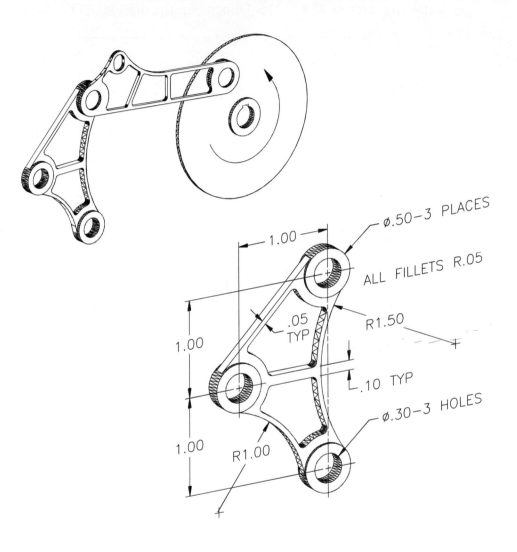

Courtesy of Gary J. Hordemann, Gonzaga University

L2 PROBLEM 37

Create and fully dimension the drawing of a master template. Use a drawing area of 22 × 17.

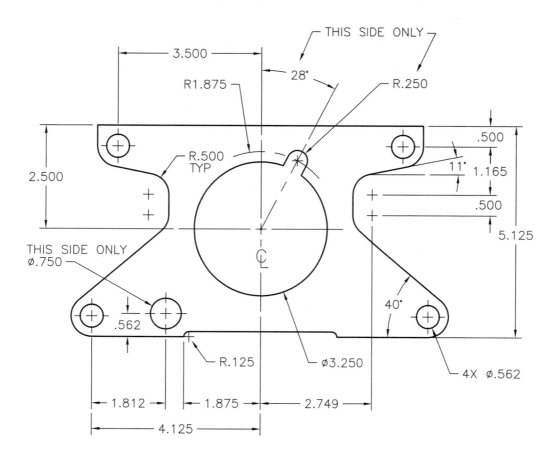

Courtesy of Steve Huycke, Lake Michigan College

L2 PROBLEM 38

Create and dimension the drawing of a steel plate.

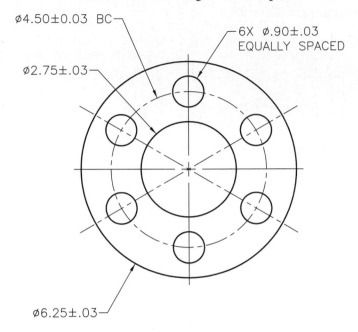

Ø4.50±0.03 BC

6X Ø.90±.03
EQUALLY SPACED

Ø2.75±.03

Ø6.25±.03

L2 PROBLEM 39

Create the drawing of a rocker arm. Dimension the drawing using decimal units.

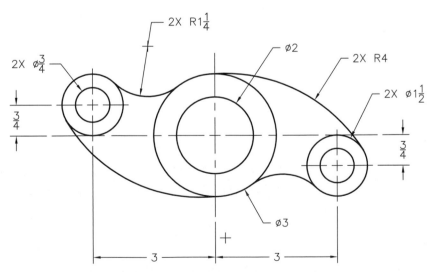

2X R1$\frac{1}{4}$

2X Ø$\frac{3}{4}$

Ø2

2X R4

2X Ø1$\frac{1}{2}$

$\frac{3}{4}$

$\frac{3}{4}$

Ø3

3

3

L2 PROBLEM 40

Create and fully dimension the drawing of a Geneva plate. Use a drawing area of 22 × 17.

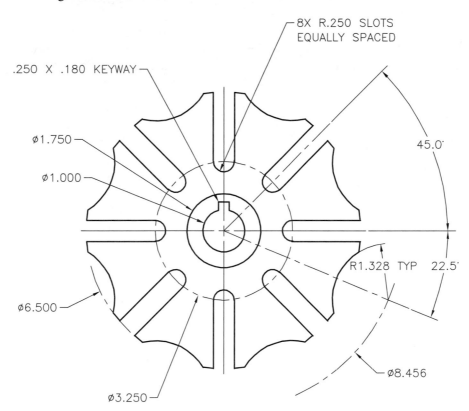

8X R.250 SLOTS
EQUALLY SPACED

.250 X .180 KEYWAY

ø1.750

ø1.000

45.0°

ø6.500

R1.328 TYP   22.5°

ø8.456

ø3.250

L2 PROBLEM 41

Create and fully dimension the drawing of a cassette reel. Use a drawing area of 280 × 215.

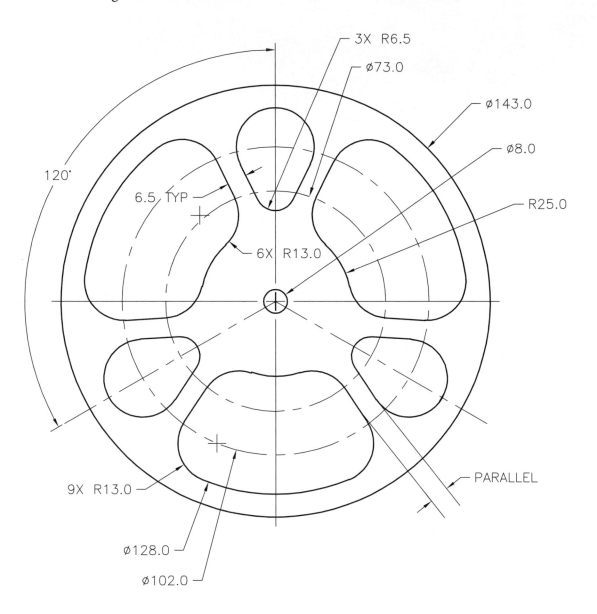

Adapted from the textbook **Drafting Fundamentals** *by Scott, Foy, and Schwendau*

L2 PROBLEM 42

Create and fully dimension the drawing of a slotted wheel. Use a drawing area of 17 × 11.

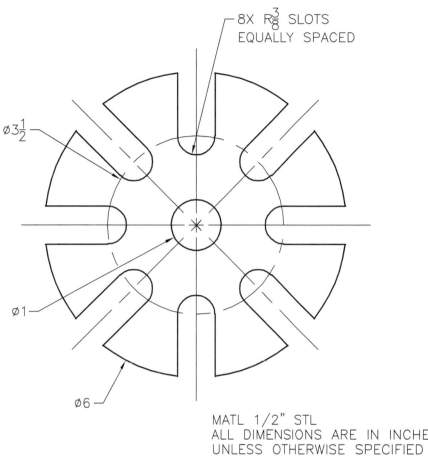

8X R$\frac{3}{8}$ SLOTS
EQUALLY SPACED

$\emptyset 3\frac{1}{2}$

$\emptyset 1$

$\emptyset 6$

MATL 1/2" STL
ALL DIMENSIONS ARE IN INCHES
UNLESS OTHERWISE SPECIFIED

Adapted from the textbook Drafting Fundamentals *by Scott, Foy, and Schwendau*

L2 PROBLEM 43

Create and dimension the drawing of a gasket.

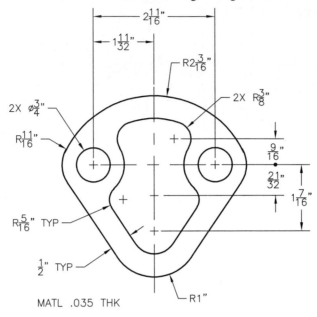

MATL .035 THK

Adapted from the textbook Drafting Fundamentals *by Scott, Foy, and Schwendau*

L2 PROBLEM 44

Create and dimension the drawing of an 18-tooth cutter. Use a drawing area of 17 × 11.

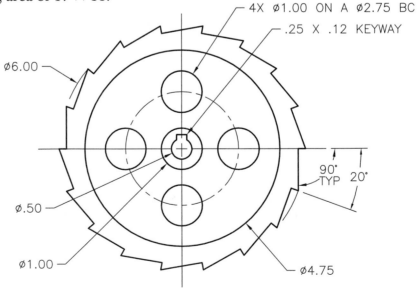

L2 **PROBLEM 45**

Create and dimension the top and front orthographic views of the slotted shaft.

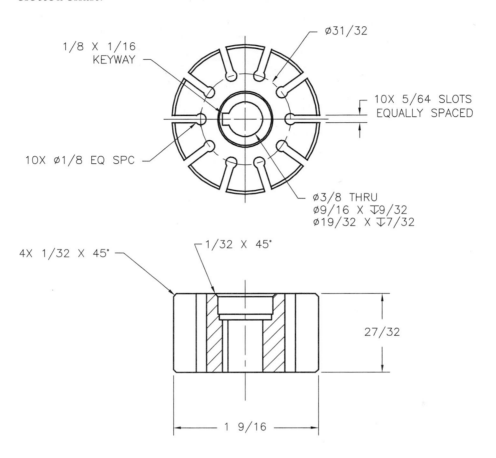

Ø31/32

1/8 X 1/16 KEYWAY

10X 5/64 SLOTS EQUALLY SPACED

10X Ø1/8 EQ SPC

Ø3/8 THRU
Ø9/16 X ↧9/32
Ø19/32 X ↧7/32

4X 1/32 X 45°

1/32 X 45°

27/32

1 9/16

Adapted from the textbook Drafting Fundamentals *by Scott, Foy, and Schwendau*

L2 PROBLEM 46

Create and dimension the drawing of an adjustable bracket.

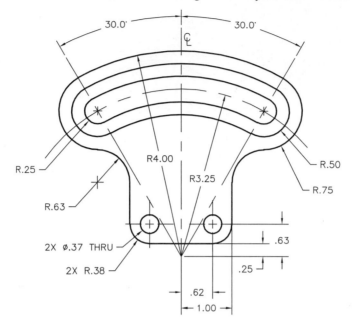

L2 PROBLEM 47

Create and dimension the drawing of a basketball backboard. Use a scale of 1″ = 1′.

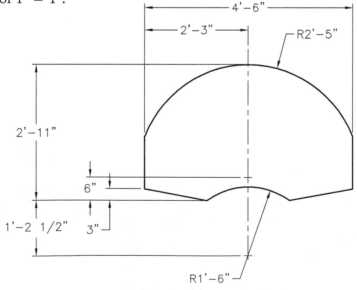

Problems 46 and 47 adapted from the textbook Drafting
Fundamentals *by Scott, Foy, and Schwendau*

L2 PROBLEM 48

Create and dimension a complete drawing of the wall plaque.

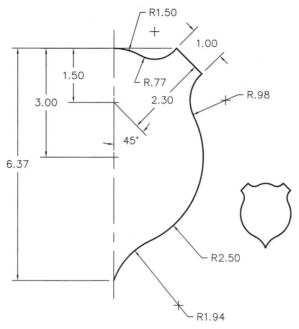

Courtesy of Mark Schwendau, Kishwaukee College

L2 PROBLEM 49

Create the electrical schematic drawing. Create blocks of each of the components and then insert them as needed.

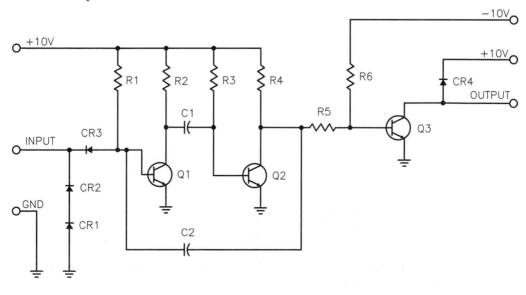

L2 PROBLEM 50

Create and dimension the spacer. Use a drawing area of 17 × 11.

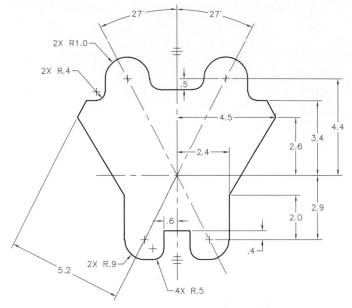

Courtesy of Mark Schwendau, Kishwaukee College

L2 PROBLEM 51

Create a single view of the spacer. Dimension the drawing and add a note indicating the thickness of the spacer.

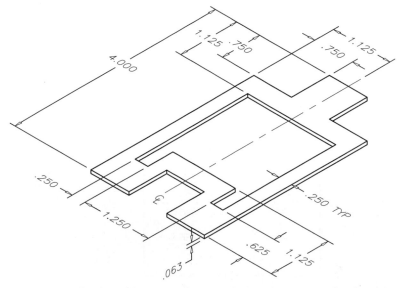

Adapted from the textbook Drafting Fundamentals *by Scott, Foy, and Schwendau*

L2 PROBLEM 52

Create the front and right-side views of the centering bushing. Include dimensions.

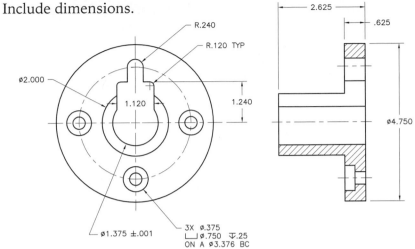

L2 PROBLEM 53

Create and fully dimension the views of the rod support. Determine the drawing area on your own, based on space requirements.

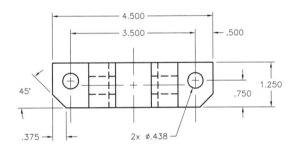

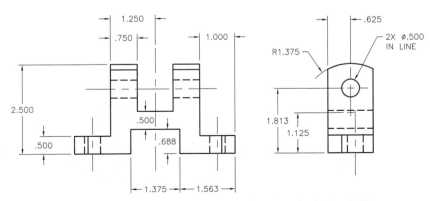

Problems 52 and 53 courtesy of Steve Huycke, Lake Michigan College

L2 PROBLEM 54

Create a drawing of the template and dimension it as shown.
Determine the drawing area based on space requirements.

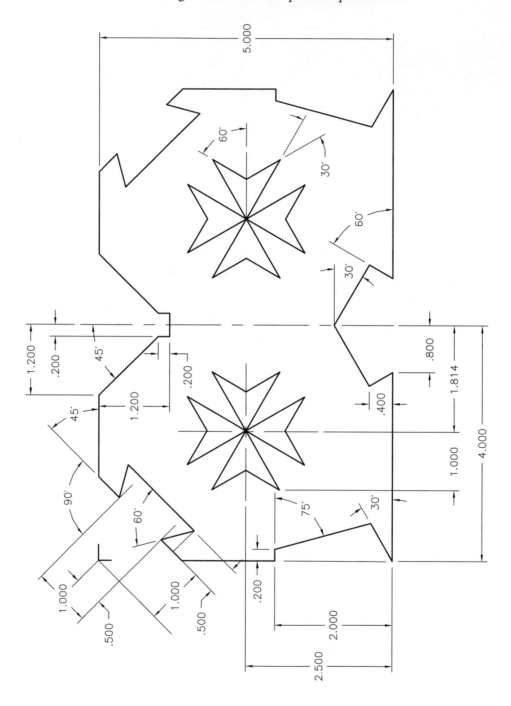

Courtesy of Steve Huycke, Lake Michigan College

L2 PROBLEM 55

Create the top view and front full section view of the arm. Be sure to include the detail.

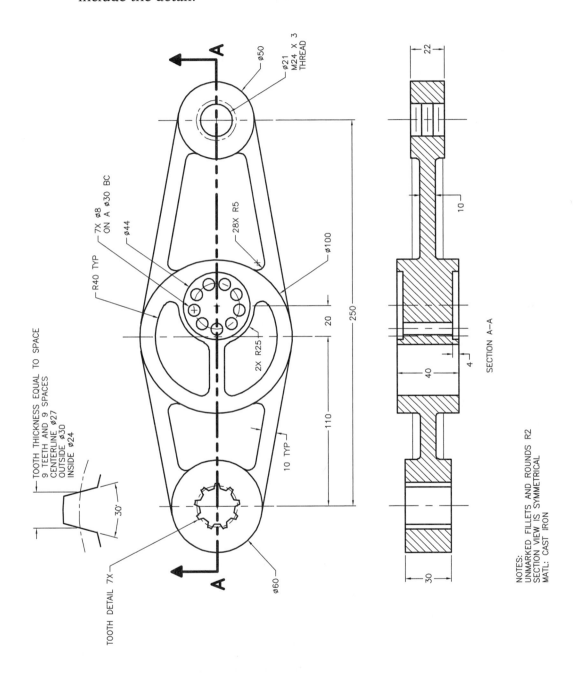

Courtesy of Alan Fitzell, Central Peel Secondary School

L2 PROBLEM 56

Create and dimension the irregular curve. Determine the drawing area based on space requirements.

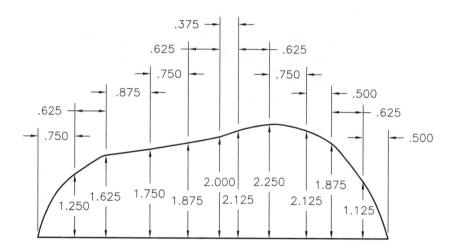

L2 PROBLEM 57

Create and dimension the irregular curve. Determine the drawing area based on space requirements.

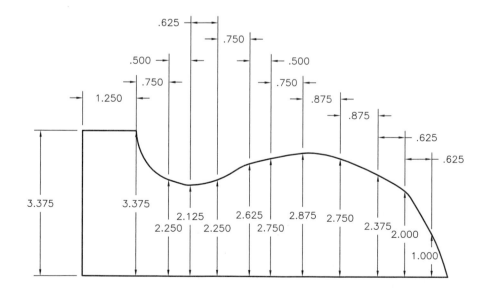

L2 PROBLEM 58

Create and fully dimension the two views of the crank handle using decimal inches to three decimal places. Estimate all dimensions.

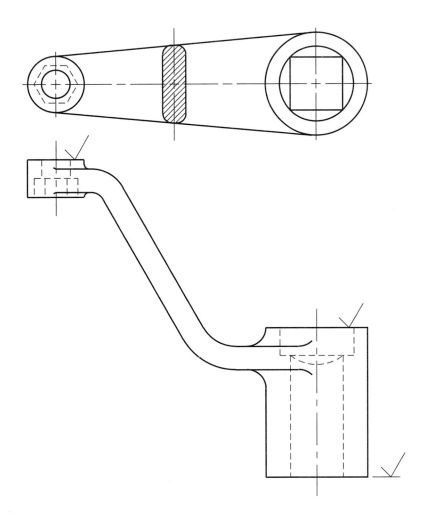

Courtesy of Joseph K. Yabu, San Jose State University

L2 PROBLEM 59

Create and fully dimension the wheel cover.

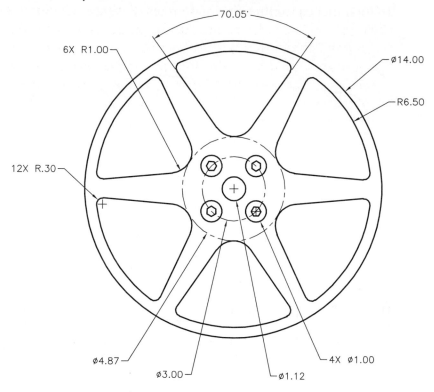

L2 PROBLEM 60

Create drawings of the wheel covers. Estimate all dimensions.

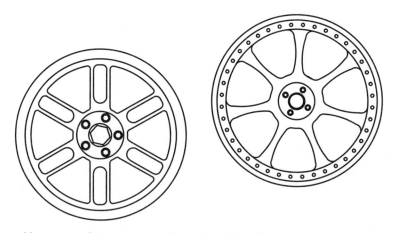

Problems 59 and 60 courtesy of Joseph K. Yabu, San Jose State University

L2 PROBLEM 61

In a steel frame building, a bracket is to be connected to a column as shown. The front and right-side views of the structural detail are given below. The W 14 × 26 wide flange beam is to be bolted to a 9 × 5 × ³⁄₈ plate which, in turn, is to be welded to the W 8 × 35 wide flange beam. The dimensions for the two beams are taken from the American Institute of Steel Construction's *Manual of Steel Construction*. Draw the two views.

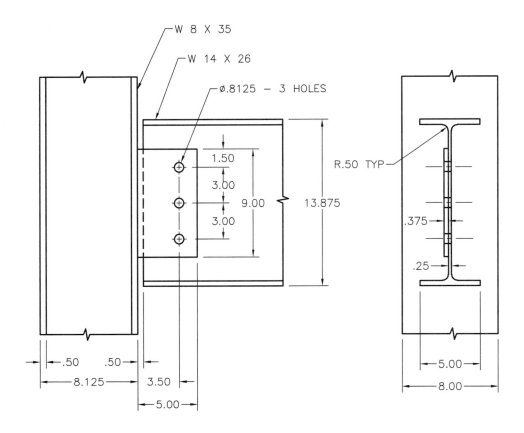

Courtesy of Gary J. Hordemann, Gonzaga University

L2 PROBLEM 62

Draw the top and front views of the barrier block. Study the top view and figure out how to minimize drawing and maximize the use of replicating commands such as COPY, MIRROR, and ARRAY.

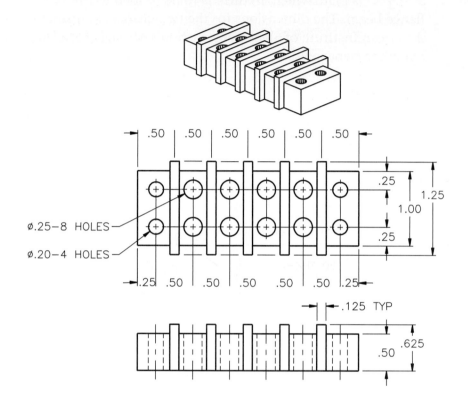

ø.25−8 HOLES

ø.20−4 HOLES

.50 | .50 | .50 | .50 | .50 | .50

.25   1.25   1.00   .25

.25 .50 .50 .50 .50 .50 .25

.125 TYP

.625   .50

Courtesy of Gary J. Hordemann, Gonzaga University

L2 PROBLEM 63

The front view of the link shown below is amazingly easy to draw if you draw the large circle first; then draw long horizontal and vertical lines snapped to the center of the circle. Offset the lines to locate the centers of the two smaller circles; if necessary, use a zero-distance chamfer or zero-radius fillet to extend the lines until they meet. Insert the two small circles, snapped to the intersections of the lines. Offset again to construct the lines composing the lower left corner and then fillet that corner. Draw a line snapped tangent to the upper small circle and the large circle; use the FILLET command to insert the arcs joining the small and large circles. Finally, insert the octagon snapped to the center of the large circle.

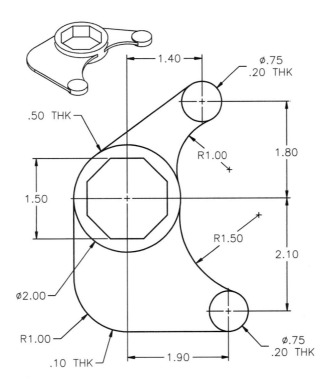

Courtesy of Gary J. Hordemann, Gonzaga University

L2 PROBLEM 64

Draw the top view of the stanchion shown below.

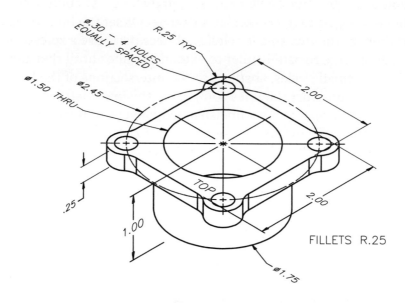

L2 PROBLEM 65

A drawing of part of the 20 scale from an engineer's scale is shown below. Using the ARRAY command, draw the 20 scale. Then draw 30 and 50 scales.

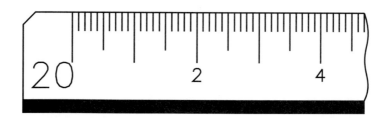

Problems 64 and 65 courtesy of Gary J. Hordemann, Gonzaga University

 PROBLEM 66

Draw the front view of the swivel link shown below. Draw half the object using LINE, CIRCLE, OFFSET, and FILLET. Construct the external fillets using circles and lines snapped to tangents. Then use BREAK to trim the circles to arcs. Use MIRROR to complete the other half.

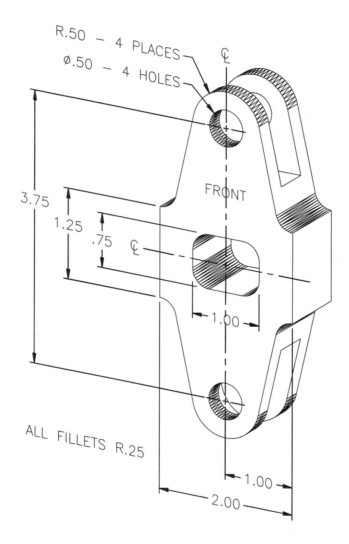

Courtesy of Gary J. Hordemann, Gonzaga University

L2 PROBLEM 67

The footprints for two cordless mice on a mouse docking (recharging) station are shown below. Draw the two footprints *without* using the FILLET command by drawing circles for all the rounded ends. Use the TTR option of the CIRCLE command for the R3.50 arc. Then use TRIM to create the final objects. Would it be easier to use filleted lines? Try it by drawing the outlines again, this time without using the CIRCLE command.

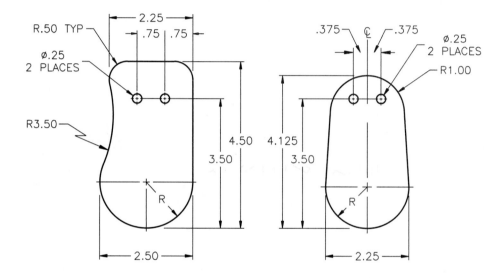

Courtesy of Gary J. Hordemann, Gonzaga University

L2 PROBLEM 68

Draw the front view of the splined separator block shown below. A suggested approach: Draw two circles bounding the spline teeth. Draw a horizontal line through the center and offset it to form the boundary of one tooth. Trim the circles and lines to form one tooth and array it about the center. Using the OFFSET command, create 1/12 of the object. Use MIRROR to create 1/6 of the object, and then array it about the center to finish it. When you have finished the drawing, use SCALE to reduce the spline teeth so that the outer diameter is 12. Use STRETCH to narrow the outer gaps from 8 to 6.

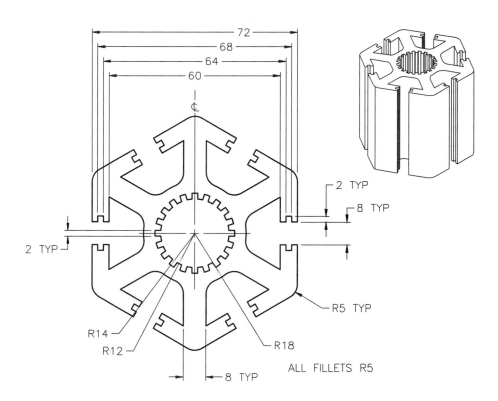

Courtesy of Gary J. Hordemann, Gonzaga University

L2 PROBLEM 69

Create and dimension the two views of the cam shown below. The spacing between the dots is .125.

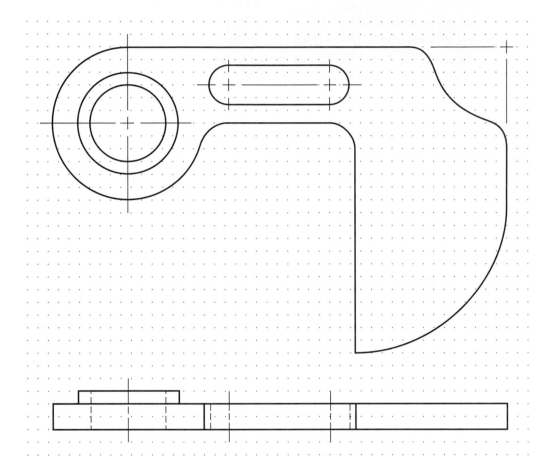

Courtesy of Gary J. Hordemann, Gonzaga University

L2 PROBLEM 70

Create a new text style called Inside with the following settings:

- Dimension line spacing: .30
- Dimension line color: ByLayer
- Extension line extension: .06
- Extension line origin offset: .05
- Extension line color: ByLayer
- Arrow size: .12
- Fit: Text only
- Unit linear precision: 0.00
- Unit angular precision: 0
- Suppress leading zeros
- Text style: ROMANS (font romans.shx)
- Text height: .12
- Text gap: .06
- Text color: Blue

Using the new dimension style, draw and dimension the two views of the clock blank shown below. The spacing between dots is .125.

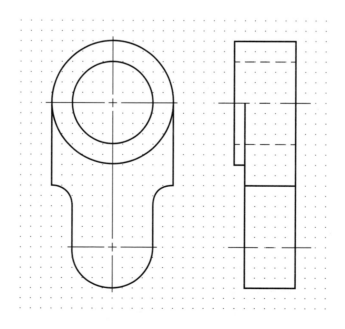

Courtesy of Gary J. Hordemann, Gonzaga University

L2 PROBLEM 71

Draw each of the three thread representations shown below and write them to files using the WBLOCK command. The external threads are most easily drawn using the ARRAY command. The distance between dots in the grid is not important, but .125 should work well.

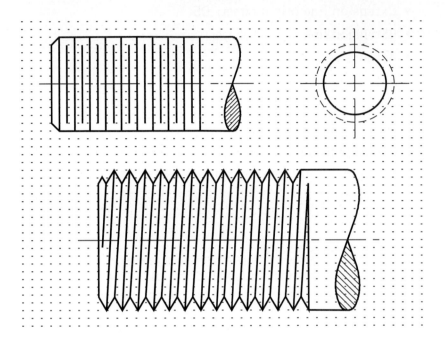

Courtesy of Gary J. Hordemann, Gonzaga University

L2 PROBLEM 72

Imagine an object composed of:

  a. A cube measuring 2″ on each side.

  b. A four-sided rectangular pyramid with a height of 1″. The pyramid sits on top of the cube and the four edges of its base align with the four edges of the cube's top surface.

  c. A hole, 1½″ in diameter, centered in the front cube face and having a depth of ½″.

  d. Another hole, coaxial with the first but 1″ in diameter, extending through the object.

Three ways of drawing this object are shown below. They are called *obliques*. Simply put, they are created by drawing one face at its full size and shape with the receding lines at an angle of 45°. The *cabinet* oblique has half-size receding lines; the *cavalier* oblique has full-size receding lines. The receding lines in *general* obliques vary; those on the one shown here are 0.707-size lines. Draw the three obliques.

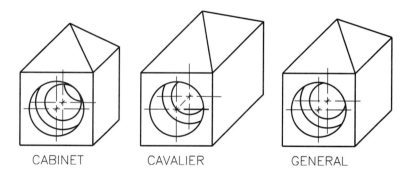

CABINET   CAVALIER   GENERAL

Courtesy of Gary J. Hordemann, Gonzaga University

L2 PROBLEM 73

Use 3D faces and a sphere to create the object shown below. Then use solid primitives with the UNION, SUBTRACT, and/or PRESSPULL commands to create a completely solid version. The large cube is 2 units on each side and the small cube is 1 unit on each side. The sphere has a diameter of .5. The total height of the object, including the sphere, is 4 units.

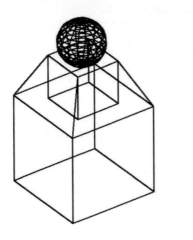 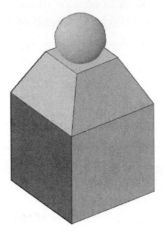

L2 PROBLEM 74

Create the chess pawn piece as a solid model. Use a polyline and the REVOLVE command. Estimate all dimensions. Then assign a material to the pawn and render it. Save both the model and the rendered image.

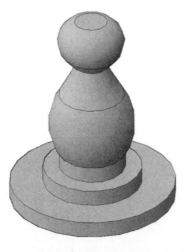

Problems 73 and 74 courtesy of Carole J. Williams, South Harrison High School

L2 PROBLEM 75

Create the floor plan for the first story of a residence. Determine the drawing area according to space requirements. Estimate the dimensions that are not given.

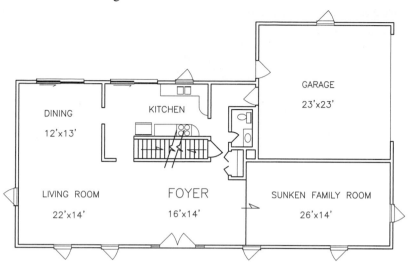

Courtesy of Mark Schwendau, Kishwaukee College

L2 PROBLEM 76

Create the wall section.

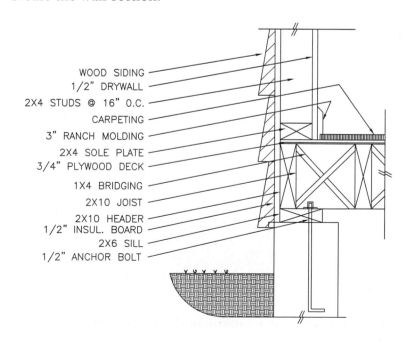

L2 PROBLEM 77

Create a dimension style named Structural that uses feet and inches, oblique arrowheads, and text placed above the dimension lines. Draw and dimension the foundation detail shown below. The hatch patterns are Earth, ANSI37, and AR-CONC.

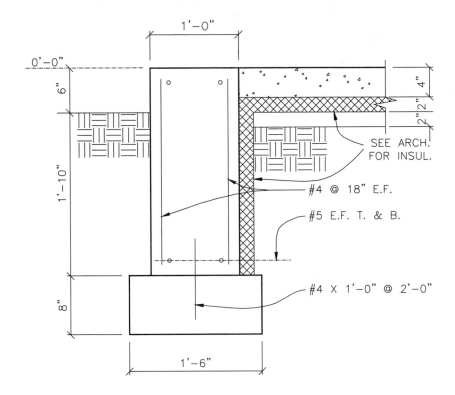

Courtesy of Gary J. Hordemann, Gonzaga University

L2 PROBLEM 78

Create and dimension an isometric view of the base.

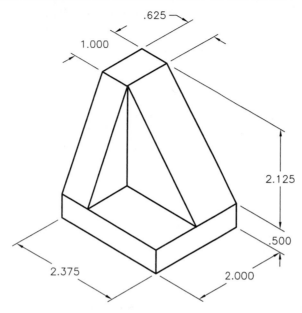

L2 PROBLEM 79

Create and dimension an isometric view of the wedge block.

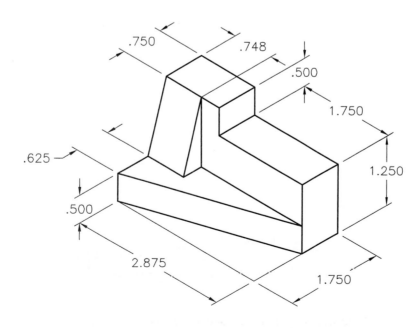

L3 PROBLEM 80

Create solid models of the table and vase. Estimate all dimensions. Display the model in the Conceptual visual style.

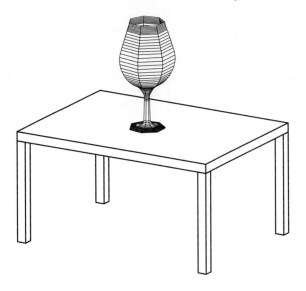

L3 PROBLEM 81

Create the top and side views of the fighter aircraft. Estimate all dimensions.

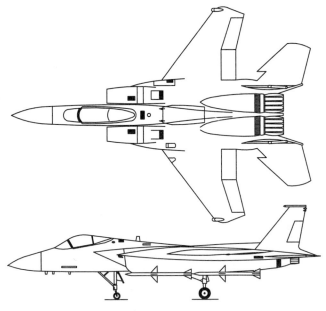

Problem 81 courtesy of Matt Melliere and Ronald Weseloh, Red Bud High School

L3 PROBLEM 82

Create the landscape drawing as shown. Create blocks of the trees and shrubs and insert them as shown. Estimate all dimensions.

Courtesy of Mills Brothers Landscape and Nursery, Inc.

L3 PROBLEM 83

Create the design elevation. Create blocks of the trees and shrubs and insert them as shown. Estimate all dimensions.

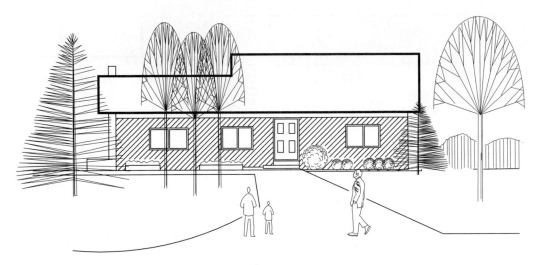

Courtesy of Mills Brothers Landscape and Nursery, Inc.

L3 PROBLEM 84

Create a solid model of the bracket. Estimate all dimensions. Display the model in the Conceptual visual style.

L3 PROBLEM 85

Create a solid model of the 90° link. Display the model in the
Conceptual visual style.

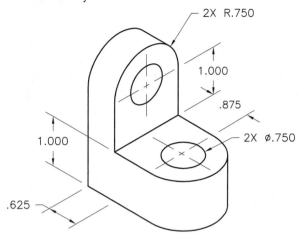

L3 PROBLEM 86

Draw the front and right-side views of the gear. Use the detail
drawing to create the teeth accurately. Hatch as indicated using
the ANSI31 pattern.

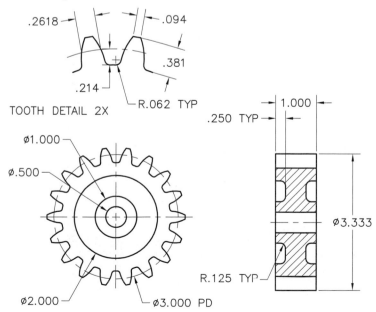

Problem 86 courtesy of Steve Huycke, Lake Michigan College

L3 PROBLEM 87

Create and fully dimension the isometric view of the strip block. Use a drawing area of 17 × 11.

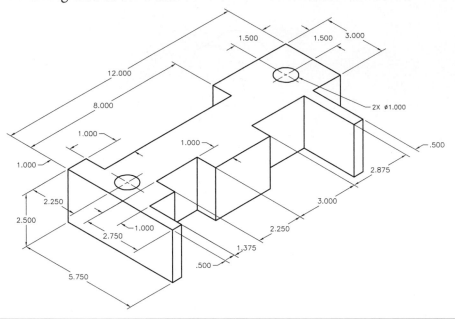

L3 PROBLEM 88

Create and dimension the drawing of a wrench. The units are millimeters. Determine the drawing area based on the dimensions of the wrench.

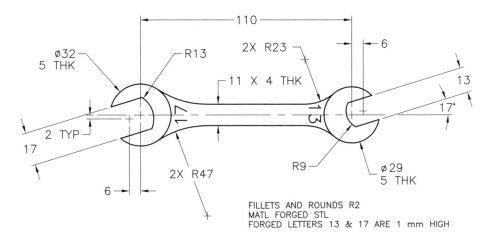

Adapted from the textbook Drafting Fundamentals *by Scott, Foy, and Schwendau*

L3 PROBLEM 89

Create the front and right-side views of the fitting. Then construct the auxiliary view. Dimension the front and right-side views.

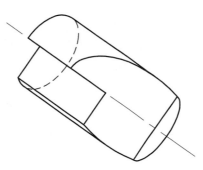

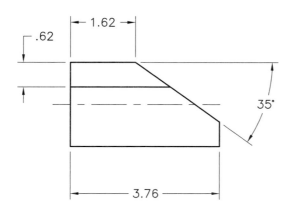

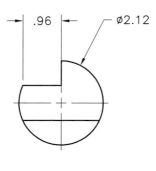

Courtesy of John F. Kirk, Kirk & Associates

L3 PROBLEM 90

Create the front and right-side views of the block. Then construct the auxiliary view. Dimension the front and right-side views.

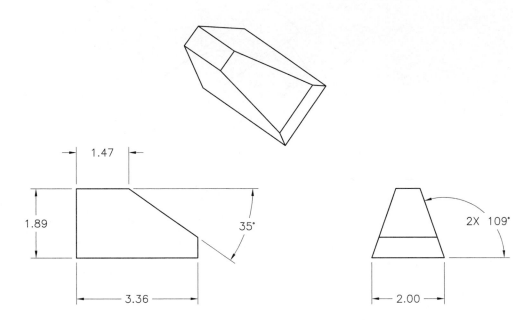

L3 PROBLEM 91

Create the front and right-side views of the casting. The right-side view should include a half section. Fully dimension both views.

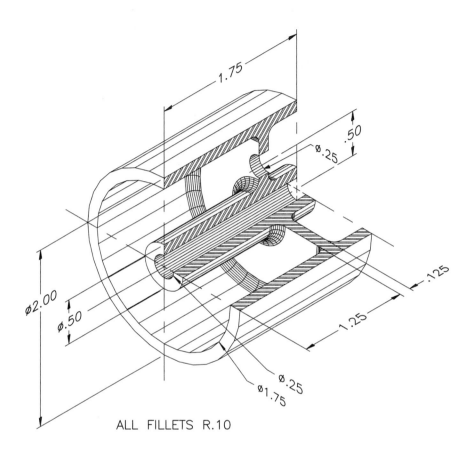

ALL FILLETS R.10

Courtesy of Gary J. Hordemann, Gonzaga University

L3 PROBLEM 92

Create the top and front views of the alternator bracket. Use a drawing area of 17 × 11.

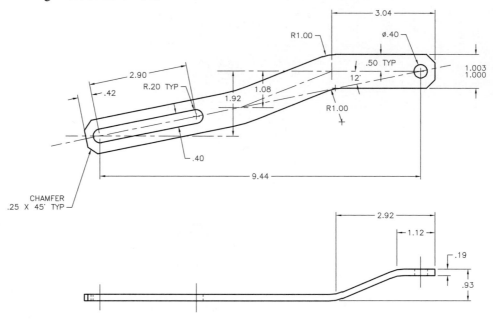

ALL FILLETS 1.00 UNLESS OTHERWISE SPECIFIED

L3 PROBLEM 93

Create the necessary orthographic views of the angle block and fully dimension the views. Add an isometric view in the upper right area of the drawing.

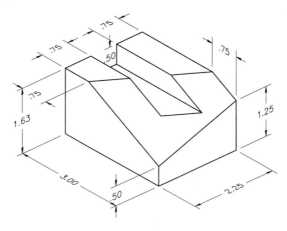

Problems 92 and 93 adapted from the textbook Drafting Fundamentals *by Scott, Foy, and Schwendau*

L3 PROBLEM 94

Create a solid model of the swivel stop. First extrude the basic profile.
Then use PRESSPULL to create the holes. Display the model in the
Conceptual visual style.

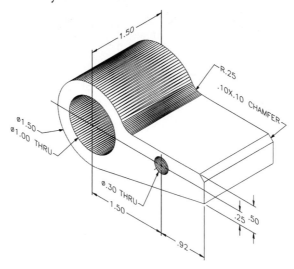

L3 PROBLEM 95

Create a solid model of the impeller. Display the model in the
Conceptual visual style.

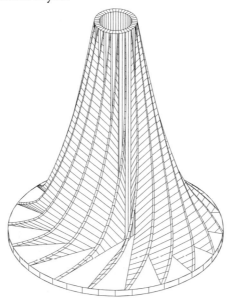

Problems 94 and 95 courtesy of Gary J. Hordemann, Gonzaga University

L3 PROBLEM 96

Create a solid model of the block.

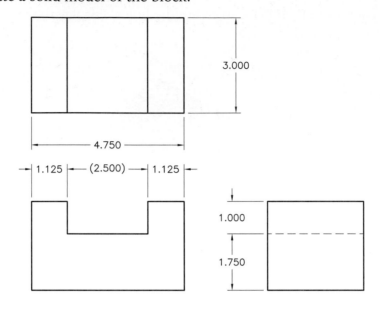

L3 PROBLEM 97

Create a solid model of the block. Then produce orthographic views from the solid model.

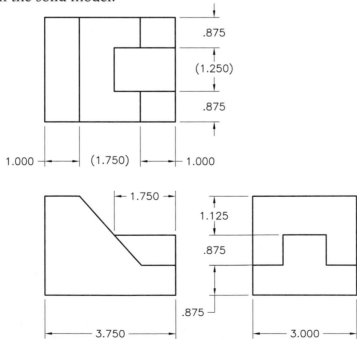

L3 PROBLEM 98

Create the views of the 45° elbow as shown and fully dimension them. Determine the drawing area based on space requirements.

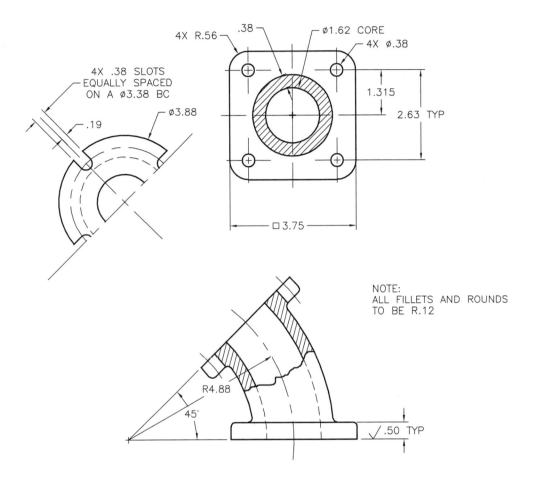

NOTE:
ALL FILLETS AND ROUNDS
TO BE R.12

Courtesy of Steve Huycke, Lake Michigan College

L3 PROBLEM 99

Create the front and full section views of the sprocket and fully dimension them. Units are in millimeters. Determine the drawing area based on space requirements.

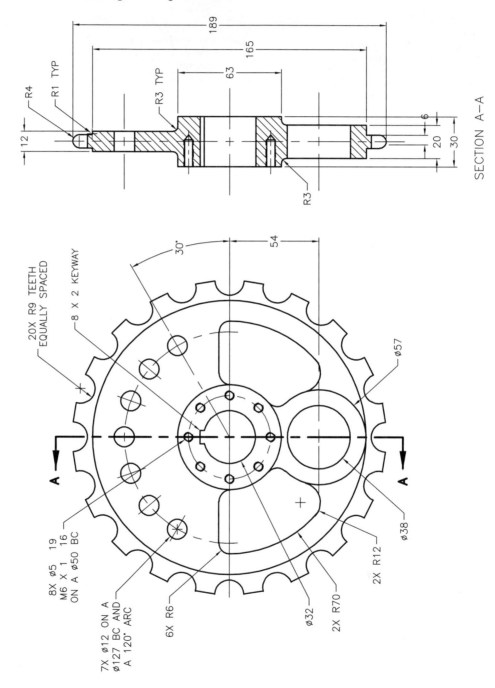

Courtesy of Alan Fitzell, Central Peel Secondary School

L3 PROBLEM 100

Create the front and full section views of the tool and dimension them. Also include the detail as shown. Determine the drawing area based on space requirements. Units are in millimeters.

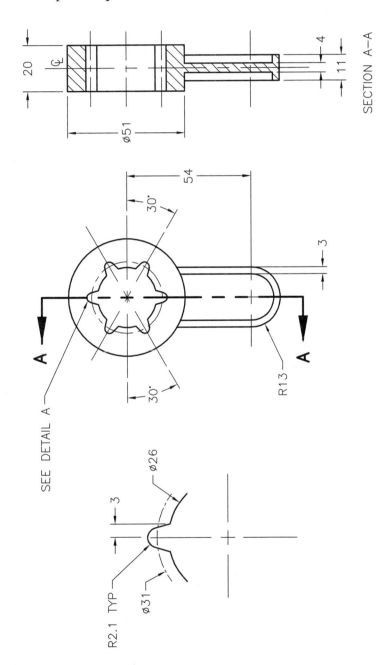

Courtesy of Julie H. Wickert, Austin Community College

L3 **PROBLEM 101**

Create the entire set of drawings of the control bracket. Determine the drawing area based on space requirements.

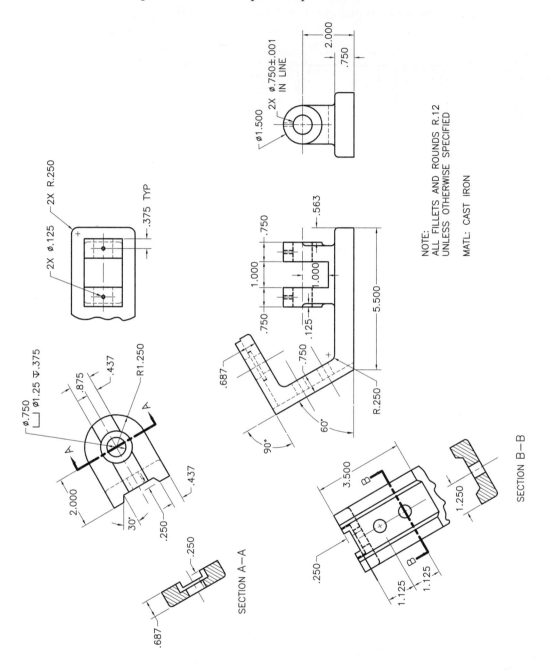

Courtesy of Steve Huycke, Lake Michigan College

L3 PROBLEM 102

Create the drawing of the fork lift as well as the bill of materials.
Estimate all dimensions.

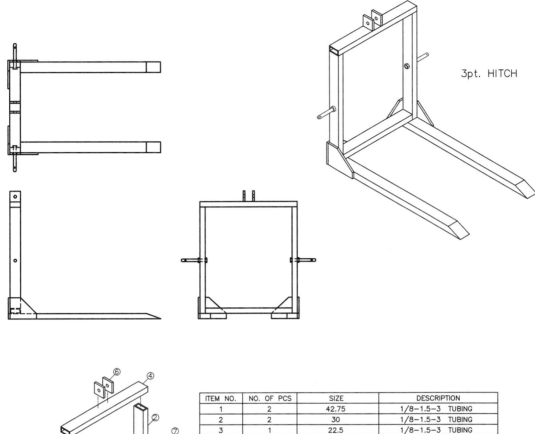

3pt. HITCH

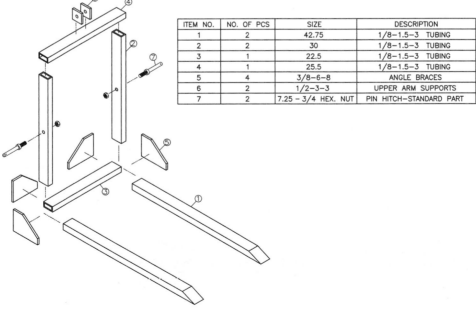

ITEM NO.	NO. OF PCS	SIZE	DESCRIPTION
1	2	42.75	1/8–1.5–3 TUBING
2	2	30	1/8–1.5–3 TUBING
3	1	22.5	1/8–1.5–3 TUBING
4	1	25.5	1/8–1.5–3 TUBING
5	4	3/8–6–8	ANGLE BRACES
6	2	1/2–3–3	UPPER ARM SUPPORTS
7	2	7.25 – 3/4 HEX. NUT	PIN HITCH–STANDARD PART

Courtesy of Craig Pelate and Ron Weseloh, Red Bud High School

L3 PROBLEM 103

Create the top, front, right-side, and isometric views of the picnic table. Do research to find appropriate sizes and dimensions.

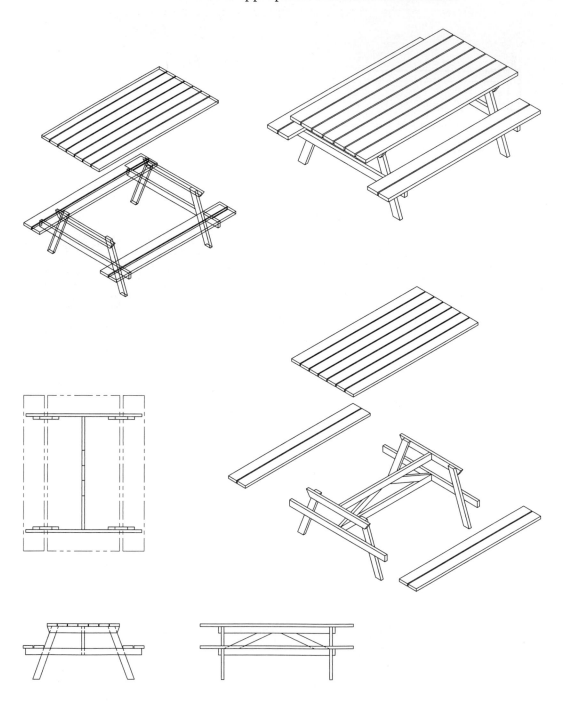

Courtesy of Dan Cowell and Ron Weseloh, Red Bud High School

L3 PROBLEM 104

Create the front and side views of the structural bracket and dimension them as shown. Also include section and isometric views. Determine the drawing area based on space requirements.

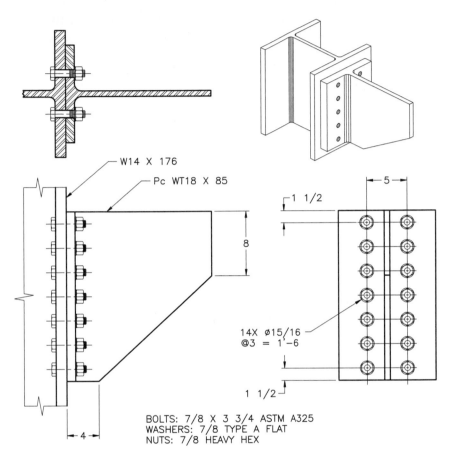

W14 X 176

Pc WT18 X 85

8

1 1/2

5

14X Ø15/16
@3 = 1'-6

1 1/2

4

BOLTS: 7/8 X 3 3/4 ASTM A325
WASHERS: 7/8 TYPE A FLAT
NUTS: 7/8 HEAVY HEX

FOR DIMENSIONS, SEE: MANUAL OF STEEL CONSTRUCTION,
AMERICAN INSTITUTE OF STEEL CONSTRUCTION

Courtesy of Gary J. Hordemann, Gonzaga University

L3 PROBLEM 105

Draw the top view of the spacer shown below. Assume that all of the ribs have a width of .2 and are symmetrical about the center lines. Try this approach: Create the hexagon and offset it. Add one set of circles and array them. Draw lines joining the centers of the circles and polygons. Then offset the lines, fillet them, and array them. *Note:* The fillets are between lines and circles and between lines and the outer polygon, *not* between lines.

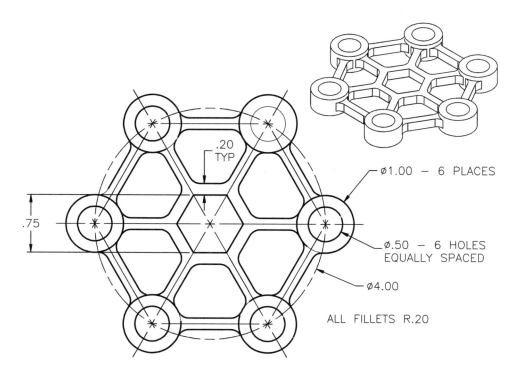

.20
TYP

Ø1.00 — 6 PLACES

.75

Ø.50 — 6 HOLES
EQUALLY SPACED

Ø4.00

ALL FILLETS R.20

Courtesy of Gary J. Hordemann, Gonzaga University

L3 PROBLEM 106

Create a solid model of the bracket using the dimensions shown in the orthographic views below. View the model in the 3D Wireframe visual style. Create orthographic views from the model.

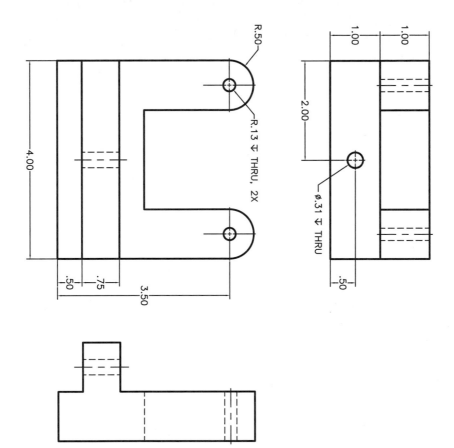

Courtesy of Mark R. Stevenson

L3 PROBLEM 107

Create and dimension a solid model of the locking sleeve according to the dimensions shown below. Display the model in the Conceptual visual style.

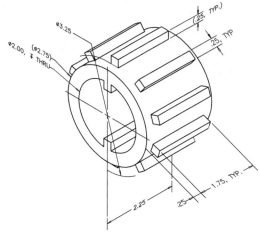

NOTES:
1.) 10 EVENLY SPACED OUTSIDE TEETH.
2.) 2 EVENLY SPACED INSIDE TEETH (.25X.25X2.25)

Courtesy of Mark R. Stevenson

L3 PROBLEM 108

Create a solid model of the roller tube block. The overall dimensions of the block are 2 × 2 × 3. The hole on the top surface has a depth of 1, and the holes on the sides are thru holes. All of the holes have a diameter of 1. Display the finished model in the Conceptual visual style. Rotate the model to view it from several viewpoints.

Model courtesy of Gary J. Hordemann, Gonzaga University

L3 PROBLEM 109

Using the REVOLVE command, create a solid model of the V-belt pulley shown below. Note that by revolving around the axis, you can create a hole without using the SUBTRACT or PRESSPULL command.

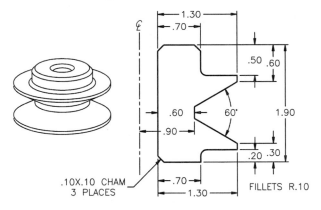

Courtesy of Gary J. Hordemann, Gonzaga University

L3 PROBLEM 110

Create top, front, and isometric views of the fence. Determine the drawing area based on space requirements. Dimension the top and front views.

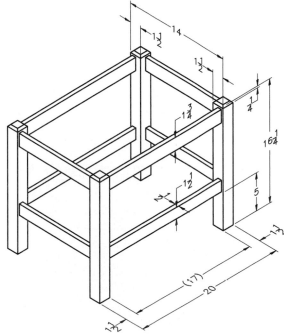

Adapted from the textbook Drafting Fundamentals *by Scott, Foy, and Schwendau*

L3 PROBLEM 111

Create a solid model of the surveyor's transit. Estimate all dimensions. Display the model in the Conceptual visual style.

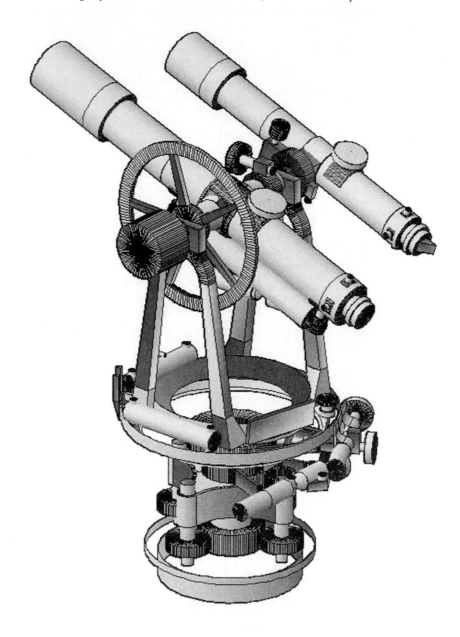

Model courtesy of Riley Clark

L3 PROBLEM 112

Create the architectural elevation of the cathedral. Estimate all dimensions.

Courtesy of David Sala, Forsgren Associates, from World Atlas of Architecture
(G.K. Hall and Company)

L3 PROBLEM 113

Create the wall section.

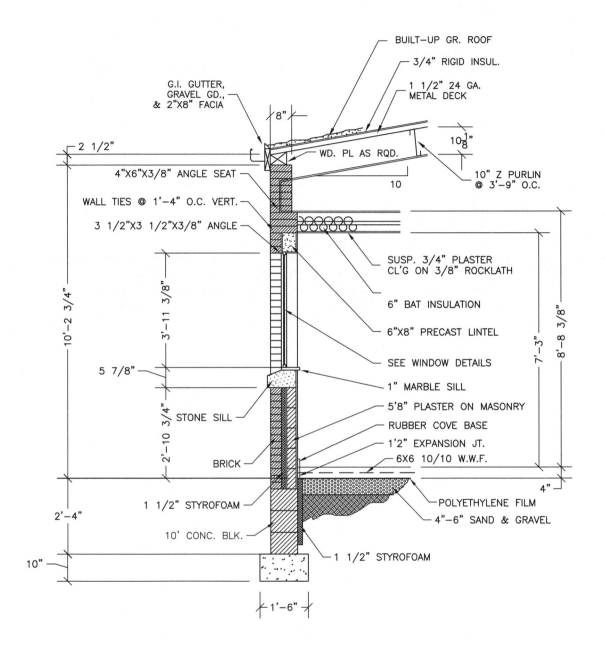

BUILT–UP GR. ROOF

3/4" RIGID INSUL.

1 1/2" 24 GA. METAL DECK

G.I. GUTTER, GRAVEL GD., & 2"X8" FACIA

8"

10 1/8"

2 1/2"

WD. PL AS RQD.

4"X6"X3/8" ANGLE SEAT

10" Z PURLIN @ 3'–9" O.C.

10

WALL TIES @ 1'–4" O.C. VERT.

3 1/2"X3 1/2"X3/8" ANGLE

SUSP. 3/4" PLASTER CL'G ON 3/8" ROCKLATH

6" BAT INSULATION

6"X8" PRECAST LINTEL

10'–2 3/4"

3'–11 3/8"

SEE WINDOW DETAILS

5 7/8"

1" MARBLE SILL

5'8" PLASTER ON MASONRY

STONE SILL

RUBBER COVE BASE

2'–10 3/4"

1'2" EXPANSION JT.

6X6 10/10 W.W.F.

BRICK

7'–3"

8'–8 3/8"

4"

1 1/2" STYROFOAM

POLYETHYLENE FILM

4"–6" SAND & GRAVEL

10' CONC. BLK.

2'–4"

1 1/2" STYROFOAM

10"

1'–6"

Adapted from a drawing by Paul Driscoll

L3 PROBLEM 114

Create the front and left-side architectural elevation drawings.
Estimate all dimensions.

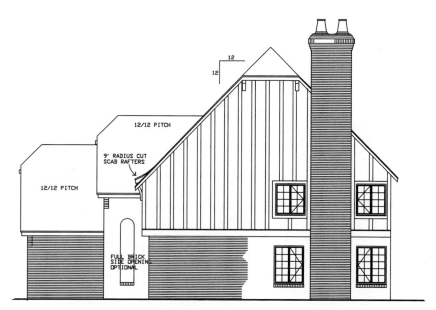

Courtesy of Rodger A. Brooks, Architect

L3 PROBLEM 115

Create the back and right-side architectural elevation drawings.
Estimate all dimensions.

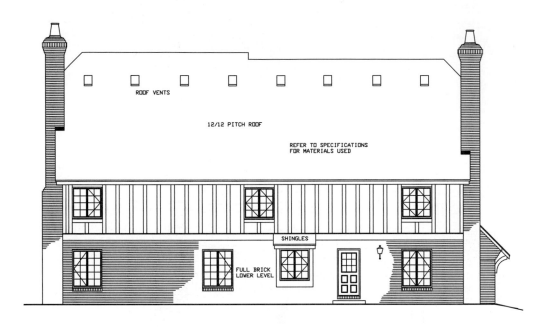

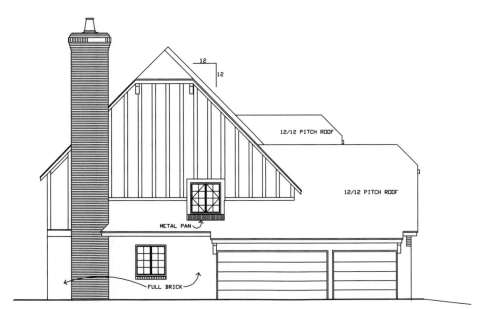

Courtesy of Rodger A. Brooks, Architect

L3 PROBLEM 116

Create the architectural elevation drawing. Estimate all dimensions.

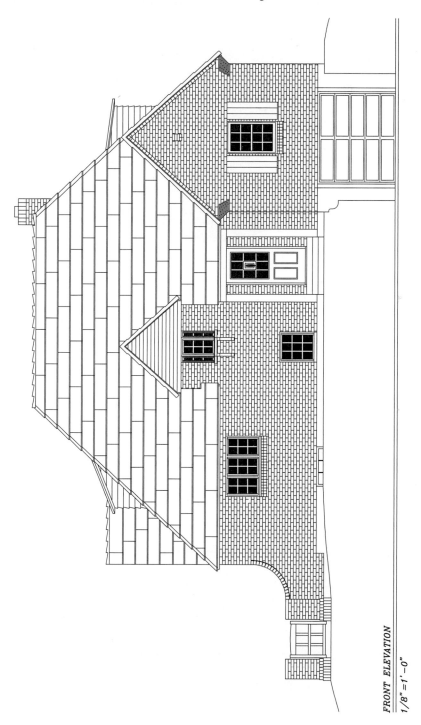

FRONT ELEVATION
1/8" = 1'-0"

Courtesy of Gary J. Hordemann, Gonzaga University

L3 PROBLEM 117

Create a solid model of the sprocket guard. Estimate all dimensions. Display the model in the Conceptual visual style.

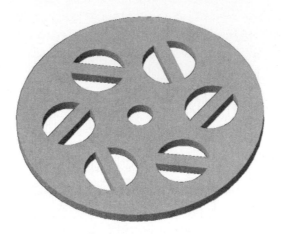

L3 PROBLEM 118

Create solid models of the parts and arrange as an exploded assembly. Estimate all dimensions. Display the assembly using the Conceptual visual style.

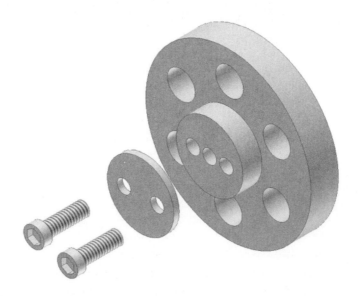

Appendix A
Options Dialog Box

AutoCAD's Options dialog box permits you to customize many of AutoCAD's settings. Also, the dialog box enables you to add pointing devices and printers to your AutoCAD system. Select the Options... button on the bottom of the Menu Browser to display the Options dialog box, as shown in Fig. A-1.

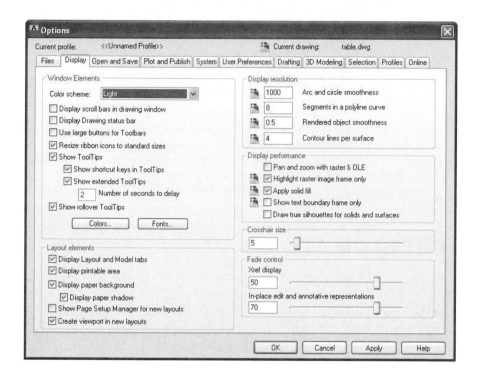

Fig. A-1

Files

The Files tab specifies the folders in which AutoCAD searches for text fonts, drawings to insert, linetypes, hatch patterns, menus, plug-ins, and other files that are not located in the current folder. It also specifies user-defined settings such as the custom dictionary to use for spell checking.

Display

This tab customizes the AutoCAD window. For instance, you can specify the display of scroll bars in the window. It allows you to set the colors for most items in the user interface, including the backgrounds for all workspaces,

Continued

Command line background and text, and AutoTrack™ vector. You can also control the size of the crosshairs and the resolution of arcs and circles.

Open and Save

This tab controls the options related to opening and saving files. Examples include the default file type when saving an AutoCAD drawing and whether a thumbnail preview image appears in the Select File dialog box. The settings on this tab also allow you to set the number of minutes between automatic saves.

Plot and Publish

The Plot and Publish tab deals with settings and controls related to plotting and printing devices. You can set the default output device and add and configure a new device. You can also add or edit plot style tables.

System

Use this tab to control miscellaneous system settings, such as the display of 3D graphics. This is also where you change the current pointing device.

User Preferences

This tab helps you optimize the way you work in AutoCAD. For example, you can customize how the right-click features in AutoCAD behave. Also, you can indicate the source and target drawing units when dragging and dropping drawing content. You can also set lineweights and edit the scale list from this tab.

Drafting

The Drafting tab controls many editing options. For instance, you can make adjustments to the AutoSnap features, such as changing the color and size of AutoSnap markers. Here, you can also change AutoTrack settings, including whether to display polar tracking vectors. You can also customize settings for light and camera glyphs.

3D Modeling

This tab allows you to control the appearance of the 3D crosshairs and the 2D and 3D UCS icons. You can also control the number of isolines used in models, customize walk, fly, and animation settings, and control the appearance of the ViewCube and UCS Icon.

Selection

The Selection tab controls settings related to methods of object selection. Examples include turning noun/verb selection on or off and adjusting the size of the pickbox and grip boxes. You can also set the color of selected and unselected grip boxes.

Profiles

This tab controls the use of profiles. In this context, a profile is a user-defined configuration that allows each user to set up and save various personal preferences, such as which toolbars are present and where they appear on the screen. Profiles allow several people to share a single computer without having to change the settings individually each time they use it.

If no profiles have been defined for a computer, you can define one by picking the Add to List... button. A dialog box appears that allows you to name the profile and give it a description. Note that creating a new profile does not automatically select it as the current profile. You must pick the Set Current button to set the active profile.

When more than one profile exists, any changes you make to AutoCAD's settings are automatically saved to the current profile. To change the current profile, display the Options dialog box, pick the Profiles tab, select the profile you want to make current, and pick the Set Current button.

Online

The Online tab allows users to set options for working online with Autodesk 360. It also provides access to documents that users store on their cloud accounts. To access the features of the Online tab, users must have an Autodesk 360 account.

Appendix B
Managing AutoCAD

Organizing an AutoCAD installation requires several considerations. For example, you should store files in specific folders. Name the files and folders using standard naming conventions. If you don't, you may not be able to find the files later. It is also very important that you back up regularly. If you overlook these and other system management tasks, your productivity may decline. Also, you may lose work, causing problems that you could otherwise avoid.

AutoCAD users and managers alike will find the following information helpful. If you manage an AutoCAD system with care, you should never have to create the same object twice. This comes only with careful planning and cooperation among those who will be using the system.

Bear in mind that the process of implementing, managing, and expanding an AutoCAD system evolves over time. As you and others become more familiar with AutoCAD and the importance of managing files and folders, refer back to this appendix.

System Manager

One person (or department) within the organization should have the responsibility for managing the AutoCAD systems and overseeing their use. This person should be the resident CAD champion and should answer questions and provide directions to other users of the systems. The manager should oversee the software, documentation, and hardware and should work with the users to establish procedural standards for use with the systems.

Management Considerations

The system manager should consider questions like those listed below when installing and organizing AutoCAD. The questions are intended to help guide your thinking, from a management perspective, as you become familiar with the various parts of AutoCAD.

- How can I best categorize the files so that each folder does not grow to more than 200 files total? Folders with a larger number of files become difficult to navigate.
- If I plan to install three or more AutoCAD stations, should I centralize the storage of user-created files and plotting by using a network and file server? A network and a centralized file system are recommended.

After AutoCAD is in place and you are familiar with it, you will create many new files. The following questions address the efficiency with which you create and store these files.

- Are there template files (or existing drawings) on file that may serve as a starting point for new drawings? It is a good practice to benefit from past work.

- Where should new drawing files be stored, what should they be named, and how can users locate them easily? Develop organizational standards that are clear and easy to follow.

- Are there predefined libraries of symbols and details that I can use while I develop a drawing? Do any of the symbol libraries that come with AutoCAD meet my needs?

- Is a custom pull-down menu or toolbar available that lends itself to my drawing application?

- Are AutoLISP routines or other programs available that would help me perform certain drafting operations more easily?

If you feel uncomfortable about your answers to these questions, there is probably room for improvement. The following discussion is provided to help you organize and manage your CAD systems more effectively.

Generally, the following discussion applies to almost all AutoCAD users and files. However, there are inevitable differences among users (backgrounds and interests, for example), drawing applications, and the specific hardware and software. Take these differences into consideration.

Software/Documentation

The AutoCAD system manager spends considerable time organizing and documenting files, establishing rules and guidelines, and tracking new software and hardware developments.

File Management—Know where files are located and the purpose of each. Understand which ones are AutoCAD system files and which are not. Create a system for backing up files and folders, and back up regularly. Emphasize this to all users. Delete "junk" files.

Continued

Template Files—Create an easy-to-use system for the development, storage, and retrieval of AutoCAD template files. Allow for ongoing development and improvement of each template. Store the template files in a folder dedicated to templates so that they are accessible by other users.

Document the contents of each template file by printing the drawing status information, layers, text styles, linetype scale, and other relevant information. On the first page of this information, write the name of the template file, its location, sheet size, and plot scale. Keep this information in a three-ring binder for future reference to other users.

User Drawing Files—Store these in separate folders. Place drawing objects on the proper layers. Assign standard colors, linetypes, and lineweights to the standard layers. Make a backup copy of each drawing and store it on a separate disk or backup system. Plot the drawings most likely to be used by others and store them in a three-ring binder for future reference.

Symbol Libraries—Develop a system for ongoing library development. (See Chapter 34 for details on creating symbol libraries.) Plot each symbol library drawing file, and place the library drawings on the wall near the system or in a binder. Encourage all users to contribute to the libraries.

Menu Files—Develop, set up, and make available custom toolbars and pull-down menus. Store the menus in a folder dedicated to menus so that they will be accessible to others.

AutoCAD Upgrades—Handle the acquisition and installation of AutoCAD software upgrades. Inform users of the new features and changes contained in the new software. Coordinate training.

AutoCAD Third-Party Software—Handle the acquisition and installation of third-party software developed for specific applications and utility purposes. Inform users of its availability and use.

Hardware

Oversee the use and maintenance of the hardware components that make up the system. Consider hardware upgrades as user and software requirements change.

Procedural Standards

Develop clear, practical standards in the organization to minimize inconsistency and confusion. Each template file should have a standard set of drawing layers, with a specific color and linetype dedicated to each layer. For example, you may reserve a layer called Dimension, with color green, a continuous linetype, and a lineweight of .3 mm, for all dimensions. This will avoid confusion and improve consistency within your organization. Develop similar standards for other AutoCAD-related practices.

AutoCAD Standards Checker

AutoCAD offers a method of checking arrowhead size, center marks, zero suppression, text height, and other items for organizational standards. The following steps outline this method.

1. Open a drawing file that you would consider a model example of how drawings should appear at your organization.
2. Using **Save As...**, save the drawing in AutoCAD Drawing Standard (DWS) format.
3. Open or create a new drawing and then open the **CAD Standards** toolbar.
4. Pick the **Configure...** button.
5. In the Configure Standards dialog box, pick the **Add Standards File** button (the one containing the plus sign) and open the DWS file you saved in Step 2.
6. Pick the **Check Standards...** button.

AutoCAD audits the current drawing against the DWS file and displays the results in the Check Standards dialog box.

7. Pick the **Settings...** button, check the **Automatically fix non-standards properties** check box, and pick **OK**.
8. Pick the **Next** button repeatedly until the Checking Complete box appears.

AutoCAD fixes the current drawing to conform to the standards contained in the DWS file.

9. Close the Check Complete and Check Standards dialog boxes.

Appendix C
Drawing Area Guidelines

Scale Type	Sheet Size	Approx. Plotting Area	Scale	Drawing Area
Architect's Scale	A: 12″ × 9″	10″ × 8″	$1/8″ = 1'$	96′ × 72′
	B: 18″ × 12″	16″ × 11″	$1/2″ = 1'$	36′ × 24′
	C: 24″ × 18″	22″ × 16″	$1/4″ = 1'$	96′ × 72′
	D: 36″ × 24″	34″ × 22″	$3' = 1'$	12′ × 8′
	E: 48″ × 36″	46″ × 34″	$3' = 1'$	16′ × 12′
Civil Engineer's Scale	A: 12″ × 9″	10″ × 8″	1″ = 200′	2400′ × 1800′
	B: 18″ × 12″	16″ × 11″	1″ = 50′	900′ × 600′
	C: 24″ × 18″	22″ × 16″	1″ = 10′	240′ × 180′
	D: 36″ × 24″	34″ × 22″	1″ = 300′	10,800′ × 7200′
	E: 48″ × 36″	46″ × 34″	1″ = 20′	960′ × 720′
Mechanical Engineer's Scale	A: 11″ × 8″	9″ × 7″	1″ = 2″	22″ × 17″
	B: 17″ × 11″	15″ × 10″	2″ = 1″	8.5″ × 5.5″
	C: 22″ × 17″	20″ × 15″	1″ = 1″	22″ × 17″
	D: 34″ × 22″	32″ × 20″	1″ = 1.5″	51″ × 33″
	E: 44″ × 34″	42″ × 32″	4″ = 1″	11″ × 8.5″
Metric Scale	A: 279 mm × 216 mm	229 mm × 178 mm	1 mm = 5 mm	1395 × 1080
	B: 432 mm × 279 mm	381 mm × 254 mm	1 mm = 20 mm	8640 × 5580
	C: 55.9 cm × 43.2 cm	50.8 cm × 38.1 cm	1 cm = 10 cm	559 × 432
	D: 86.4 cm × 55.9 cm	81.3 cm × 50.8 cm	2 cm = 1 cm	43.2 × 27.95
	E: 111.8 cm × 86.4 cm	106.7 cm × 81.3 cm	1 cm = 2 cm	237.6 × 172.8

Note: 1″ = 25.4 mm

Appendix D
Dimensioning Symbols

Geometric Characteristic Symbols

Type of Tolerance	Symbol	Name
Location	◎	Concentricity
	⊕	Position
	⚌	Symmetry
Orientation	∠	Angularity
	//	Parallelism
	⊥	Perpendicularity
Form	/○/	Cylindricity
	▱	Flatness
	○	Circularity (roundness)
	—	Straightness
Profile	⌒	Profile of a line
	⌓	Profile of a surface
Runout	↗	Circular runout
	↗↗	Total runout
Supplementary	Ⓜ	Maximum material condition (MMC)
	Ⓛ	Least material condition (LMC)
	Ⓟ	Projected tolerance zone

Basic Dimensioning Symbols

Symbol	Type of Dimension	Symbol	Type of Dimension
Ø	Diameter	⌵	Countersink
R	Radius	⌴	Counterbore/Spotface
SR	Spherical radius	⤓	Deep
SØ	Spherical diameter	×	Places, times, or by
( )	Reference		

Appendix E
AutoCAD Fonts

AutoCAD provides several standard fonts, which have file extensions of SHX. You can use the STYLE command to apply expansion, compression, or obliquing to any of these fonts, thereby tailoring the characters to your needs. (See Chapter 19 for details on the STYLE command.) You can draw characters of any desired height using any of the fonts.

Examples of some of the fonts supplied with AutoCAD are listed in this appendix, along with samples of their appearance. With the exception of monotxt.shx (not included in the samples below), each font's characters are proportionately spaced. Hence, the space needed for the letter "i," for example, is narrower than that needed for the letter "m."

Each font resides in a separate file. This is the "compiled" form of the font, for direct use by AutoCAD. Examples of standard and TrueType fonts are shown in this appendix.

TrueType Fonts

AutoCAD uses several TrueType fonts and font "families" (groups of related fonts). In the TrueType fonts, each font in a family has its own font file. The examples below are shown in outline form, as they appear by default in

TrueType Fonts

Name	Appearance
Arial Narrow	ABCDEFGHIJKLMNOPQRSTUVWXYZ abcdefghijklmnopqrstuvwxyz0123456789
Dutch801 Rm BT	*ABCDEFGHIJKLMNOPQRSTUVWXYZ abcdefghijklmnopqrstuvwxyz0123456789*
Lucida Console	ABCDEFGHIJKLMNOPQRSTUVWXYZ abcdefghijklmnopqrstuvwxyz0123456789
Swis721 BdOul BT	ABCDEFGHIJKLMNOPQRSTUVWXYZ abcdefghijklmnopqrstuvwxyz0123456789
UniversalMath1 BT	ΑΒΨΔΕΦΓΗΙΞΚΛΜΝΟΠΘΡΣΤΘΩ6ΧΥΖ αβψδεφγηιξκλμνοπϑρστθωφχυζ″ + − × ÷ = ± ∓°ʹ
Wingdings	[symbol characters]

Standard Fonts

Name	Appearance
txt.shx	ABCDEFGHIJKLMNOPQRSTUVWXYZ abcdefghijklmnopqrstuvwxyz
romans.shx	ABCDEFGHIJKLMNOPQRSTUVWXYZ abcdefghijklmnopqrstuvwxyz
romand.shx	**ABCDEFGHIJKLMNOPQRSTUVWXYZ** **abcdefghijklmnopqrstuvwxyz0123456789**
italicc.shx	*ABCDEFGHIJKLMNOPQRSTUVWXYZ* *abcdefghijklmnopqrstuvwxyz0123456789*
scripts.shx	*ABCDEFGHIJKLMNOPQRSTUVWXYZ* *abcdefghijklmnopqrstuvwxyz0123456789*
gothice.shx	**ABCDEFGHIJKLMNOPQRSTUVWXYZ** **abcdefghijklmnopqrstuvwxyz0123456789**
syastro.shx	☉☿♀⊕♂♃♄♅♆♇☌⚹✳︎☊♈︎♉︎♊︎♋︎♌︎♍︎♎︎♏︎♐︎♑︎≈ ✳"⊂∪⊃∈∈→↑←↓∂∇^`´˘ℵ§†‡℈℔®©
symusic.shx	·‚♪♩○●♯♮♭▬─×⌐𝄞℮:𝄀•─⌐⌐∸⌐▽ ·‚♪♩○●♯♮♭▬─⅟⌐𝄢℮:℔☉☿♀⊕♂♃♄♅♆♇

AutoCAD. To display solid characters, set the TEXTFILL system variable to 1. You may also change the print quality by changing the value of the TEXTQLTY system variable.

Standard Fonts

AutoCAD's standard SHX fonts include both text and symbol files that have been created as AutoCAD shapes. You can change the appearance of these fonts by expanding, compressing, or slanting the characters.

Glossary of Terms

Note: Numbers in parentheses are chapter numbers.

A

absolute points Specific coordinate points entered directly (without using relative positioning).

additive manufacturing A process of joining liquids, powders, filaments, or sheet materials to form parts from 3D model data in plastics, metals, ceramics, or composite materials, usually layer upon layer as opposed to subtractive manufacturing methodologies.

alignment grid A non-printing grid that can be set at any desired spacing in either Standard or Isometric drawing mode.

alignment paths Temporary, non-printing lines that appear at predefined or customizable angles to help you create objects more accurately.

alternate units A second set of distances shown inside brackets on some dimensioned drawings.

aperture box A small box that defines an area around the center of the crosshairs within which an object or point will be selected when you press the pick (left) button.

arcball An object that displays on the screen when 3D Orbit is active; it allows the user to manipulate the drawing view in realtime.

area The number of square units needed to cover an enclosed 2D shape or surface.

array An orderly arrangement of objects. AutoCAD can create rectangular, polar (circular), and path arrays.

aspect ratio The ratio of height to width of a rectangular object or region.

associative array An array that updates automatically when its defining parameters are edited or modified.

associative dimensions Dimensions that update automatically when you change the size of objects by stretching, scaling, etc.

associative hatch A hatch object that updates automatically when its boundaries are changed; a hatch created with the BHATCH command.

attribute tag A variable that identifies each occurrence of an attribute in a drawing.

attribute values The information stored by attribute tags in specific instances of an attribute.

attributes Text data stored in blocks.

audits Examines the validity of a drawing file.

B

baseline dimensions A set of dimensions that have a common baseline to reduce measurement error that can accumulate from "stacking" dimensions.

basic dimension A dimension to which allowances and tolerances are added to obtain limits.

bisect Divide into two equal halves.

block A collection of objects that you can group to form a single object.

Boolean logic (see *Boolean mathematics*.)

Boolean mathematics The mathematical principles behind the creation of Boolean solid models.

Boolean operations AutoCAD operations such as union, subtraction, and intersection, which are based on Boolean mathematics and are used to create composite solids.

B-splines Splines created using the PEDIT or SPLINE command. (See *splines*.)

button In AutoCAD, the small areas in toolbars and palettes that, when pressed, enter an associated command or function.

C

camera The point of view that moves around the target in 3D Orbit.

Cartesian coordinate system A system used in geometry that assigns a specific coordinate pair (*x,y*) for 2D or triplet (*x,y,z*) for 3D to points in space so that each point has a unique identifier.

cascading menu A submenu in a pulldown menu that offers further tool choices.

cells The individual boxes in a table.

center lines Imaginary lines that mark the exact center of an object or feature; shown by a series of alternating long and short line segments.

chamfers Beveled edges.

circular array See *polar array*.

circumference The distance around a circle.

clipping plane A plane that slices through one or more 3D objects, causing the part of the object on one side of the plane to be omitted.

colinear Two or more line segments lying on the same infinite line.

color palette A selection of colors, similar to an artist's palette of colors.

comma-delimited file A file format that allows a user to write attributes to an ASCII text file. Entries in this file format are separated by commas.

command aliases Command abbreviations that usually consist of the first one to three letters of a command name.

Command **line** A text area in AutoCAD where you can key in commands and options and view information.

compass When the VPOINT command is active, the compass is a globe representation that allows the user to orient the drawing in 3D space.

composite solid A solid composed of two or more solid objects.

concentric Sharing a common center point.

constrain Assign limits or values for parameters in dynamic blocks.

construction lines Line objects in AutoCAD that extend infinitely in both directions.

context-sensitive A menu that changes depending on which option is selected.

context-sensitive help Help related to the currently entered command.

control panels Sections within the Dashboard, which appears in the 3D Modeling workspace, that hold related buttons and controls.

control points Points on or near a curve that exert a "pull," influencing the shape of the curve.

coordinate pair A combination of one *x* value and one *y* value that specifies a unique point in the 2D coordinate system.

coordinates Sets of *x,y* (for 2D drawings) or *x,y,z* (for 3D models) values used to describe specific locations in space.

copy and paste A feature that allows you to copy objects from one drawing and insert them into another.

crosshairs A special cursor that consists of two intersecting lines and a pickbox; used to select objects and pick points in the drawing area.

crosshatch (see *hatch.*)

D

data extraction The process of gathering attribute data and placing it into an AutoCAD table or an electronic file that can be read by a computer program.

datum A surface, edge, or point that is assumed to be exact.

datum dimensions (see *ordinate dimensions.*)

delta In mathematics, a Greek symbol used to signify change.

deviation tolerancing A method of tolerancing that allows the user to set the upper and lower tolerances separately (to different values).

dialog box A window that provides information and permits you to make selections and enter information.

diameter The length of a line that extends from one side of a circle to the other, passing through its center.

dimension line The part of a dimension that typically contains arrowheads at its ends and shows the extent of the area being dimensioned.

dimension styles Saved sets of dimension settings that control the format and appearance of dimensions and leaders in a drawing.

dimensioning A system of describing the size and location of features on a drawing.

docked toolbar A toolbar that is attached to the top or side of the drawing area.

donuts Thick-walled or solid circles created using the DONUT command.

double-clicking Positioning the pointer on an object or button and pressing the pick button twice.

drawing area The part of the AutoCAD screen in which the drawing or model appears. Also refers to the boundaries for constructing the drawing, which should correspond to both the drawing scale and the sheet size.

drawing database Information associated with each AutoCAD file that defines the objects in the drawing and contains specific numerical information about individual objects in the file and their locations.

dynamic blocks Blocks created with certain changeable characteristics, or parameters, such as length, width, rotation, or location.

dynamic input Entering commands and options at the cursor.

dynamic UCS A feature that automatically moves the current UCS to a selected face of a solid model, temporarily making that face the work plane. Dynamic UCS allows you to create objects in 3D without constantly changing UCSs.

E

elevations Z heights in three-dimensional space.

ellipse A regular oval shape that has two centers of equal radius.

entity An individual predefined element in AutoCAD; the smallest element that you can add to or erase from a drawing. Also called an *object*.

extension lines The lines in a dimension that extend from the object to the dimension line.

extrude To add depth to a 2D region or closed polyline.

extrusion thickness Thickness in the Z axis; can be set using the THICKNESS variable or by applying the EXTRUDE command.

F

feature control frames The frames used to hold geometric characteristic symbols and their corresponding tolerances.

file attributes The characteristics of a file, such as its size, type, author, and the date and time it was created and last modified.

fillets Rounded inside corners.

filtering Including or excluding items from a selection set using specific criteria.

fixed attributes Attributes whose values you define when you first create a regular (nondynamic) block.

floating toolbar A toolbar that displays as a window in the drawing area.

font Distinctive set of characters consisting of letters, numbers, punctuation marks, and symbols.

formula An expression that performs an automatic mathematical equation in an AutoCAD table.

freeze In AutoCAD, to remove a layer from view without turning it off or deleting it.

full section A section that extends all the way through a part or object. A view of a 3D model that slices all the way through the part, showing the interior of the part as it appears at that slice or cross section.

G

geometric characteristic symbols Symbols that are used to specify form and position tolerances on drawings.

geometric dimensioning and tolerancing (GD&T) Standardized format for dimensioning and showing variations (called *tolerances*) in the manufacturing process. Tolerances specify the largest variation allowable for a given dimension. GD&T uses geometric characteristic symbols and feature control frames that follow industry standard practices. GD&T standards are defined in the American National Standards Institute (ANSI) Y14.5M Dimensioning/ Tolerancing handbook.

gizmo An icon that simplifies the section process for 3D commands.

glyph In AutoCAD, an icon that has a specified location in the drawing area and represents one of AutoCAD's tools, such as a spotlight or camera.

grips Small boxes that appear at key points on an object when you select the object without first entering a command. Grips allow you to perform several basic operations such as moving, copying, or changing the shape or size of an object.

ground plane The XY plane in 3D space, on which all Z values = 0.

group A named set of objects. A circle with a line through it, for instance, could become a group. Unlike blocks, groups cannot be saved as files.

H

hatch A repetitive pattern of lines or symbols that shows a related area of a drawing.

helix An open 3D spiral.

hidden line removal The deletion of tessellation lines that would not be visible if you were viewing the 3D model as a real object.

hidden lines Lines that would not be visible if you were looking at an object without "see-through" capability; shown as dashed or broken lines. Dashed or broken lines used to show lines that would not be visible if you were looking at an actual object. Lines removed in a plot because they would not be visible if you were looking at the actual object.

I

icon The small picture on a button in AutoCAD that helps identify the function of the button.

image tiles Selectable squares displayed in some dialog boxes that contain thumbnail images of tools, procedures, or options from which the user can choose.

in-place text editor The text editor that allows you to edit dynamic text and mtext in place on the screen.

integer Any positive or negative whole number or 0.

interference The overlap of two or more objects that occupy the same area in 3D space.

island An area within an object, such as a hole, that is not hatched.

isometric drawing A type of 2D drawing that gives a 3D appearance.

isometric planes Three imaginary drawing planes spaced at 120° from each other; these planes are used to draw isometric views.

J

justify Align multiple lines of text at a specific point (left, right, center, etc.) relative to the point of insertion.

L

landscape A paper orientation in which the wider edge of the paper is at the top and bottom of the sheet.

layers Similar to transparent overlays in manual drawing; allow AutoCAD users to display and hide information to clarify the drawing.

layout An arrangement of one or more views of an object or assembly on a single sheet.

leaders Lines with an arrowhead at one end that point out a particular feature in a drawing.

limits The size of the drawing area, which can be specified using the LIMITS command or by using a Startup wizard.

limits tolerancing A method of tolerancing in which only the upper and lower limits of variation for a dimension are shown.

linear dimensions Vertical and horizontal dimensions and dimensions that are aligned with an edge of the object being dimensioned.

live sectioning A setting you can specify when creating a section to provide realtime views while manipulating the grips that control where a section slices through a 3D model.

lofting Creating a new solid model or surface by specifying a series of 2D cross sections.

M

material condition symbols Symbols used in geometric dimensioning and tolerancing to modify the geometric tolerance in relation to the produced size or location of the feature.

Maximum Material Condition (MMC) A notation specifying that a feature such as a hole or shaft is at its maximum size or contains its maximum amount of material.

menu bar A row of menu items located below the title bar.

mesh Unlike a solid primitive, a mesh is not closed volume. However, mesh primitives can be modified more easily and more extensively than solid primitives and can be converted to a solid at any time.

mesh model Created using a mesh modeling operation.

mirror To produce a mirror image of an object or set of objects.

mirror line The axis or line about which objects are mirrored using the MIRROR command.

mline (see *multiline*.)

model space The drawing mode in AutoCAD in which most drawing is done.

mtext A multiple-line text object created using AutoCAD's text editor.

multiline A line object in AutoCAD that consists of up to 16 parallel lines created simultaneously and viewed by AutoCAD as a single object.

multiview drawing A drawing that describes a three-dimensional object completely using two or more two-dimensional views.

N

notes Text that refers to an entire drawing, rather than one specific feature. (See also *specifications*.)

noun/verb selection An object selection method in which you select the object first, then enter a command (verb) to perform an operation on the object.

O

object A predefined element; the smallest element that you can add to or erase from a drawing. Also called an *entity*.

object snap tracking A feature that, when used in conjunction with object snap, provides alignment paths that help the user produce objects at precise positions and angles.

object snaps Magnet-like aids in AutoCAD that allow the user to "snap" to endpoints, midpoints, and other specific points easily and accurately.

offsetting Creating a new object (such as a line or circle) at a specific distance from an existing line or circle.

opacity In AutoCAD, the value of the VSFACEOPACITY system variable. A value of 100 makes objects opaque; a value of 0 makes them transparent.

ordinate dimensions Dimensions that use a datum, or reference dimension, to help avoid confusion that could result from "stacking" dimensions; also called *datum dimensions*.

origin The point at which the axes cross in a coordinate system.

ortho A mode in AutoCAD that forces all lines to be perfectly vertical or horizontal.

orthogonal Drawn at right angles.

orthographic projection A projection of views at right angles to each other in a multiview drawing.

P

palette A special toolbar that contains a large grouping of buttons, tools, or information.

panning A feature that allows the user to move the viewing window around the drawing without changing the zoom magnification.

paper space The space, or drawing mode, in AutoCAD from which multiple views can be arranged and plotted on the same sheet.

parallel projection The standard projection mode in AutoCAD, in which objects are shown in parallel regardless of perspective.

parameters Characteristics in a dynamic block that can be changed after the block is inserted into a drawing.

parametric shape editing Editing an object's shape by activating grips and entering a new size in the text box that appears.

perimeter The distance around a two-dimensional shape or surface.

perpendicular Lines, polylines, or other linear or curved objects that meet at a precise 90° angle.

perspective projection Projection mode in which parts of an object that are farther away appear smaller, the same as in a real-life photograph.

phantom lines Lines drawn using a thick lineweight, alternating two short dashes with one long dash; used for cutting planes in sectional views.

pickbox A cursor that consists of a small box that allows you to pick objects on the screen.

pictorial representation A drawing of an object that offers a realistic view, compared to orthographic projections.

plane An imaginary flat surface that can be defined by any three points in space or by a coplanar object.

plot scale The scale at which a drawing is plotted to fit on a drawing sheet. The plot scale has no effect on the sizes of the objects in the drawing file.

plot style A collection of property settings saved in a plot style table.

plotters Originally, output devices that could handle very large sheets of paper; now used interchangeably with *printers*.

polar array An array in which objects are arranged radially around a center point. Also called a *circular array*.

polar method A relative method of entering points in which the user can produce lines at specific angles around a central point.

polar tracking method A method of entering points in which AutoCAD displays alignment paths at prespecified angles to help the user place the points at precise positions and angles.

polyline A line object in AutoCAD that consists of a connected sequence of line and arc segments and is treated by AutoCAD as a single object.

polysolid A solid primitive that is similar to a two-dimensional polyline, but is three-dimensional.

portrait A paper orientation in which the narrower edge of the paper is at the top and bottom of the sheet.

primary units The units that appear by default when you add dimensions to a drawing in AutoCAD.

primitives The basic building blocks of solid modeling, including the box, sphere, cone, cylinder, torus, wedge, pyramid, and polysolid.

printers Originally small, tabletop output devices; now used interchangeably with *plotters*.

projected tolerance zone Controls the height of the extended portion of a perpendicular part.

pull-down menus Menus whose names appear across the top of the screen, from which you can perform many drawing, editing, and file manipulation tasks.

purge In AutoCAD, selectively deleting blocks or other elements that are not currently in use in a drawing.

Q

quadrant A quarter of a circle, defined by the points at exactly 0°, 90°, 180°, and 270°.

quadrant points The points at 0°, 90°, 180°, and 270° on a circle.

R

radial Describes a feature on which every point is the same distance from an imaginary center point.

radius The length of a line extending from the center to the periphery of an arc or circle.

rays Line objects in AutoCAD that extend infinitely in one direction from a specified point.

realtime zooming A method of changing the level of magnification by picking the Realtime Zoom button and moving the pointing device.

rectangular array An array in which the objects are arranged in rows and columns.

regular polygon A polygon in which all the sides are of equal length.

relative method A method of entering points that allows the user to base the new points on the position of a point that has already been defined.

rendered An object that has shading applied.

rendering A display of an object that combines visual styles, lighting, materials, and shadows to create a photorealistic image of a solid model.

resolution The number of pixels per inch on a display system; this determines the amount of detail you can see on the screen.

revolve Turn a 2D object about an axis to create a 3D model.

roll In 3D Orbit, moving the view around an axis at the center of the arcball.

rotating Changing the angle of an object in the drawing without changing its scale.

rounds Curved or rounded outside corners on an object.

rubber-band effect The stretching effect when a command is active that lets you see the line, circle, etc., "grow" as you move the pointing device.

running object snap modes/running object snaps Object snaps that have been preset to function automatically as you work in AutoCAD.

S

scale Increasing or decreasing the overall size of an object without changing the proportions of the parts to each other. A means of reducing or enlarging a representation of an actual object or part that is too small or too large to be shown on a drawing sheet.

screen regeneration Recalculation of each vector, in a drawing.

section drawings Drawings used to show the interior of a part.

section plane indicator A transparent plane that appears to show where a section will be taken in a 3D model.

selectable In AutoCAD groups, *selectable* means that when you select one of the group's members (*i.e.*, one of the circles), AutoCAD selects all members in the group.

selection previewing AutoCAD's automatic highlighting of objects as you run the cursor over them.

selection set All of the objects that are currently highlighted, or selected, for a command or operation.

selection window A window created by picking diagonally opposite corners of an imaginary box at the Select objects prompt.

shortcut menu A context-sensitive menu that appears when you right-click; shortcut menus contain several of the commands and functions the user is most likely to need in the current drawing situation.

snap grid An invisible, user-controlled grid that restricts cursor movement to points that lie on the grid.

solid model In AutoCAD, the type of model produced when a modeling operation results in a closed (watertight) volume.

solid modeling Computer modeling of three-dimensional volumes of solid parts. CAD solid modeling programs that take advantage of constructive solid geometry (CSG) techniques use primitives, such as cylinders, cones, and spheres, and Boolean operations to construct shapes. CAD programs that use the boundary representation (B-Rep) solid modeling technique store geometry directly using a mathematical representation of each surface boundary. Some CAD programs use both CSG and B-Rep modeling techniques.

special characters Characters that are not normally available on the keyboard or in a basic text font.

specifications Text that provides information about sizes, shapes, and surface finishes that apply to specific portions of an object or part.

spline leader Similar to a regular leader, except that the leader line consists of a spline curve, offering more flexibility with its shape.

splines Curves that use sampling points to approximate mathematical functions that define a complex curve.

status bar The line at the bottom of the AutoCAD screen that displays various types of status information about modes and settings, including snap, grid, the ortho mode, polar tracking, object snap, object tracking, and lineweight.

subobject A face, edge, or corner of a solid that can be edited directly using grip tools.

surface controls Feature control frames not associated with a specific dimension that refer to a surface specification regardless of feature size.

surface model In AutoCAD, the type of model produced when a modeling operation results in an open 3D surface.

sweeping Moving a 2D profile along a path curve to create a 3D model. Sweeps can result in solids or surfaces, depending on whether the profile is open (surface) or closed (solid).

symbol library Drawing file that contains a series of blocks for use in other drawings.

symmetrical deviation (see *symmetrical tolerancing*.)

symmetrical tolerancing A method of tolerancing in which the upper and lower limits of a plus/minus tolerance are equal.

system variable Similar to a command, except that it holds temporary settings and values instead of performing a specific function.

T

table A set of rows and columns that contain text.

tangent Circles, arcs, or other objects that meet at a single point.

template file A file that contains predefined characteristics and is used to create new drawings with those characteristics. A file that defines text format for attribute extraction.

tessellations A collection of planer facets on a curved surface that approximate curvature.

3D faces Three- or four-sided planar meshes that have no width.

thumbnail A small picture of a drawing that gives you a quick view of the graphic content.

title block A portion of a drawing that is set aside to give important information about the drawing, the drafter, the company, and so on.

tolerance A designation that specifies the largest variation allowable for a given dimension.

toolbar A collection of related buttons that can be displayed or hidden by the user.

tooltips One or more words that appear when you position the pointer over a button for a second or longer. The purpose of tooltips is to help you understand the purpose or function of the button.

traces Thick lines commonly used on printed circuit boards; also, lines drawn using AutoCAD's TRACE command.

transparent zooms Zooms that the user forces to occur while another command is in progress.

2D representation A single profile view of an object seen typically from the top, front, or side.

U

UCS icon An icon located in the lower left corner of the drawing area that shows the orientation of the axes in the current coordinate system.

user coordinate systems (UCSs) Custom coordinate systems that a user can define for use in creating 3D objects and models.

V

value set The set of specified values to which a constrained parameter is limited in dynamic blocks.

variable attributes Attributes whose value you can change as you insert the block.

vector A line segment defined by its endpoints or by a starting point and direction in 3D space.

verb/noun selection The traditional method of entering a command in AutoCAD, in which the user enters a command and then selects the object or objects on which the command should operate.

vertex Any endpoint of an individual line or arc segment in a polyline, polygon, or other segmented object.

vertex grip A corner grip on a solid model that has corners, such as a box or pyramid.

view resolution The accuracy with which curved lines appear on the screen.

viewport A portion of the drawing area that a user can define to show a specific view of a drawing.

visual style The visual representation of objects in the drawing area, including the way the edges are displayed and the way the object is shaded. The display style of a 3D model.

W

wireframe A "stick" representation of a 3D model whose shape is defined by a series of connected lines and curves. A representation of a 3D object in which lines and curves are used to define its boundaries.

wizard A series of dialog boxes that steps you through a sequence of steps to perform a procedure.

workspace A complete user interface, including the location and appearance of the drawing area and all menus, buttons, tool palettes, and toolbars.

world coordinate system (WCS) AutoCAD's name for the Cartesian coordinate system. (See *Cartesian coordinate system*.) AutoCAD's default coordinate system, in which the X axis is horizontal on the screen, the Y axis is vertical, and the Z axis is perpendicular to the XY plane (the plane defined by the X and Y axes).

X

xlines A type of construction lines created by the XLINE command that extend infinitely in both directions.

xref An ordinary drawing file that has been attached to the current (base) drawing for reference.

XYZ octant The area in 3D space where the *x*, *y*, and *z* coordinates are greater than 0.

Z

Z axis The axis in a 3D coordinate system on which depth is measured.

Index

and folder navigation, 100–102

maintenance, 104–106

moving, 103

navigation, 100–102

new folders, 103

permanently damaged, 106

preview, 5

recovering, 106

renaming, 103

repairing, 104–106

viewing details, 101–102

File attributes, 101

File maintenance, 100–106

diagnosing, 104–105

file and folder navigation, 100–102

finding drawing files, 102–104

recovering, 104–106

repairing, 104–106

Fill, 184

FILL command, 184

FILLET command, 199

Fillets, 196, 199–200

Filtering, 316

Find and replace, 279

Fit tolerance, 189

Fixed attributes, 477–481

FLATSHOT command, 621

Floating toolbar, AutoCAD Classic, 25–26

Folder. *See also* File

creating, 103

deleting, 104

mechanical sample, 5

sample, 5

Font, 265, 270, 752–753

Formatting text. *See* Text

Formula, in tables. *See* Tables

Fraction format, 394

Freeze, layers, 314–315

From Model Space button, 625

Full navigation wheel, 519–520

Full section, 246

G

GD&T (Geometric dimensioning and tolerancing), 403, 407–412. *See also* Tolerancing

datum modifiers, 411–412

datum references, 411–412

feature control frames, 407

material condition symbols, 410–411

maximum material condition (MMC), 411

projected tolerance zone, 412

surface controls, 408–409

symbols, 407, 409–410

Geometric characteristic symbols, 407, 409–410. *See also* GD&T (Geometric dimensioning and tolerancing)

Geometric dimensioning and tolerancing (GD&T). *See* GD&T (Geometric dimensioning and tolerancing)

Geometry calculator, 421–422

Gizmo, 552

Glossary of terms, 754–766

Glyph, 635

GRID command, 145

Grid display button, 145

Grips, 74–78

and commands, 78

copying objects, 76

edge, 76

editing shape, 75

editing size, 75

feature modification, 76–77

modifying objects, 76–77

and options dialog box, 77

Ground plane, 516

Group, 434–438. *See also* Object grouping

changing name, 436

changing properties, 435–437

creating, 434–435

making selectable, 436–437

reordering objects in, 438

GROUPEDIT command, 435–436

Guides, 96

H

Hatch, 246. *See also* Hatching

Hatch and gradient dialog box, 248

Hatch pattern palette, 248

HATCHEDIT command, 250

Hatching, 245–252

adding and removing islands, 250–251

boundary, 246–249

characteristics, changing, 251

editing, 249–251

hatch and gradient dialog box, 248

hatch pattern palette, 248

tool palettes, 251–252

Hatching and sketching, 245, 252–253. *See also* Hatching; Sketching

Helix, 574

Help, 96–100

additional sources, 99–100